Developmental Behavior Genetics

# Developmental Behavior Genetics: Neural, Biometrical, and Evolutionary Approaches

Edited by
  Martin E. Hahn
  John K. Hewitt
  Norman D. Henderson
  Robert H. Benno

New York   Oxford
OXFORD UNIVERSITY PRESS
1990

Oxford University Press

Oxford  New York  Toronto
Delhi  Bombay  Calcutta  Madras  Karachi
Petaling Jaya  Singapore  Hong Kong  Tokyo
Nairobi  Dar es Salaam  Cape Town
Melbourne  Auckland

and associated companies in
Berlin  Ibadan

Copyright © 1990 Oxford University Press, Inc.

Published by Oxford University Press, Inc.,
200 Madison Avenue, New York, New York 10016

Oxford is a registered trademark of Oxford University Press

All rights reserved. No part of this publication may be reproduced,
stored in a retrieval system, or transmitted, in any form or by any means,
electronic, mechanical, photocopying, recording, or otherwise,
without the prior permission of Oxford University Press.

Library of Congress Cataloging-in-Publication Data
Developmental behavior genetics: nerual, biometrical, and
evolutionary approaches / edited by Martin E. Hahn . . . [et al.].
    p. cm.  Includes bibliographies and index.
ISBN 0-19-505446-6
1. Behavior genetics.  2. Developmental genetics.  I. Hahn,
Martin E.
QH457.D47, 1990        89-16061
155.7—dc20  CIP

9 8 7 6 5 4 3 2 1
Printed in the United States of America
on acid-free paper

*We dedicate this book to the memory of
Ronald S. Wilson, Ph.D.,
a pioneer in Developmental Behavior Genetics*

# Foreword

How the genes operate is one of the major questions of genetic research. But before it could be answered, the problem of how biological heredity is transmitted had to be solved. The chromosome theory of heredity was rapidly established following the rediscovery of Mendel's research in 1900, and this soon led to the discovery by Morgan and his colleagues of the basic unit of heredity, the gene. What characteristics were affected by these units was still another general problem, and in the seven or eight decades since these fundamental discoveries were made, researchers have established the fact that any characteristic of a living organism, including its behavior, is subject to genetic variation. They have also discovered a great deal about primary gene action, and one of the most active current fields of research is molecular genetics. But there is still a vast uncharted territory between primary gene action and behavior, and that is the subject of this book.

What do genes do? They *organize* nonliving material into living systems. Development, whose study began as the biological science of embryology, is the study of organizational processes. Therefore, the study of development is a fundamental method of analyzing gene action. John Fuller and I (Scott and Fuller, 1965) recognized this when we began our study of genetics and the social behavior of the dog and planned our study to include changes in the behavior of five breeds of dogs from birth until one year of age. We made several discoveries which Hahn has summarized in Chapter 4 of this book.

Some of these were unexpected. We thought that the best time to study the effects of genetics would be soon after birth, when behavior still had little opportunity to be altered by experience. On the contrary, we found that the different dog breeds were most alike as newborns; that is, genetic variation in behavior *develops* postnatally, in part as a result of the timing of gene action and in part from the interaction of gene action and experience, social and otherwise.

A second was that the major social behavior patterns of the dog were little changed from those of the ancestral wolf and differed among breeds chiefly in frequency, intensity, and timing of expression rather than in form. That is, the organization of major behavior patterns is somehow resistant to genetic change (reminiscent of Maynard Smith's "evolutionarily stable strategies"), and there is as yet no obvious genetic explanation.

As Cairns and Gariépy point out in this volume, variation in the time of appearance of behavior patterns provides an avenue for genetic change through selection that avoids the problem of disruption of systems inherent in any gross change in organization. But this still does not provide a genetic basis for behavioral stability.

Other insights are sure to emerge from any developmental genetic study of behavior. The major implication of this book is that it is time to study gene action on behavior through painstaking longitudinal studies rather than following the easier path of cross-sectional studies whose results provide only indirect inferences regarding organizational change processes.

John Paul Scott
*Professor Emeritus*
*Bowling Green State University*
*Bowling Green, Ohio*

# Acknowledgments

We thank the Center for Applied Science in the School of Science at William Paterson College for support in the preparation of the manuscript for this book. We wish to thank the editorial and production staffs of Oxford University Press, particularly Shelley Reinhardt, Joan Bossert, Louise Chang, and Stan George for their thoroughly professional work and personal encouragement during the preparation of this book.

The symposium that preceded this book was sponsored by the School of Science, the Biopsychology Honors Program, the Department of Biology and the Natural Science Club, all of William Paterson College. Three individuals deserve individual mention for their roles in making the symposium run smoothly: Dean Robert Simpson, Mrs. Carol Wilderson, and Mr. John Rockman. We especially thank all the students of the Biopsychology Honors Program who gave freely of their time and energies.

# Contents

| | |
|---|---|
| *Contributors* | xiii |
| *Introduction:* Integrative Approaches to the Study of Behavior<br>Martin E. Hahn and Robert H. Benno | xv |

**Part I  The Challenge of Developmental Behavior Genetics**  3
Martin E. Hahn

1. A Contextual History of Behavior Genetics  7
   Glayde Whitney
2. Developmental Behavior Genetics: Contributions from the Louisville Twin Study  25
   Adam P. Matheny, Jr.
3. Dual Genesis and the Puzzle of Aggressive Mediation  40
   Robert B. Cairns and Jean-Louis Gariépy
4. Approaches to the Study of Genetic Influence on Developing Social Behavior  60
   Martin E. Hahn

**Part II  Genetic Approaches to the Developing Nervous System**  81
Robert H. Benno

5. Genetic Studies of Brain Development  85
   Cynthia Wimer
6. Genetic and Molecular Analysis of Neural Development and Behavior in *Drosophila*  100
   Jeffrey C. Hall, Shankar J. Kulkarni, Charalambos P. Kyriacou, Qiang Yu, and Michael Rosbash
7. Development of the Nervous System: Genetics, Epigenetics, and Phylogenetics  113
   Robert H. Benno
8. The Evolution of Brain and Body Size: Genetic and Maternal Influences  144
   Larry Leamy

## Part III  Biometrical Approaches to Evolution and Behavioral Development    161
John K. Hewitt

9   Inheritance and the Evolution of Behavioral Ontogenies    167
Stevan J. Arnold

10  Genetic Analysis as a Route to Understanding the Evolution of Animal Behavior: Examples Using the Diallel Cross    190
Norman D. Henderson

11  Changes in Genetic Control during Learning, Development, and Aging    217
John K. Hewitt

12  What Can Adoption Studies Tell Us about Cognitive Development?    236
Robin P. Corley and David W. Fulker

13  Approaches to the Quantitative Genetic Modeling of Development and Age-related Changes    266
Lindon Eaves, John K. Hewitt, Joanne Meyer and Michael Neale

## Part IV  Integration Themes and Future Directions    281
John K. Hewitt

14  Quantitative Genetic Analysis of Neurobehavioral Phenotypes    283
Norman D. Henderson

15  Genetics as a Framework for the Study of Behavioral Development    298
John K. Hewitt

16  Issues of Integration    304
Martin E. Hahn and Robert H. Benno

*Index*    311

# Contributors

**Stevan J. Arnold,** Department of Biology, University of Chicago, Chicago, Illinois 60637

**Robert H. Benno,** Department of Biology, William Paterson College, Wayne, New Jersey 07470

**Robert B. Cairns,** Department of Psychology, University of North Carolina, Chapel Hill, North Carolina 27514

**Robin P. Corley,** Institute for Behavioral Genetics, University of Colorado, Boulder, Colorado 80309

**Lindon Eaves,** Department of Human Genetics, Medical College of Virginia, Richmond, Virginia 23298

**David W. Fulker,** Institute for Behavioral Genetics, University of Colorado, Boulder, Colorado 80309

**Jean-Louis Gariépy,** Department of Psychology, University of North Carolina, Chapel Hill, North Carolina 27514

**Martin E. Hahn,** Department of Biology, William Paterson College, Wayne, New Jersey 07470

**Jeffrey C. Hall,** Department of Biology, Brandeis University, Waltham, Massachusetts 02154

**Norman D. Henderson,** Department of Psychology, Oberlin College, Oberlin, Ohio 44074

**John K. Hewitt,** Department of Human Genetics, Medical College of Virginia, Richmond, Virginia 23298

**Shankar J. Kulkarni,** Department of Biology, Brandeis University, Waltham, Massachusetts 02154

**Charalambos P. Kyriacou,** Department of Biology, Brandeis University, Waltham, Massachusetts 02154

**Larry Leamy,**[*] Department of Biology, California State University, Long Beach, California 90840

**Adam P. Matheny, Jr.,** Department of Pediatrics, University of Louisville, Louisville, Kentucky 40292

[*]Present address: Department of Biology, University of North Carolina, Charlotte, North Carolina 28223

**Joanne Meyer,** Department of Human Genetics, Medical College of Virginia, Richmond, Virginia 23298

**Michael Neale,** Department of Human Genetics, Medical College of Virginia, Richmond, Virginia 23298

**Michael Rosbash,** Department of Biology, Brandeis University, Waltham, Massachusetts 02154

**Glayde Whitney,** Department of Psychology, Florida State University, Tallahassee, Florida 32306

**Cynthia Wimer,** Beckman Research Institute of the City of Hope, Duarte, California 91010

**Qiang Yu,** Department of Biology, Brandeis University, Waltham, Massachusetts 02154

# Introduction: Integrative Approaches to the Study of Behavior

MARTIN E. HAHN AND ROBERT H. BENNO

## BACKGROUND

In 1973 Ronald Wilson published now classic data tracing the mental development of pairs of monozygotic twins from the ages of 6 to 24 months. It is clear from these data (see figure A) that most identical twin pairs are quite similar in their mental development at a given age. This finding is important, but of even greater interest is that the twins vary together in apparent spurts and plateaus of mental development. Thus, Wilson's data show that the genome is integrally involved in changes as well as in continuity of mental functioning over at least the early life of humans.

The primary mediators between genes and behaviors are individual neurons that form complex circuits in the nervous system. Understanding of the functioning of neurons and of developing neural circuits is basic to understanding of developing behavior. A series of innovative experiments in which both in vivo preparations (LeDouarin, 1980) and in vitro preparations (Furshpan, MacLeish, O'Lague, & Potter, 1976; Potter, Landis & Furshpan, 1981; Reichardt & Patterson, 1977) were used showed that neurons that normally become norepinephrine cells can be converted to acetylcholine cells when they are placed in the appropriate environment. Thus the result of interaction between the genome and the environment was an altered expression of the genome with subsequent production of a different phenotype. Although the concept of gene × environment interaction is not new, this particular interaction is defined in a specific manner. Clearly the role of the environment vis-à-vis the genome is instructional. Identifying, defining, and characterizing the types of interactions between genomes and environments will be a rewarding but difficult task in the study of the development of behavior.

Over 100 years ago in *On the Origin of Species,* Darwin wrote about natural selection and proposed a role for it in the evolution of behaviors and the evolution of ontogenies.

> It will be universally admitted that instincts are as important as corporeal structure for the welfare of each species, under its present conditions of life. Under changed conditions of life, it is at least possible that slight modifications of instinct might be profitable to a

species; and if it can be shown that instincts do vary ever so little, then I can see no difficulty in natural selection preserving and continually accumulating variations of instinct to any extent that may be profitable. (p. 209)

... so in a state of nature, natural selection will be enabled to act on and modify organic beings at any age, by the accumulation of profitable variations at that age, and by their inheritance at a corresponding age.... Natural selection may modify and adapt the larvae of an insect to a score of contingencies, wholly different from those which concern the mature insect. These modifications will no doubt affect, through the laws of correlation, the structure of the adult.... So conversely, modifications in the adult will probably often affect the structure of the larvae.... (p. 86)

Here, then, are <u>four observations</u> relevant to the study of behavior. First, the genotype is seen to produce continuity as well as change in the behavior of devel-

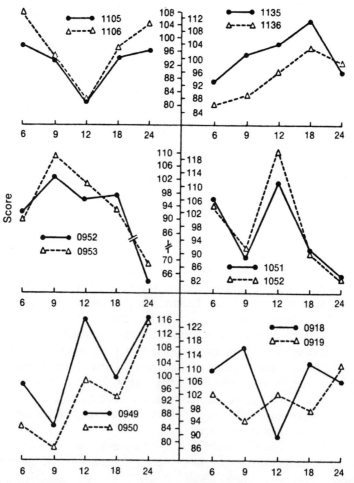

Illustration of identical twin similarities in mental development from ages 6 through 24 months. Strong similarities are observed in five of six pairs for the end point and for spurts and lags in development. (From Wilson, R. S., 1972. Copyright 1972 by the American Association for the Advancement of Science.)

# INTRODUCTION

oping organisms. Second, the environment and the genome interact at the most fundamental level to produce structure and function in the nervous system during development. Third, behavior is subject to the influences of natural selection just as are such characteristics as the color of a bird's feathers or the shape of its bill. Fourth, the forces of natural selection act not just on adults but on organisms throughout their development. These and closely related ideas are the subject of this book and are at the core of a new multidisciplinary field—developmental behavior genetics. Our book is an effort to gain a greater understanding of behavior; more specifically, we hope to facilitate study on the interactive roles of genes, environments, and development in the establishment of behavior. Furthermore, we feel that it is critical for such study to take place within the context of current evolutionary thinking.

What is the best way to study the influences of genes, environments, and development on behavior? Considering the extraordinary scope and complexity of the phenomena, it would seem that investigations that are integrative—combining approaches from genetics and the neurosciences, for example, or from genetics, developmental psychology, and evolutionary biology—would be the methods of choice. Interdisciplinary efforts entail disadvantages as well as advantages, however, and thinking about integrated, multidisciplinary approaches to the understanding of behavior led us to pose a series of practical questions.

> Is it possible to forge productive links between such research areas as the neurosciences and quantitative genetics?
> Would integrated studies be so expensive in resources that, though desirable, they could never be accomplished?
> Would a diverse group of single approaches yield more understanding of behavior than complex integrated ones?

These and similar questions brought us together to organize a symposium, the third in a series of symposia sponsored by William Paterson College in Wayne, New Jersey, with the general theme of the evolution of behavior. (The previous two were published by Hahn and Simmel, 1976, and Hahn, Jensen, and Dudek, 1979.) This symposium was entitled, "Developmental Behavior Genetics in the Evolutionary Context: Theory and Methods." Our purpose in the symposium was to bring together leading scientists in animal and human behavior, the neurosciences, quantitative genetics, and evolutionary theory to engage in face-to-face discussion in search of common ground.

## FOCUSES OF DISCUSSION

### Integration of Development

The first focus of our discussion in this book is to examine various approaches to integrating development into the study of genetic influences in behavior. Though this is not a new idea, research on the topic has been energized recently as investigators have expanded their thinking about the way in which the genome functions (see, for example, Davidson, 1986). Put simply, genes are more than blueprints. The genome is now viewed as interacting with elements of the environment in the formation of the structure of an organism. Pushing this view further, the genome is seen as fundamentally involved in the regulation of the functioning of the constructed organism. Finally, the interchanges between genome and the environment take place all during the developmental lifetime of the organism, and a description of genetic influence on behavior at one age in an organism may not be a good predictor of the genetic influence at another age. We thus think that development is more than just a new dimension of the relationship between genes and behavior—we view it as integral to understanding that relationship.

How can development be described and integrated into our understanding of behavior? Virtually all of the chapters in this book deal with development, and we believe that chapters by Matheny; Hall et al.; Corley and Fulker; and Eaves, Hewitt, Meyer, and Neale illustrate an assortment of fruitful approaches to the study of behavior integrated with development.

## Genes and the Nervous System

The second focus of this book is to consider how genes influence the development of the nervous system and, through the nervous system, behavior. Tracing the pathway from genomic expression to even the simplest behaviors is not easy, though significant progress has been made (see Hall et al., Chapter 6). Moreover, simple behaviors whose neural substrates have been well studied may hold little interest for the psychologist whose focus is human cognitive processes or problems of attention. The chapters by Benno (Chapter 7), Wimer (Chapter 5), Hall et al. (Chapter 6), and Leamy (Chapter 8) provide representative approaches to the study of genetic influence on the structure and function of the nervous system in mice and *Drosophila*. Using these two genetically well-defined animals, these investigators have begun to demonstrate the links between the gene and the nervous system and between the nervous system and behavior. Whether animal studies of this type will ultimately answer questions important to students of complex human behaviors remains to be seen, but it is our hope that the material presented in this book will stimulate further research at the interfaces of genes, the nervous system, development, and behavior.

## Evolution

Much of the research into behavior that has been carried out in the United States has ignored the principles and findings of evolution, though that seems to be changing slowly (Dewsbury, 1978). As illustrated by several chapters in this book, especially those by Arnold (Chapter 9); Cairns and Gariépy (Chapter 3); and Henderson (Chapter 10), evolutionary thinking can enter the discussion of behavior in several different ways. First, as has been well argued by Lorenz (1981), those wishing to understand any biological system are obliged to follow a sequence of steps in their approach. They must describe the system in the broadest possible terms. After this broad description is completed, the investigator is able to take the more analytical steps of identifying the parts of the system and then describing the relationships among the parts. Arnold's chapter in this book illustrates the approach advocated by Lorenz.

Second, there has long been a discussion of the relationship between ontogeny,

# INTRODUCTION

the developmental history of an organism, and phylogeny, the evolutionary history of the organism (Gould, 1977). Darwin argued that development would be subjected to the forces of natural selection and thus ontogenies would have a phylogeny. Cairns (1976) and in his chapter with Gariépy in this book reports on empirical studies of the relationship between ontogeny and phylogeny of a social behavior in mammals.

Third, a theory and method, usually called biometrical genetics (Mather & Jinks, 1982), have developed over the last 25 or so years that allow an investigator to discover the genetic architecture underlying a behavioral trait. That genetic architecture can then be related to the history of the pressures of natural selection that have impinged on that behavior. We feel that the development of behavior is especially amenable to the approaches of biometrical genetics. Henderson presents here a detailed review of the use of the biometrical method to study the behavior of animals.

## "How to"

Finally, we have focused our attention on ways to stimulate integrative, multidisciplinary approaches to the study of behavior. As a result, much of this book, especially Chapters 4, 10, and 12 through 16 deal directly with methodological and quantitative issues. It is our hope that, in part, we are producing a "how to" manual for aspects of developmental behavior genetics.

## INTEGRATED APPROACHES TO BEHAVIOR

Developmental behavior genetics, as seen in this book, covers a range of topics in animals and humans and can be productively articulated with a number of other disciplines. In the inaugural issue of the journal *Behavior Genetics,* Vandenberg and DeFries (1970) listed the disciplines to which behavior genetics could relate; they included animal behavior, anthropology, demography, ecology, ethology, evolution, political science, psychology, psychiatry, and sociology. Behavior genetics, using an integrated developmental approach, could relate to these disciplines and others as well. A number of recent publications illustrate the successful integration of behavior genetics and other disciplines. Especially important among these are Robert Plomin's *Development, Genetics and Psychology* (1986); and a dedicated issue of the journal *Child Development* edited by Plomin (Plomin, 1983); a special, dedicated issue of the journal *Behavior Genetics,* entitled "Multivariate Behavior Genetics and Development" (DeFries & Fulker, 1986); and the report of a symposium, "The Inter-face of Life-History, Evolution, Whole-Organism Ontogeny, and Quantitative Genetics," edited by Stearns (1983).

These successes have led one developmental behavior geneticist (Scarr, 1987) to suggest that the field of behavior genetics has served a useful function and could now be absorbed into related and larger disciplines—for example, developmental psychology. Other views are available. It could easily be argued that the success of behavior genetics in explaining individual differences in human characteristics, such as intelligence and personality, has placed a discipline like developmental psychology on the road to reduction by behavior genetics.

We take a more heuristic view than either of these assessments and are convinced

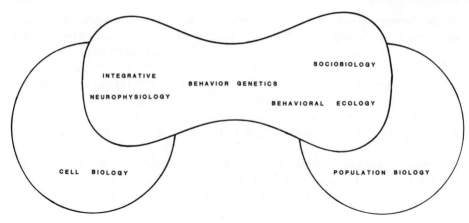

View of the integrated behavioral sciences. (Modified from Wilson, E. O., 1975.)

that developmental behavior genetics, as defined by the chapters that follow, has a clear identity and can serve as a bridge between the population sciences on the one hand and the molecular sciences on the other. This is a view with some similarities to the model proposed by Wilson (1975) for the structure of behavioral and related sciences in the year 2000. Wilson, however, neglected the genetics of behavior in his schema, and so we now offer an amended illustration, indicating our view of the integrated behavior sciences in the year 2000. It is in this spirit that we have written this book.

## REFERENCES

Cairns, R. (1976). The ontogeny and phylogeny of social interactions. In M. Hahn & E. Simmel (Eds.), *Communicative Behavior and Evolution*. New York: Academic Press.
Darwin, C. (1964). *On the Origin of Species: A Facsimile of the First Edition*. Cambridge, MA: Harvard University Press.
Davidson, E. H. (1986). *Gene Activity in Early Development* (3rd ed.). Orlando, FL: Academic Press.
DeFries, J., & Fulker, D. (Eds.). (1986). Multivariate Behavior Genetics and Development [Special Issue]. *Behavior Genetics, 16* (1).
Dewsbury, D. A. (1978). *Comparative Animal Behavior*. New York: McGraw-Hill.
Furshpan, E. J., MacLeish, P. R., O'Lague, P. H., & Potter, D. D. (1976). Chemical transmission between rat sympathetic neurons and cardiac myocytes developing in microcultures: Evidence for cholinergic, adrenergic and dual-function neurons. *Proceedings of the National Academy of Science USA, 73*, 4225–4229.
Gould, S. J. (1977). *Ontogeny and Phylogeny*. Cambridge, MA: Belknap Press.
Hahn, M. E., & Simmel, E. C. (Eds.). (1976). *Communicative Behavior and Evolution*. New York: Academic Press.
Hahn, M. E., Jensen, C., & Dudek, B. (Eds.). (1979). *Development and Evolution of Brain Size: Behavioral Implications*. New York: Academic Press.
LeDouarin, N. M. (1980). The ontogeny of the neural crest in avian embryo chimeras. *Nature, 286*, 663–669.
Lorenz, K. (1981). *The Foundations of Ethology*. New York: Simon and Schuster.

Mather, K., & Jinks, J. L. (1982). *Biometrical Genetics* (3rd ed.). London, Chapman and Hall.
Plomin, R. (Ed.). (1983). Developmental Behavior Genetics [Special Issue]. *Child Development, 54.*
Plomin, R. (1986). *Development, Genetics and Psychology.* Hillsdale, NJ: Lawrence Erlbaum.
Potter, D. D., Landis, S. C., & Furshpan, E. J. (1981). Adrenergic-cholinergic dual function in cultured sympathetic neurons of the rat. In K. Elliot & G. Lawrenson (Eds.), *Development of the Autonomic Nervous System,* CIBA Foundation Symposium #83 (pp. 122–138). London: Pitman Books.
Reichardt, L. F. & Patterson, P. H. (1977). Neurotransmitter synthesis and uptake by isolated sympathetic neurons in microcultures. *Nature, 270,* 147–151.
Scarr, S. (1987). Three cheers for behavior genetics: Winning the war and losing our identity. *Behavior Genetics, 17* (3), 219–228.
Stearns, S. C. (Ed.). (1983). The Inter-face of Life-History Evolution, Whole-Organism Ontogeny, and Quantitative Genetics [Special Issue]. *American Zoologist, 23* (3–4).
Vandenberg, S., & DeFries, J. (1970). Our hopes for behavior genetics. *Behavior Genetics, 1* (1), 1–2.
Wilson, E. O. (1975). *Sociobiology: The New Synthesis.* Cambridge, MA: Belknap Press.
Wilson, R. S. (1972). Twins: Early mental development. *Science, 175,* 914–917.

# Developmental Behavior Genetics

# I

# THE CHALLENGE OF DEVELOPMENTAL BEHAVIOR GENETICS

MARTIN E. HAHN

Behavior genetics, like other interdisciplines, has had a difficult birth. Glayde Whitney in Chapter 1 details the stepwise progression of the discipline from its roots in the ideas of Darwin and Galton through Mendel's theory of particulate inheritance, the beginnings of biometrical theory, and neo-Darwinism to the publication of the first textbook in the field, *Behavior Genetics* (Fuller & Thompson, 1960). The field finally emerged following powerful demonstrations of the influence of genes on behavior by genetic selections for behaviors in animals and in twin studies in humans. The publication of *Behavior Genetics,* the formation of the Behavior Genetics Association in 1970, and publication of the first issue of the journal *Behavior Genetics,* also in 1970, formally launched the discipline. Progress in the field has been slow, however, in part because each step forward seems rife with controversy. The eugenics movement, the nature–nurture debate, and the issue of race and IQ, all have posed difficult questions. Where does behavior genetics fit today in the array of disciplines involving behavior? E. O. Wilson, in his classic text *Sociobiology* (1975), projected the relationships among the behavioral sciences in the year 2000. Interestingly, though sociobiology assumes that genes underlie behavior, Wilson did not include behavior genetics in his flowchart of disciplines. I argue here that behavior genetics—with the integration of development, and touching the neurosciences, and cell biology on the one hand and population biology on the other—provides the necessary interdisciplinary bridge that Wilson needed for his view of the year 2000.

What is the promise of behavior genetics as the core discipline for the study of behavior? The major goal of behavior genetics has been to further understanding of the etiology of individual differences in behavior. This main goal has been subdivided into such general questions as: How do genes influence behavior? How do environments influence the expression of genotypes? Behavior geneticists can provide the integrated knowledge required to answer these questions. For example, how is personality or cognitive ability formed and why do such characteristics often differ substantially from child to child of the same parents living in the same family? Scarr and McCartney (1983) proposed an interesting theory that relates genotype and environment to the question of differences in cognitive and personality traits between children. They argued that the genotype propels a child effectively to "choose" particular interpersonal and cognitive environments by behaving in ways that increase the frequency of some situations and decrease the frequency of others. A baby that pulls away when held is less likely to be cuddled by its mother than one that snuggles.

There are other questions and other approaches. For example, in a biometrical approach, behavior geneticists provide data that speak to the relationship between behavior and evolution. With methods that allow the substitution of one gene for another in a genotype, behavior geneticists produce a greater understanding of complex "gene to the nervous system to behavior pathways." The evidence is in hand—genes clearly are involved in the construction and regulation of behaviors. The question has shifted from *whether* genes influence behavior to *how* genes influence behavior. Even this question is too simplistic. Not only do genes influence behavior, but clearly environments interact with genotypes in the construction of behavioral pathways. Furthermore, genes and environments interact over the life span of organisms to produce continuities and changes in behavior over that life span. Once the study of behavior development over a life span is incorporated into the field of behavior genetics, behavior geneticists will be in a position to answer the most intriguing and potent questions about behavior. That is the promise of developmental behavior genetics. The first part of this book provides the historical context of behavior genetics, discusses some difficult methodological issues, and describes model research programs for human and animal studies in developmental behavior genetics. These chapters begin to fulfill the promise.

In the lead chapter, Whitney chronicles the development of the field of behavior genetics. Trained as a behavior geneticist and serving as an associate editor to the journal *Behavior Genetics* for many years, Whitney is positioned to ask an incisive question. Noting that there has been very slow progress in understanding the relationships among development, evolution, and behavior, Whitney asks why so little is known since Darwin recognized and wrote about those relationships over 100 years ago. The reasons are several, but the dominant ones appear to be political. Darwin agonized as he toppled man from a reserved niche and placed him in the natural lineage. Darwin's half-cousin Galton suggested that eugenics, the advocacy of "good" genes, be applied to man. After 1900, geneticists split on the role of individual characters versus biometrical characters in inheritance and natural selection, and this split is still evident today. With the behaviorist revolution in American psychology, the nature–nurture debate pitted behaviorists, armed with environmental theories, against European ethologists, who opted for the innate control of behavior. Whitney provides us with some insight into why progress in behavior genetics has not been swift. He projects a bright future, however, because the tools of biometrical genetics and molecular biology are now in place to probe genetic and environmental influences on the behavior of developing organisms.

Adam Matheny also presents a history. His is more limited in scope than Whitney's as he details the events that established the Louisville Twin Study as a classic in developmental behavior genetics. The study was begun in 1958-1959 as a long-term longitudinal examination of the physical and behavioral development of twins. I describe the study as classic because of its long duration (almost 30 years), its size (500 pairs of twins), and especially because of the care the investigators use in measurement. From a developmental behavior genetics perspective, Matheny highlights two pervasive findings of the study. First, with increasing age, monozygotic twins come to resemble each other more, while dizygotic twins come to resemble each other less. Second, spurts and lags are observed in the development of twins, with MZ twins exhibiting a remarkably high concordance for developmental events. Evidence provided by Matheny shows that these two important findings hold for physical, cognitive, and temperament development. This evidence indicates the pervasive influence of genes acting along the dimension of age to produce continuities and discontinuities in physical characteristics and behaviors.

Speculation on the possible relationships between ontogeny and phylogeny has a long and sometimes confusing history (Gould, 1977). From a classic interdisciplinary research program, Robert Cairns and Jean-Louis Gariépy report on their investigations into the relationship between ontogeny and phylogeny in social behavior of the mouse. There is a fundamental parallel between the twin studies reported by Matheny and those reported by Cairns and Gariépy—both studies employed the basic paradigm of a successful developmental study—the longitudinal approach. The basis of Cairns and Gariépy's chapter is a successful and rapid genetic selection for aggression in male mice. Follow-up studies showed the selected lines to be differentially aggressive in an age-dependent manner. Large line differences in young adult males were followed by much more modest differences in males of a more advanced age. The search for a mediator of the selection differential and age-dependent effects led to evidence that selection operated on the developmental timetable of elements of aggressive behavior. Having obtained evidence on the relationship between ontogeny and phylogeny for this social behavior in the mouse, Cairns and Gariépy appropriately entitled their chapter "Dual Genesis and the Puzzle of Aggressive Mediation."

In the final chapter of this part, Martin Hahn pursues methodological questions inherent to developmental behavior genetics at large, but particularly those issues relevant to the study of social behaviors. This chapter has as its basis an earlier article on issues in the study of genetic influences on social behavior (Fuller & Hahn, 1976), but it expands on that paper by discussing definitional issues, such as: What is social behavior and what is a social relationship? Central to the chapter is a proposed agenda for research in the genetics of developing social behavior. Agenda items range from the question of how social relationships are established to the question of how individual differences in the establishment of social relationships are related to Darwinian fitness. Hahn describes four model research studies on the development of social behavior in dogs, mice, and humans that deal with various questions from the proposed agenda. Hahn points out that research in developmental social behavior genetics will be difficult and time consuming. His hope is that the discussion of issues as well as the presentation of model studies will assist in energizing and guiding the way to further progress.

## REFERENCES

Fuller, J. L. & Hahn, M. E. (1976). Issues in the genetics of social behavior. *Behavior Genetics, 6,* 391–406.
Fuller, J. L. & Thompson, W. R. (1960). *Behavior Genetics.* New York: John Wiley.
Gould, S. J. (1977) *Ontogeny and Phylogeny.* Cambridge, MA.: Belknap Press.
Scarr, S., & McCartney, K. (1983). How people make their own environments: A theory of genotype–environment effects. *Child Development, 54,* 424–435.
Wilson, E. O. (1975). *Sociobiology: The New Synthesis.* Cambridge, MA: Belknap Press.

# 1
# A Contextual History of Behavior Genetics

## GLAYDE WHITNEY

Broadly considered, most if not all of behavior genetics is implicitly concerned with developmental questions. This is explicitly stated by Fuller and Thompson (1978) in the preface to their *Foundations of Behavior Genetics:* "In our view the major goal of behavior genetics is to increase our understanding of the etiology of individual and group differences" (p. *v*). Thus, a history of developmental behavior genetics is the history of behavior genetics itself. As such it is the joint history of the interactions and amalgamations of evolution, genetics, and psychology. Obviously a brief chapter cannot do justice to the rich history of three very active branches of science in all their intricacies and interactions. The intent of this chapter is much more limited. In the framework of a volume devoted to developmental behavior genetics in an evolutionary context, a fascinating question with a historical perspective is: Why after so long do we know so little? If genetics and evolution are important to an understanding of developmental psychology, then why, after about a century of investigation, are we still just beginning? Insight into this question and current issues necessitates a historical perspective. Thus the intent of this chapter is to provide background about how we come to be where we are today. For this limited purpose, major historical features can be divided into four epochs:

- the pre-Mendelian era (1859–1900) of Darwinian evolution by natural selection and the concomitant beginnings of scientific psychology
- Era of rapid growth of experimental Mendelian genetics and experimental psychology (1900–1930) characterized by an intense "nature–nurture" debate and with a partial eclipse of Darwinian evolution
- Era (1930–1960) during which the dominant themes included separation between a heavily environmental determinist psychology and a developing evolutionary genetics and behavioral biology
- Present epoch (beginning about 1960) of uneven and sometimes uneasy rapproachment and synergism among evolution, genetics, and psychology, of which the present volume is representative

## PRE-MENDELIAN ERA: 1859–1900

Considerations of organic evolution as it is known today started with a book by Charles Darwin published in 1859 under the lengthy title *On the Origin of Species by Means of Natural Selection, or the Preservation of Favoured Races in the Struggle for Life*. There had been many suggestions by others that came close to Darwin's evolutionary theory, but the precursors were inadequate for various reasons. Most of them fell short since they could not account for how evolution occurred, that is, what the mechanism might be. Darwin had the essential insight of "natural selection" as the mechanism.

In the years 1837–1838 Darwin developed his basic theory and he wrote unpublished abstracts of it in 1842 and 1844. After about 20 years of collecting evidence and examples, he finally published *The Origin of Species* in 1859, after Alfred Russell Wallace had independently also discovered natural selection. Traditionally Darwin gets most of the credit because he both (1) had the idea and (2) amassed overwhelming supporting evidence. Many people have pointed out that the idea of natural selection is elegant and profound and basically a very simple notion. Darwin's great genius was to have had the original thoughts in the intellectual climate of his day, and then to have amassed evidence relevant to the theory.

On July 1, 1858, brief statements of the theory by A. R. Wallace and Charles Darwin were read at a meeting of the Linnean Society (Darwin & Wallace, 1859). Interestingly, Darwin recounted in his autobiography that those initial presentations

> excited very little attention, and the only published notice . . . was by Professor Haughton . . . whose verdict was that all that was new in them was false, and what was true was old. This shows how necessary it is that any new view should be explained at considerable length in order to arouse public attention. (F. Darwin, 1958, p. 44)

In *The Origin* Darwin did not dwell on implications for humans. His main reference to humans was a brief paragraph near the end:

> In the distant future I see open fields for far more important researches. Psychology will be based on a new foundation, that of the necessary acquirement of each mental power and capacity by gradation. Light will be thrown on the origin of man and his history. (p. 458)

Darwin's avoidance of discussion of humans in *The Origin* was a calculated decision, to minimize initially the predictable acrimonious debate. Of course, human evolution remains a lively topic today. Darwin's view was (F. Darwin, 1958):

> As soon as I had become, in the year 1837 or 1838, convinced that species were mutable productions, I could not avoid the belief that man must come under the same law. Accordingly I collected notes on the subject for my own satisfaction, and not for a long time with any intention of publishing. Although in the *Origin of Species* the derivation of any particular species is never discussed, yet I thought it best, in order that no honourable man should accuse me of concealing my views, to add that by the work "light would be thrown on the origin of man and his history." It would have been useless, and injurious to the success of the book to have paraded, without giving any evidence, my conviction with respect to his origin.
>
> But when I found that many naturalists fully accepted the doctrine of the evolution of species, it seemed to me advisable to work up such notes as I possessed, and to publish a special treatise on the origin of man. (p. 49)

Darwin's *The Descent of Man and Selection in Relation to Sex* was first published in 1871.

Although *The Origin* had avoided issues of humans, behavioral involvement in evolution loomed large. Indeed, Chapter 7, entitled "Instinct," dealt almost exclusively with behavioral evolution, and the importance of behavior in natural selection generally was emphasized throughout the book. Many accounts in twentieth-century psychology sources suggest that early Darwinian evolution dealt with physical and not behavioral evolution. This suggestion could hardly be further from the truth. Mind–body, structure–function, brain–thought, and other sorts of Cartesian dualism, wherein bodies are subject to evolution while minds are molded by experience—although important impediments in more recent psychology—were not a part of Darwin's legacy.

Darwin's half-cousin, Francis Galton (same grandfather—Erasmus Darwin—different grandmother), although already an eminent scientist in his own right, was not an intimate of Charles Darwin and thus first became aware of Darwinian natural selection following publication of *The Origin*. Galton immediately saw many of the implications for humans and psychology and devoted most of the remainder of his scientific career (Galton was 37 years old in 1859; he lived until 1911) to human evolution, heredity, and behavior. Within a decade Galton had published landmarks (Galton, 1869) dealing with human behavioral heredity, thus establishing himself as the "father of behavior genetics." Of course, in common with most nineteenth-century evolutionists, Galton was not aware of Mendelian genetics. To deal precisely and quantitatively with the "laws of heredity," he originated much of biometrics and invented many of the important statistical approaches still in use, including the notion of regression and correlation (Stigler, 1986). The importance of twins and the distinction between monozygotic and dizygotic twinning in allowing a partial disentangling of the effects of heredity from the effects of common environment was first emphasized by Galton (Galton, 1875), who essentially initiated approaches and concepts that have matured as among the central themes of developmental behavior genetics.

Unfortunately Galton's approach was plagued from the beginning by attempts to combine the science with applications for the benefit of humanity. In 1883 Galton labled the new science *eugenics*. Eugenic was "a brief word to express the science of improving stock," and eugenic questions were "questions bearing on what is termed in Greek *eugenes*, namely, good in stock, hereditarily endowed with noble qualities" (Bajema, 1976). The two interrelated aspects of eugenics—the scientific endeavor and the social/philosophical area of interest—developed together across the late nineteenth century and into the twentieth century. Most of the scientific parts of eugenics are now within what are usually considered to be general genetics and biometrical genetics. The applied humanitarian goal of eugenics was summarized by Galton in 1908:

> Man is gifted with pity and other kindly feelings; he has also the power of preventing many kinds of suffering. I conceive it to fall well within his province to replace Natural Selection by other processes that are more merciful and not less effective.... Natural Selection rests upon excessive production and wholesale distruction; Eugenics on bringing no more individuals into the world than can be properly cared for, and those only of the best stock. (p. 323)

On the science side, eugenics amassed great quantities of data and statistics as investigators tried to understand the laws of heredity, but it was always somewhat controversial with regard to applications. A fundamental problem for Galton and for other nineteenth-century scientists was that they did not know how heredity worked. From the perspective of present knowledge it is easy to fault Galton for not properly appreciating modern distinctions between inherited and acquired influences on individual differences. However, in their pre-Mendelian context his endeavors were actually quite sophisticated. To illustrate the murky (by current standards) status of nineteenth-century appreciation of the laws of heredity we probably cannot do better than to quote from Galton's eminent half-cousin:

> I may take this opportunity of remarking that my critics frequently assume that I attribute all changes of corporeal structure and mental power exclusively to the natural selection of such variations as are often called spontaneous; whereas, even in the first edition of the "Origin of Species," I distinctly stated that great weight must be attributed to the inherited effects of use and disuse, with respect both to the body and mind. I also attributed some amount of modification to the direct and prolonged action of changed conditions of life. Some allowance, too, must be made for occasional reversions of structure; nor must we forget what I have called "correlated" growth, meaning, thereby, that various parts of the organisation are in some unknown manner so connected, that when one part varies, so do the others; and if variations in the one are accumulated by selection, other parts will be modified. (Darwin, 1874/1972, pp. *v, vi*)

As evolution and studies of heredity were progressing during the latter half of the nineteenth century, psychology as an experimental science also began, but from quite separate roots. From centuries-old philosophical questions combined with doses of experimental physics and sensory physiology there emerged an experimental approach to mental philosophy that was called psychology. Wundt is often credited as the founder of psychology, presumably for labeling his endeavors *experimental psychology* in 1879 (Boring, 1957).

Initially, and in important respects to the present day, experimental psychology was mainly concerned with normative questions. That is, typological approaches to modal or average effects in which variation is more often than not viewed simply as error variance to be minimized or eliminated by experimental methods (Stigler, 1986). But of course it is variation, real variation, that provides the raw material for natural selection and that is the subject of most genetic research. James M. K. Cattell is credited with being one of the first influential scientists to combine experimental psychology with an appreciation of individual differences, traveling as a student from Wundt's operation in Germany to Galton's base in England.

From his perspective as an evolutionary theorist, Ernst Mayr suggested that displacement of typological thinking by population thinking may be the greatest conceptual revolution that has taken place in the history of biology (Mayr, 1963, 1982). Such a conceptual revolution may indeed be taking place, but it is far from complete. It has been repeatedly pointed out that many of the recurrent controversies that touch on theory and research in evolutionary and developmental behavior genetics have a dimension that involves typological (modal) versus populational (variational) viewpoints.

By around the turn of the century, evolutionary views were important in some areas of psychology. On the one hand there developed an animal comparative psychology represented, for instance, by the work of Romanes (1970) that included evolutionary notions and has frequently been criticized as being overly anthropomorphically anecdotal. A functional neurology also existed, for which quite a few data concerning innate reflexes and nervous system function were amassed. Some data apparently were on what has turned out to be the right track (e.g., Broca's area), while certain major contemporary developments succumbed to the lure of too-early applied generalizations and degenerated into the applied pseudoscience of phrenology (Boring, 1957; Hilgard, 1987). With regard to the psychology of humans, the work of Galton and Darwin was very important. An evolutionarily grounded theoretical structure involving faculties of the mind and invoking evolved instinctual motives and thought patterns reached a quite high state of development. McDougall's *An Introduction to Social Psychology* (1908) is a marvelous presentation of this tradition, still worth reading. It is as unlike modern social psychology as anything in psychology could be. The era prior to 1900 thus included the initiation and an elaboration of scientific evolution, evolutionary psychology, and individual differences studied with an eye toward natural and artificial selection.

## ERA OF GROWTH OF MENDELIAN GENETICS: 1900-1930

Around the turn of the century, two fundamentally important developments began that were to profoundly impact twentieth-century psychology, evolution, and behavior genetics. One was the origin of experimental approaches to understanding behavioral development which, as learning and conditioning, came to dominate theoretical psychology at least in the United States. The other was, of course, the rediscovery of Mendel's work and the establishment of the generality of Mendelian particulate inheritance. Both of these developments contributed to a turning away of psychology from heredity and evolutionary considerations.

The story of the serendipitous discovery of conditioning by the eminent physiologist Pavlov is well known to students of psychology. Essentially Pavlov was using a nervous system reflex, known to be innate, as part of his experimental procedures to study physiological variables influencing digestion. Accidentally he discovered that such innate, inherited evolutionarily given rigid reflexes were in fact subject to modification through experience. His subsequent research into the conditions under which reflexes were modifiable by learning, and the research of others along this same line, originated the tradition of classic or Pavlovian conditioning (Pavlov, 1927).

At about the same time that Pavlov was initiating classic conditioning, E. L. Thorndike began his laboratory approach to learning using animals as subjects (Thorndike, 1898). Thorndike's efforts originated what became the tradition of instrumental conditioning, with its emphasis on rewards and reinforcement in molding behavior. Both classical and instrumental conditioning held great appeal to investigators interested in application of manipulative scientific methods to psychology. That is, methods were emphasized that involved direct testing of causal notions through manipulation of independent variables in controlled experiments.

The "rediscovery" in 1900 of Mendel's epoch-setting work of the 1860s has fascinated historians and theoreticians ever since. Why did Mendel's work lay dormant after 1865, only to be appreciated and to experience explosively widespread application after 1900? Whatever the reasons (see Dunn, 1965), the theory of particulate inheritance with Mendel's laws and genes (so named by Bateson in 1906) was widely investigated after 1900. Like learning and conditioning, the study of genetics with animals and plants was eminently suited to manipulative experimental design. Genetics attracted the attention of many investigators who preferred controlled experimentation to computation of correlations among existing dependent variables.

Especially in England, a rift soon developed between biometricians and geneticists that was to have a profound influence for decades. To oversimplify a bit, the biometricians following Galton were primarily interested in quantitative measurement of continuously variable phenotypes. Regression and correlation techniques were the primary statistical tools. Biometricians knew that phenotypic variation for many important traits was essentially continuous, and they recognized the suitability of such continuous variation for gradualism in classic Darwinian natural selection. Genetics, dealing as it did with discrete qualitative types, often extreme or abnormal, was obviously concerned with abnormalities of little importance to natural selection.

The early geneticists, on the other hand, thought they knew how heredity worked, and furthermore they could do real experiments. Heredity was discrete, particulate, and not continuous as the biometricians would have it. It followed that the gradualism inherent in the Darwinian concept of evolution by natural selection was wrong. Evolution could occur in a saltatory manner through better adaptation created by new mutations.

The debates became more heated and polemic as time went on. The basis for resolution of the apparent incompatibility between Mendelian particulate inheritance and biometrical continuous phenotypic variation was contained in speculative suppositions in Mendel's original paper. Both East (1910) and Nilsson-Ehle (1908) early developed the basis for multiple gene theory underlying quantitative variation. In 1918, R. A. Fisher laid the basis for quantitative genetics and a complete reconciliation of the biometricians and geneticists with his classic paper entitled "The correlation between relatives on the supposition of Mendelian inheritance" (Fisher, 1918). Interestingly, Fisher had difflculty finding a publisher for that paper (Kevles, 1985). Although the technical and theoretical base for a reconciliation thus existed by about 1910, a widely accepted reconciliation between genetics, biometrics, and evolution was delayed until the 1930s. Indeed, this debate between genetics and biometrics was so fundamental that during the decade of the 1920s acceptance of Darwinian evolution among scientists was at its lowest point ever. Only after 1930 and into the 1940s did evolution by natural selection of genetic variants gradually become widely accepted (Mayr & Provine, 1980).

Contributing to the heat of the polemics were contemporary developments in eugenics. In addition to scientific interest, by the early decades of the twentieth century eugenics had become a popular, and exceedingly controversial, social and political movement. As so often happens, the sociopolitical proselytizers and their opponents outpaced the scientific data (Kevles, 1985).

Support for eugenic notions did not neatly covary with the biometrician-geneticist dimension among scientists. Eminent biometricians such as Galton and Karl

Pearson were among the staunch eugenicists. So were many individuals from the genetics camp. In America, Charles Davenport was a well-known geneticist who supported eugenic social policy. Davenport, along with other early geneticists, is often held up to modern-day ridicule for his credulity in seeing simple Mendelian segregation behind many phenotypic variations of relevance to social interests. In their enthusiasm, some early geneticists did err in the direction of overinterpreting data as favoring single-locus models. Moreover, it is probably important to emphasize that a continuing source of difficulty for geneticists is a tendency to name a locus on the basis of acceptance of a null hypothesis. As Wahlsten (1982) has discussed, it is too easy to fail to reject a single-locus model and then name a nonexistent locus from data that are too weak to differentiate between any number of possible alternative hypotheses. Be that as it may, the "findings" of some early geneticists fed the fires of a sociopolitical eugenics movement.

Then, as now, scientists in a particular discipline were not of a single opinion regarding social application of their subject. Some geneticists were among the outspoken critics of genetically based eugenic proposals. And probably equally important, propagandists could seize upon simple portions of complex papers to further their cause. As an example, in 1917 the eminent geneticist R. C. Punnett presented the now-familiar argument against negative eugenics by pointing out the futility of selection against rare recessive alleles (Punnett, 1917). Punnett's paper was actually a call for research aimed at improving differential diagnosis of heterozygote carriers of recessive feeblemindedness. He mentioned that complete dominance is rare in animal and plant genetics, and suggested that Binet–Simon or other tests might exhibit a lower mentality of heterozygotes (Punnett, 1917). Punnett drew attention to an earlier article by East concerning "hidden feeblemindedness" due to heterozygotes for recessive deleterious alleles. Punnett mentioned Hardy's discussion of the now well-known Hardy-Weinberg equilibrium and, adopting a random breeding model with complete selection against a rare recessive allele, demonstrated how ineffective negative eugenics would be to decrease the incidence of feeblemindedness caused by such an allele. It is probably safe to surmise that all current students of behavior genetics have been led through this genetically based illustration of the futility, and by implication nonsense, of negative eugenics as a social ameliorative tool. However, no less astute a geneticist than R. A. Fisher countered that Punnett's paper had "led to a widespread misapprehension of the effectiveness of selection, either by segregation or sterilization, in purging a population of its feebleminded" (Fisher, 1924, p. 114). Fisher lamented that Punnett had "inadvertently supplied material for anti-eugenic propaganda" (p. 114). He went on to illustrate that by changing two assumptions (random mating and the equation of feeblemindedness with a Mendelian recessive) one could come to quite different conclusions. On the basis of other, perhaps reasonable, assumptions Fisher suggested over a 30 percent reduction in the incidence of feeblemindedness as a result of just one generation of selection. "This is perhaps about what might be expected from an effective policy of segregation, and it is of a magnitude which no one with a care for his country's future can afford to ignore" (Fisher, 1924, p. 115).

During these first decades of the century when lively debates were going on between geneticists and biometricians, eugenicists and antieugenicists, other related and wide-ranging issues flared. Reliable mental testing began and psychometrics became established as an important specialty area of psychology. Following the success of Binet and Simon in France, Terman applied Binet's techniques to develop an American test and in 1916 came up with the "Stanford Revision of the Binet Scale," better known as the Stanford-Binet test. During World War I the army alpha and beta tests were administered to thousands, Terman's IQ test was widely administered, and the "IQ controversy" was off and running. Popular arguments about the validity of mental testing became heated around questions of ethnic/racial differences at a time when high-volume immigration was a volatile political issue. In the popular press, effective polemicists such as Walter Lippmann attacked mental testing. Lippmann was factually wrong and flailing against invented strawmen, according to Terman (Hilgard, 1987). However, at least in the short term, cautious reasoned science, with its uncertainties and caveats, is almost certain to come out second best in public debates not bounded by the canons of science. Recent critics of behavior genetics and mental testing have claimed that early hereditarian mental testers played an important role in influencing passage of restrictive immigration legislation (Gould, 1981; Kamin, 1974). Other authors have maintained that those claims are incorrect (Davis, 1983; Snyderman & Herrnstein, 1983).

With even the basic science approaches to heredity, genetics, biometrics, and evolution in acrimonious disarray, the stage was set for extremist posturing by theoreticians who preferred to purge from their emerging disciplines all taint of eugenic hereditarian notions. Sociology pulled out of the group of evolutionarily based sociobiological disciplines, as did cultural anthropology. Freeman (1983) has provided a very readable account of this era in the formation of cultural anthropology. Under the leadership of Franz Boas, who was by training and outlook non-Darwinian and antieugenics, doctrinnaire polemics proceeded from heredity–environment interactionism to militant environmental determinism for all things cultural.

One of psychology's preeminent polemicists thus found fertile ground for his methodological and theoretical views. From "Psychology as the behaviorist views it" (Watson, 1913) to *Behaviorism* (Watson, 1930), J. B. Watson developed a position that was to be so influential as to lead mainstream psychology to redefine itself from a science of mind or consciousness to a science of behavior. Similarly, because of this influence we now have "behavior genetics" rather than "psychogenetics" or another categorical label.

Watson's methodological emphasis was to attack introspective techniques as inadequate for development of a scientific data base. Instead he advocated measurement of behaviors as the only "objective" route to a public, and therefore scientifically adequate, data base. Experimental methods were emphasized, with the conditioned reflex as a theoretical hinge pin. Conditioning and learned habits were the theoretical elements of the new system, while instincts, faculties of the mind, and associated hereditarian notions were ridiculed as unbelievable products of the earlier nonscientific introspective methods. Watsonian behaviorism became extremely environmentalistic and dualistic. The body (structure) might reflect genetic variation, but behavior (function) was determined by experience.

During these turbulent times, developmental psychology, or child psychology, was also emerging as an identifiable specialty. Essentially from its beginnings it too has been plagued by a confusion of science with sociopolitical preferences. In 1909 G. Stanley Hall announced plans for an institute at Clark University that would conduct science *and* social activism for children. The first child research center actually materialized at Iowa in 1917 (Scarr, Weinberg, & Levine, 1986). Watson conducted research on learning in child development (Watson & Morgan, 1917; Watson & Raynor, 1920) and promoted a learning theory approach to child rearing (Watson, 1928). At about the same time, Gesell was emphasizing the importance of maturation in child development and exploited the method of co-twin control to obtain data concerning the relative importance of maturation and specific training in development (Gesell & Thompson, 1929).

The decades prior to 1930 can thus be characterized as years of substantial theoretical ferment around the ages-old nature–nurture questions. The newly emerging scientific domains of genetics and psychology were individually primitive and jointly uncoordinated. The potential importance of these disciplines for human affairs was obvious to many, but in different ways. During these decades many behavior genetics research endeavors began, which are excellently chronicled elsewhere (e.g., Fuller & Thompson, 1960; McClearn, 1963).

## ERA OF SEPARATION BETWEEN BEHAVIORIST PSYCHOLOGY AND DEVELOPMENTAL BEHAVIOR GENETICS: 1930–1960

The year 1930 is a convenient, if somewhat arbitrary, date at which to set the turning point in the history of developmental behavior genetics. Around 1930 landmark publications appeared that were to be influential in both psychology and evolutionary genetics. With regard to psychology, it was the 1930 revision of Watson's *Behaviorism* that greatly influenced many "behaviorists" and social scientists (Hilgard, 1987). Genetics along with eugenics was effectively deleted from mainstream theory and research in psychology proper. Developmental causal questions in many aspects of psychology dealt with earlier and earlier ontogeny and formative learning experiences. Although individual differences might be traced to ever earlier stages of development, birth became the starting point for many psychologists. Early postnatal experience assumed great theoretical importance because conception (genetic differences) tended not to have a place in learning-dominated child psychology.

Political developments through the 1930s and 1940s contributed to this trend. As national socialism in Germany introduced programs involving first sterilization and euthanasia and then mass killings, in other countries acceptable public and scientific opinion went in the opposite direction. Considerations of genetic bases of individual and group differences in human and animal behavior tended to be received with an assortment of responses that ranged from impolite to insensitive to outrageous violations of taboo. Even today it is not unusual for the epithet "Nazi" to be hurled at any public discussant of behavior genetics. In the Soviet Union under "Lysenkoism," which lasted from the late 1940s through the middle 1960s, all study of genetics, from agricultural applications through psychological investigations, was officially banned (Medvedev, 1971).

In Western psychology this era from 1930 to around 1960 was a period of "great theories." Leaders attempted to construct general theories, which started from a parsimoniously small set of learning postulates and ended in deductive construction of causal frameworks for essentially all of psychology. Lacking an adequate inductive base and ignoring or minimizing genetics and evolution, these grand theories were of course failures. It is in many senses ironic that just as behavioristic psychology was purged of genetics and evolution, these domains were entering an era of rapid development.

The "neo-Darwinian" consolidation, or the "synthetic theory of evolution," is often dated to about 1930 because of the near-simultaneous publication of three independent theoretical syntheses: Fisher's *The Genetical Theory of Natural Selection* (1930), Haldane's *The Causes of Evolution* (1932), and Wright's *Evolution in Mendelian Populations* (1931). Widespread acceptance did not occur overnight, however. Fundamentally important later works included Dobzhansky's *Genetics and the Origin of Species* (1937), Huxley's *Evolution, The Modern Synthesis* (1942), which introduced the term "evolutionary synthesis" (Mayr, 1980), and Simpson's *Tempo and Mode in Evolution (1944)*. Neo-Darwinism consists essentially of a synthesis of genetics (providing the variation) with natural selection (providing the main cause of directional change).

During this time, and progressively to the present, new findings in basic genetics and evolution have accumulated and seem to form an ever-expanding base of solid empirical information and slightly shifting integrative theory. This is in contrast to the status of data and theory in environmental determinist psychology, where changing fads seemed to repeatedly displace out-of-style approaches with a lamentable lack of overall progress.

Many important developments occurring during this phase of course have continuities with earlier developments and exhibit ramifications for present study. Among empirical discoveries, Folling's initial recognition (1934) of a specific physiological condition associated with some familial cases of severe mental retardation was profoundly important. Combined with Penrose's pedigree analyses establishing a Mendelian recessive pattern of inheritance (1935) and Garrod's earlier (1909) notions concerning inborn errors of metabolism, the phenomenon of phenylketonuria provided a model for many later discovered, genetically determined causes of severe mental anomaly.

Chromosomal cytology came into its own and, by 1960, was beginning to have an impact on theory in human developmental psychology. Knowledge of the relationship between chromosomes and nuclear Mendelian genes developed early (Sutton, 1902, 1903). But detailed information concerning mammalian, including human, chromosomes was slow to accumulate because of technical problems. Only in 1956 was the normal human karyotype established as consisting of 46 chromosomes. By 1959 aneuploidy had been established as being causally related to some developmental anomalies with both behavioral and physiological/morphological symptoms. Down's syndrome, as well as Kleinfelter's syndrome and Turner's syndrome, was attributed to aneuploidy (Vogel & Motulsky, 1979). Molecular genetics, including the biochemistry of chromosomes, developed during this phase as well. Important landmarks of cumulative knowledge include the identification of DNA as hereditary material (Avery, MacLeod, & McCarty, 1944), the architecture of the double-helix model involving structure and replication (Watson & Crick, 1953), and discovery of the triplet code (Nirenberg & Matthei, 1961).

Advances in theory and methods of biometrical or quantitative genetics occurred that were to be fundamentally important in application to psychological phenotypes.

These developments actually began coalescing by the 1920s and led, as population genetics, to the neo-Darwinian revolution of the 1930s. As commonly understood, the population genetics that formed the basis for the synthetic theory of evolution dealt mainly with the properties of single Mendelian loci influencing traits in "populations" larger than families. Quantitative genetics also deals with populations but goes further by dealing with multiple loci (polygenic theory) and continuously variable phenotypes. As already mentioned, Galton and his followers had attempted to deduce the laws of heredity from correlational data of metric traits. Fisher (1918), in establishing the theoretical basis of quantitative genetics, essentially reversed the approach: From the properties of Mendelian genetics he deduced the correlational structure of metric phenotypes. After Fisher, Wright, and Haldane, many investigators further refined polygenic theory–quantitative genetics and, often in applied plant and animal breeding contexts, developed both methods and concepts. Thus Lush (1940) introduced the technical notion of heritability, while Hazel (1943) developed an approach to genetic correlations. Mather (1949) published his *Biometrical Genetics,* and the first edition of Falconer's very influential text, *Introduction to Quantitative Genetics,* was published in 1960.

A wide variety of conceptual clarifications gradually emerged, which were important in setting the stages for empirical advances in developmental behavior genetics. One recurrent difficulty in the developmental context is the dual application of the notion of heritable influence. On the one hand, in a typological mode, a trait was said to be likely heritable if it is resistant to environmental modification and if it is phenotypically relatively invariant among individuals. Maturational stages in development, whether Gesselian or Piagetian, are often traits of this sort. Such traits are deeply developmentally canalized in Waddington's sense. On the other hand, in a variational view, a trait is said to be heritable if individual differences are based to some extent on genetic differences among the indivduals. The latter meaning of heritable, being genetic and dealing with causes of variation, is of course fundamental to developmental behavior genetics. The former meaning, being phenotypic and typological, in the hands of nongenetic developmentalists leant itself to the claim that the nature-versus-nurture problem was an unanswerable nonquestion. After all, both nature and nurture are inextricably interwoven and necessary for the development of an individual. The second meaning, the populational-variational view, led to the behavioral genetic approach of investigation of causes of individual differences, including study not only of nature versus nurture but also of the nature of nurture effects. That is, with the populational view investigators escaped from the behavioristic learning-versus-genetics issue by investigating the genetics of learning itself.

Another important conceptual clarification was Tinbergen's emphasis on the mutual legitimacy of different approaches and different interests in the investigation of any biological phenomenon, including behavior. He first enumerated the three major problems of (1) causation, (2) adaptiveness, and (3) evolution (Tinbergen, 1951), and later separated causation into immediate aspects (physiological, environmental stimuli, etc.) and developmental-ontogenetic aspects (Tinbergen, 1963). Although not a unique insight, the clear delineation of different questions or problems, of differential degrees of interest to different theoreticians and investigators, did much to end the not very useful arguments between investigators with different interests. The distinction between proximate (immediate causation) and ultimate (adaptiveness, selective advantage) questions and their mutual interest is another version of this recognition of theoretical breadth and tolerance (Mayr, 1982). Rather than having sterile arguments at cross-purposes, many investigators at the present time are of the view that a satisfactory "understanding" of any behavioral phenomenon will include an integration of knowledge from each of the four perspectives. This, of course, is the spirit of this book, which addresses genetic approaches to the four questions in an integrative manner.

It has been emphasized here that genetics was not an important aspect of mainstream behaviorist psychology during this time frame. However, although not considered important (or polite) by many, behavior genetics investigations and theorizing did not cease altogether. A steady stream of research from mavericks such as J. Paul Scott and John Fuller emphasized the importance of genetics to an understanding of behavioral development (e.g., Fuller, 1948, 1951, 1954; Scott, 1936, 1942, 1943, 1949; Scott & Fuller, 1951). Reviews noteworthy for their position and influence include Woodworth's (1941) statement of the joint importance of heredity and environment and Hall's (1951) chapter entitled "The genetics of behavior." This era in the history of behavior genetics can be said to have ended, and the current era of behavior genetics as an identified specialty area to have begun, in 1960. The landmark event was the publication of Fuller and Thompson's (1960) *Behavior Genetics,* a compendium of previous work with over 800 references.

## THE MODERN EPOCH: 1960-PRESENT

Following Fuller and Thompson (1960), behavior genetics became an identified specialty area that, in a decade, spawned both a dedicated journal (*Behavior Genetics,* volume 1 in 1970) and a scientific organization (Behavior Genetics Association, first annual meeting in 1970). The field is an odd hybrid that acquired a coherent identity mainly in contrast to the previous eras' relative neglect of genetic influences on behavioral phenotypes. The discipline rather quickly passed through a phase of straightforward descriptive reports from both animal and human research. Many of us who are fascinated by the etiology of individual differences continue to be interested in reports of yet more phenotypes that display genetically influenced variation. However, given that the general (or universal) point has been made (contrary as it may be to environmental determinist theory and philosophy), such further demonstrations in themselves are now akin to stamp collecting and form but the beginning of scientifically interesting endeavors. Behavior genetics as a coherent specialty area of scientific investigation may soon disappear as a consequence of its success. Analogous to a widespread successful species in phylogenesis, it is giving rise to a number of separate lineages, each thriving in a different conceptual ecological framework. The genus characteristic is primarily an application of genetic methods to problems with a behavioral dimension. Behavior genetics investigations with both human and animal subjects are addressing in various combinations each of the "questions" or approaches delineated by Tinbergen (1963). Scarr and Kidd (1983) provided a recent overview specifically in the developmental context.

Substantial advances have occurred during this most recent era in both theory and methods. With regard to evolution, W. D. Hamilton (1964) codifed the concept of inclusive fitness as an adjunct to the population genetics mathematical definition of Darwinian (individual) fitness in the context of discussions of group and kin selection. With an emphasis on behavioral evolution, Hamilton's theoretical developments contributed to the emergence of so-called sociobiology (Wilson, 1975). Sociobiology, according to some of its first-generation practitioners, is nothing more nor less than the application of neo-Darwinian evolutionary concepts to the domains of behavior, including social behavior (Whitney, 1979). There was much that was not new to behavior geneticists in sociobiology; even the label had a long history. "Sociobiological systematics" was part of a title used by J. P. Scott in 1950. Beginning in 1952, Scott was chairman of a "Committee on Animal Behavior and Sociobiology" within the Ecological Society of America. From that committee developed first a Section of Animal Behavior and Sociobiology and then the Animal Behavior Society (Guhl & Schein, 1976). Although general statements have tended to be quite typological at a species level, there are obvious interconnections between behavior genetics and recent sociobiology (Fuller, 1978; Rushton, 1985; Rushton, Littlefield, & Lumsden, 1986; Thiessen, 1979).

A perennial problem with evolutionary hypotheses concerning past or present selection pressures is the ease of positing, and difficulty of testing, various scenarios. This difficulty had contributed to earlier "behaviorists" turning away from evolutionary notions. More recently and to the present time, many investigators of both human and animal behavior genetics disagree with various sociobiological suggestions. Many scientifically fertile differences exist even among current sociobiological approaches (e.g., compare Alexander, 1979, and Lumsden & Wilson, 1981). Sociobiology as an approach to understanding human and animal behavior also received very strong polemic attacks from the same camp of sociopolitical and philosophical environmental determinists that traditionally attacks behavior genetics endeavors. Indeed, during this most recent era, behavior genetics has once again become a focus of contention. The current round of philosophical discord arose largely in the context of world wide civil rights issues involving questions related to race. Political philosophers pursuing "equality" unfortunately early tied their movements to a typological framework espousing either an absence of group differences or an environmental determination of any phenotypic group differences that might be proven to exist. Hence the entire enterprise of behavior genetics was anathema in their view. Among practitioners of behavior genetics the responses have covered a wide range, from advocacy and practice of research addressing questions of the etiology of race differences (e.g., Jensen, 1973; Scarr, 1981) through a practice of explicit voluntary censorship to attacking any juxtaposition of behavior genetics with questions of racial group differences (Hirsch, 1981). In spite of the polemics, behavior genetics has experienced remarkable progress, at least in areas other than those explicitly tied to race. This recent progress has been primarily through applications of genetic, molecular physiological, biometric, and general technological advances to behavioral domains.

Vogel and Motulsky (1979) suggested that the development of human cytogenetics after 1956 has the characteristics of a scientific revolution in the Kuhnian sense. Chromosome banding techniques became available, which allowed individual recognition of differentiated portions of each human chromosome. By the 1970s cytogenetics had become the most popular branch of human clinical genetics (Vogel & Motulsky, 1979). As banding methods have rapidly become more sophisticated, their widespread application is suggesting ever more causal links to behavioral variants.

A combination of newly developed molecular techniques with traditional Mendelian analyses of extended pedigrees is rapidly extending knowledge of phenotypes for which patterns of single-locus segregation can be discerned. As an example, markers established through restriction enzyme site polymorphisms, combined with somatic cell hybridization techniques in application with extended pedigrees, led to chromosomal localization of a marker for Huntington's disease (Gusella et al., 1983). A similar approach may have recently discovered a locus on chromosome 11 causal to some cases of affective disorder (Egeland et al., 1987).

Biometrical genetic approaches to quantitative phenotypes have entered a new era of sophisticated application. The advent and ever-widening availability of computers since about the 1960s has for the first time rendered feasible large-scale complex modeling. Univariate analytical models are being extended to, or are giving way to, multivariate techniques. The methods of path coefficients was developed in 1918–1921 by Sewell Wright in the context of population genetics, but only since the mid 1960s has it been widely employed in other disciplines (Li, 1977). Recently in the behavior genetics context, path analytical techniques have been extended from univariate to bivariate and general multivariate models (Vogler & DeFries, 1986).

Multivariate analyses with regard to human behavior genetics were first reported by Loehlin (1965) and Vandenberg (1965). Such endeavors quickly advanced in sophistication, especially since the early 1970s, with contributions to multivariate modeling by Eaves and others (DeFries & Fulker, 1986a). At present the forefront of biometrical behavior genetics is represented by large-scale multivariate investigations of developmental questions, in the context of carefully designed longitudinal studies. In addition to this book, such endeavors have recently been presented in dedicated issues of the journals *Child Development* (Plomin, 1983), and *Behavior Genetics* (DeFries & Fulker, 1986b).

This chapter began with the rhetorical question: Why after so long do we know so little? Now in a historical context a general answer is perhaps obvious. On the one hand, knowledge concerning developmental behavior genetics, and in an evolutionary context, has depended on acquisition of theory and methods in basic genetics, biometrics, and evolution. Many of these enabling developments have become available only rather recently. On the other hand, and antedating current methods, philosophical positions relevant to nature–nurture questions in combination with sociopolitical movements have tended to retard free and open discussion and scientific investigation of behavior genetics. We can only hope, along with many other commentators (e.g. Davis, 1975; Koshland, 1987), that realistic appraisal of complex phenomena based on data will supplant overly simplistic and frankly incorrect notions.

## ACKNOWLEDGMENT

Preparation supported in part by Grant NS15560 from the National Institute of Neurological and Communicative Disorders and Stroke.

## REFERENCES

Alexander, R. (1979). *Darwinism and Human Affairs.* Seattle: University of Washington Press.
Avery, O. T., MacLeod, C. M., & McCarty, M. (1944). Studies on the chemical nature of the substance inducing transformation of pneumococcal types. *Journal of Experimental Medicine, 79,* 137–158.
Bajema, C. J. (Ed.). (1976). *Eugenics Then and Now.* Stroudsburg, PA: Dowden, Hutchinson & Ross.
Boring, E. G. (1957). *A History of Experimental Psychology* (2nd ed). New York: Appleton-Century-Crofts.
Darwin, C. (1859). *On the Origin of Species by Means of Natural Selection, or the Preservation of Favoured Races in the Struggle for Life.* London: John Murray. (Reprint of 1st ed., 1979, New York: Avenel Books)
Darwin, C. (1874/1972). Preface, *The Descent of Man and Selection in Relation to Sex* (2nd ed., dated September 1874). New York: Appleton (1896); New York: AMS Press, 1972.
Darwin, C., & Wallace, A. (1859). On the tendency of species to form varieties; and on the perpetuation of varieties and species by natural means of selection. *Journal of the Proceedings of the Linnean Society (Zoology), 3,* 45–62.
Darwin, F. (Ed.). (1958). *The Autobiography of Charles Darwin and Selected Letters.* New York: Dover. (Original work published 1892)
Davis B. D. (1975). Social determinism and behavioral genetics. *Science, 189.* 1049.
Davis, B. D. (1983). Neo-Lysenkoism, IQ, and the press. *The Public Interest, 73,* 41–59.
DeFries, J. C., & Fulker, D. W. (1986a). Multivariate behavioral genetics and development: An overview. *Behavior Genetics, 16,* 1–10.
DeFries, J. C., & Fulker, D. W. (Eds.) (1986b). Symposium on multivariate behavioral genetics and development. *Behavior Genetics, 16,* 1–235.
Dobzhansky, T. (1937). *Genetics and the Origin of Species.* New York: Columbia University Press.
Dunn, L. C. (1965). *A Short History of Genetics.* New York: McGraw-Hill.
East, E. M. (1910). A Mendelian interpretation of variation that is apparently continuous. *American Naturalist, 44,* 65–82.
Egeland, J. A., Gerhard, D. S., Pauls, D. L., Sussex, J. N., Kidd, K. K., Allen, C. R., Hostetter, A. M., & Housman, D. E. (1987). Bipolar affective disorders linked to DNA markers on chromosome 11. *Nature, 325,* 783–787.
Falconer, D. S. (1960). *Introduction to Quantitative Genetics.* New York: Ronald Press.
Fisher, R. A. (1918). The correlation between relatives on the supposition of Mendelian inheritance. *Transactions of the Royal Society of Edinburgh, 52,* 399–433.
Fisher, R. A. (1924). The elimination of mental defect. *Eugenics Review, 16,* 114–116.
Fisher, R. A. (1930). *The Genetical Theory of Natural Selection.* London: Oxford University Press.
Folling, A. (1934). Uber Ausscheidung von Phenylbrenztraubensaure in den Harn als Stoffwechselanomalie in Verbindung mit Imbezzillitat. *Zeitschrift für Physiologische Chemie, 227,* 169–176. Quoted in Knox, W. E. (1972). Phenylketonuria. In J. B. Stanbury, J. B. Wyngaarden, & D. S. Fredrickson (Eds.), *The Metabolic Basis of Inherited Disease.* New York: McGraw-Hill (pp. 266–295).
Freeman, D. (1983). *Margaret Mead and Samoa: The Making and Unmaking of an Anthropological Myth.* Cambridge MA: Harvard University Press.
Fuller, J. L. (1948). Individual differences in the reactivity of dogs. *Journal of Comparative and Physiological Psychology, 41,* 339–347.

Fuller, J. L. (1951). Gene mechanisms and behavior. *American Naturalist, 85,* 145–157.
Fuller, J. L. (1954). *Nature and Nurture: A Modern Synthesis.* New York: Doubleday.
Fuller, J. L. (1978). Genes, brains, and behavior. In M. S. Gregory, A. Silvers, & D. Sutch (Eds.), *Sociobiology, and Human Nature.* San Francisco: Jossey-Bass (pp. 98–115).
Fuller, J. L., & Thompson, W. R. (1960). *Behavior Genetics.* New York: John Wiley.
Fuller, J. L., & Thompson, W. R. (1978). *Foundations of Behavior Genetics.* St. Louis: C. V. Mosby.
Galton, F. (1869). *Hereditary Genius: An Inquiry Into its Laws and Consequences.* London: MacMillan.
Galton, F. (1875). The history of twins as a criterion of the relative powers of nature and nurture. *Journal of the Anthropological Institute, 6,* 391–406.
Galton, F. (1908). *Memories of My Life.* London: Methuen (3rd ed., April 1909).
Garrod, A. E. (1909). *Inborn Errors of Metabolism.* London: Oxford University Press.
Gesell, A. L., & Thompson, H. (1929). Learning and growth in identical infant twins: An experimental study by the method of co-twin control. *Genetic Psychology Monographs, 6,* 5–124.
Gould, S. J. (1981). *The Mismeasure of Man.* New York: Norton.
Guhl, A. M., & Schein, M. W. (1976). The Animal Behavior Society: Its Early History and Activities (pamphlet). Morgantown, WV: M. W. Schein, A.B.S. Society Historian, pp. 1–66.
Gusella, J. F., Wexler, N. S., Conneally, P. M., Naylor, S. L., Anderson, M. A., Tanzi, R. E., Watkins, P. C., Ottina, K., Wallace, M. R., Sakaguchi, A. Y., Young, A. B., Shoulson, I., Bonilla, E., & Martin, J. B. (1983). A polymorphic DNA marker genetically linked to Huntington's disease. *Nature, 306,* 234–238.
Haldane, J. B. S. (1932). *The Causes of Evolution.* London: Longmans, Green.
Hall, C. S. (1951). The genetics of behavior. In S. S. Stevens (Ed.), *Handbook of Experimental Psychology.* New York: Wiley (pp. 304–329).
Hamilton, W. D. (1964). The genetical theory of social behaviour, I, II. *Journal of Theoretical Biology, 7,* 1–52.
Hazel, L. N. (1943). The genetic basis for constructing selection indexes. *Genetics, 28,* 476–490.
Hilgard, E. R. (1987). *Psychology in America: A Historical Survey.* New York: Harcourt Brace Jovanovich.
Hirsch, J. (1981). To "unfrock the charlatans." *Sage Race Relations Abstracts, 6,* 1–65.
Huxley, J. S. (1942). *Evolution, The Modern Synthesis,* London: Allen & Unwin.
Jensen, A. R. (1973). *Educability and group differences.* New York: Harper & Row.
Kamin, L. J. (1974). *The Science and Politics of I.Q.* Potomac, MD: Lawrence Erlbaum.
Kevles, D. J. (1985). *In the Name of Eugenics: Genetics and the uses of human heredity.* New York: Alfred A. Knopf.
Koshland, D. E. (1987). Nature, nurture, and behavior. *Science, 235,* 1445.
Li, C. C. (1977). *Path Analysis—a Primer.* Pacific Grove, CA: Boxwood Press.
Loehlin, J. C. (1965). A heredity-environment analysis of personality inventory data. In S. G. Vandenberg (Ed.), *Methods and Goals in Human Behavior Genetics.* New York: Academic Press (pp. 163–170).
Lumsden, C., & Wilson, E. O. (1981). *Genes, Mind, and Culture.* Cambridge, MA: Harvard University Press.
Lush, J. L. (1940). Intra-sire correlations or regressions of offspring on dam as a method of estimating heritability of characteristics. *Thirty-third Annual Proceedings of the American Society of Animal Production,* pp. 293–301.
Mather, K. (1949). *Biometrical genetics: The study of continuous variation.* London: Methuen.

Mayr, E. (1963). *Animal Species and Evolution.* Cambridge, MA: Belknap-Harvard University Press.

Mayr, E. (1980). Prologue: Some thoughts on the history of the evolutionary synthesis. In E. Mayr & W. B. Provine (Eds.), *The Evolutionary Synthesis.* Cambridge, MA: Harvard University Press (pp. 1–48).

Mayr, E. (1982). *The Growth of Biological Thought.* Cambridge, MA: Belknap-Harvard University Press.

Mayr, E., & Provine, W. B. (Eds.). (1980). *The Evolutionary Synthesis. Perspectives on the Unification of Biology.* Cambridge, MA: Harvard University Press.

McClearn, G. E. (1963). The inheritance of behavior. In L. Postman (Ed.). *Psychology in the Making. Histories of Selected Research Problems.* New York: Alfred A. Knopf (pp. 144–252).

McDougall, W. (1908). *An introduction to social psychology.* London: Methuen.

Medvedev, Z. A. (1971). *The rise and fall of T. D. Lysenko* (I. M. Lerner, Trans.). Garden City, NY: Anchor-Doubleday. (Original work published 1969)

Nilsson-Ehle, H. (1908). Einige Ergebnisse von Kreuzungen bei Hafer und Weisen. *Botanische Notiser,* 1908–1909, 257–294.

Nirenberg, M. W., & Matthei, J. H. (1961). The dependence of cell-free protein synthesis in *E. coli* upon naturally occurring or synthetic polyribonucleotides. *Proceedings of the National Academy of Science, 47,* 1588–1602.

Pavlov, I. P. (1927). *Conditioned Reflexes* (G. V. Anrep, Trans.). London: Oxford University Press.

Penrose, L. S. (1935). Inheritance of phenylpyruvic amentia. *Lancet, 2,* 192–194.

Plomin, R. (1983) (Guest Ed.). Special section on developmental behavioral genetics. *Child Development, 54,* 253–435.

Punnett, R. C (1917). Eliminating feeblemindedness. *Journal of Heredity, 8,* 464–465.

Romanes, G. J. (1970). *Animal Intelligence.* London: Kegan Paul, Trench. (Original work published 1882)

Rushton, J. P. (1985). Differential K theory: The sociobiology of individual and group differences. *Personality, Individual Differences, 6,* 441–452.

Rushton, J. P., Littlefield, C. H., & Lumsden, C. J. (1986). Gene-culture coevolution of complex social behavior: Human altruism and mate choice. *Proceedings of the National Academcy of Sciences, 83,* 7340–7343.

Scarr, S. (1981). *Race, Social Class and Individual Differences.* Hillsdale, NJ: Lawrence Erlbaum.

Scarr, S., & Kidd, K. K. (1983). Developmental behavior genetics. In M. M. Haith & J. J. Campos (Eds.), *Handbook of Child Psychology. Vol. II. Infancy and Developmental Psychobiology.* New York: Wiley (pp. 345–433).

Scarr, S., Weinberg, R. A., & Levine, A. (1986). *Understanding Development.* New York: Harcourt Brace Jovanovich.

Scott, J. P. (1936). Inherited behavior in *Drosophila. Proceedings of the Indiana Academy of Science, 46,* 211–216.

Scott, J. P. (1942). Genetic differences in the social behavior of inbred strains of mice. *Journal of Heredity, 33,* 11–15.

Scott, J. P. (1943). Effects of single genes on the behavior of *Drosophila. American Naturalist, 77,* 184–190.

Scott, J. P. (1949). Genetics as a tool in experimental psychological research. *American Psychologist, 4,* 526–530.

Scott, J. P. (1950). The social behavior of dogs and wolves: An illustration of sociobiological systematics. *Annals of the New York Academy of Science, 51,* 1009–1021.

Scott, J. P., and Fuller, J. L. (1951). Research on genetics and social behavior at the Jackson Laboratory, 1946-51: A progress report. *Journal of Heredity, 42,* 191-197.

Simpson, G. G. (1944). *Tempo and Mode in Evolution.* New York: Columbia University Press.

Snyderman, M., & Herrnstein, R. J. (1983). Intelligence tests and the immigration act of 1924. *American Psychologist, 38,* 986-995.

Stigler, S. M. (1986). *The History of Statistics.* Cambridge, MA: Harvard University Press.

Sutton, W. S. (1902). On the morphology of the chromosome group of *Brachystola magna. Biological Bulletin, 4,* 24-39.

Sutton, W. S. (1903). The chromosomes in heredity. *Biological Bulletin, 4,* 231-251.

Thiessen, D. (1979). Biological trends in behavior genetics. In J. R. Royce & L. P. Mos (Eds.), *Theoretical Advances in Behavior Genetics.* Germantown, MD: Sijthoff-Noordhoff (pp. 169-212).

Thorndike, E. L. (1898). Animal intelligence: An experimental study of the associative processes in animals. *Psychological Review, Monograph Supplements, 2 (#8).*

Tinbergen, N. (1951). *The Study of Instinct.* New York: Oxford University Press. (Reprinted 1974)

Tinbergen, N. (1963). On the aims and methods of ethology. *Zeitschrift für Tierpsychologie, 20,* 410-433.

Vandenberg, S. G. (1965). Multivariate analysis of twin differences. In S. G. Vandenberg (Ed.), *Methods and Goals in Human Behavior Genetics.* New York: Academic Press (pp. 29-44).

Vogel, F., & Motulsky, A. G. (1979). *Human genetics. Problems and approaches.* New York: Springer-Verlag.

Vogler, G. P., & DeFries, J. C. (1986). Multivariate path analysis of cognitive ability measures in reading-disabled and control nuclear families and twins. *Behavior Genetics, 16,* 89-106.

Wahlsten, D. (1982). Genes with incomplete penetrance and the analysis of brain development. In I. Lieblich (Ed.), *Genetics of the Brain.* New York: Elsevier Biomedical Press (pp. 367-391).

Watson, J. B. (1913). Psychology as the behaviorist views it. *Psychological Review, 20,* 158-177.

Watson, J. B. (1928). *Psychological Care of Infant and Child.* New York: Norton.

Watson, J. B. (1930). *Behaviorism* (rev. ed.). New York: Norton. (Original work published 1924)

Watson, J. B., & Morgan, J. J. B. (1917). Emotional reactions and psychological experimentation. *American Journal of Psychology, 28,* 163-174.

Watson, J. B., & Raynor, R. (1920). Conditioned emotional reactions. *Journal of Experimental Psychology, 3,* 1-14.

Watson, J. D., & Crick, F. H. C. (1953). The structure of DNA. *Cold Spring Harbor Symposium on Quantitative Biology, 23,* 123-131.

Whitney, G. (1979). The second century of the Darwinian revolution. *Contemporary Psychology, 24,* 682-685.

Wilson, E. O. (1975). *Sociobiology. The New Synthesis.* Cambridge, MA: Belknap-Harvard University Press.

Woodworth, R. S. (1941). *Heredity and Environment: A Critical Survey of Recently Published Material on Twins and Foster Children.* New York: Social Science Research Council, Bulletin No. 47.

Wright, S. (1931). Evolution in Mendelian populations. *Genetics, 16,* 97-158.

# 2
# Developmental Behavior Genetics: Contributions from the Louisville Twin Study

ADAM P. MATHENY, JR.

The Louisville Twin Study was instituted by Frank Falkner, a pediatrician who had previously organized and coordinated international studies of child development at seven university centers: London, Paris, Zurich, Brussels, Kampala, Dakar, and Stockholm. The center at the University of Louisville, like the other centers, was organized to follow the physical and behavioral development of children from infancy on. Unlike studies at the other centers, however, at Louisville the Twin Study emphasized the longitudinal study of twins as a first step in determining genetic and environmental contributions to growth and development (Falkner, 1957).

A pilot study initiated in 1957 established the core elements of the Louisville Twin Study. During 1958–1959, the first infant twin pairs were recruited into a longitudinal program that now, some 30 years later, includes an active sample of more than 500 pairs of twins.

Falkner's primary interest was in the physical development of children, as reflected by his more recent contributions (e.g., Falkner & Tanner, 1986). All of the international centers were encouraged, however, to assess the behavioral development of children as well. Therefore, even though appraisals of growth were the common connection among the longitudinal programs, each center initiated more or less systematic assessments of the unfolding behaviors of children.

The Louisville Twin Study, more than the other international centers, responded to this encouragement because of the research interests of two psychologists—Steven Vandenberg, who directed the study from 1964 until 1967, and Ronald Wilson, who directed the study from 1967 until his untimely death in 1986. Under their leadership the Louisville Twin Study continued to obtain the fundamental measures of physical growth but expanded the program to include a wide range of assessments of intelligence, personality, academic skills, and temperament.

Within the span of 30 years, the research reports issuing from the Louisville Twin Study have represented a series of analyses of a steadily increasing data base on twins' physical and behavioral development. Although the reports differ according to the age ranges concerned, the characteristics assessed, and the genetic analyses performed, the abiding preoccupation has been the behavior genetic analyses of longitudinal data. This preoccupation is the major concern of a new interdiscipline, developmental behavior genetics, that fosters an integration of behavior genetics and developmental psychology.

A chapter devoted to the Louisville Twin Study is not a "summing up." Rather it is a "summing forward," providing an overview of the main research efforts to date, yet recognizing that the longitudinal behavioral studies of twins are not completed.

The organization of this chapter is arbitrary in the sense that the chronological sequence of research reports was ignored in favor of the major concern regarding behavioral development as genetically influenced. Framed by this concern, the chapter uses the physical development of twins as a prototype for behavioral development. The longitudinal studies of twin's mental development and, more recently, temperament are then summarized to illustrate the same major concern. The unifying theme throughout is the growing recognition that there may be a genetic influence on change as well as stability of behavior throughout development. To that end, the references to published research by the Louisville Twin Study are selective, not exhaustive, and the references to other investigators' important studies involving different methods of analyses are kept to a minimum. For a wider sampling of these reports, one can find references elsewhere (e.g., Goldsmith, 1983; Plomin, 1986; Rowe, 1987).

## THE RESEARCH PROGRAM

### Physical Development

The study of twins' physical growth serves as an ideal model for indirect examination of genesystems guiding the pathways of growth. Starting with the assessment of physical status at birth, one expects physical growth to accelerate rapidly in the early years of childhood, proceed at a slower pace in middle childhood, accelerate again during adolescence, and slow to zero during the early adult years. The growth rate is not uniform for any given child, however; there are periods of acceleration and lag. Each child has a distinctive pattern—a growth signature—that is not duplicated by any other child taken at random.

These individualized patterns of growth may give the impression that the bodily processes are erratic. If, however, there is a coherent plan, and if that plan is regulated by timed genetic processes (a chronogenetic pattern), then the individualized patterns of growth should be synchronous for twins sharing exactly the same genetic makeup: Two identical twins should follow a parallel growth curve for periods of acceleration and lag. Wilson presumed that evidence for such synchronized patterns of growth would be indicative of the influences of timed gene-action systems, switching on and switching off according to a genetic plan (Wilson, *1979b*, 1981b).

The ages for and methods of measurement of physical growth have been described in detail elsewhere (Wilson, 1974a). Weight of children from 3 to 24 months of age was measured on a balance scale and of older children on a platform scale (see Table 2-1). Birth weight and height were obtained from birth certificates

**Table 2-1** Monozygotic (MZ) and dizygotic (DZ) twin pair correlations for weight and height and year-to-year stability for height

| | Correlations | | | | |
|---|---|---|---|---|---|
| | Weight | | Height | | |
| Age | MZ | SS-DZ | MZ | SS-DZ | Year to-year stability |
| Birth | .64 | .71 | .66 | .77 | |
| 3 mos | .78 | .66 | .77 | .74 | |
| 6 | .82 | .62 | .81 | .70 | |
| 9 | .83 | .55 | .83 | .64 | |
| 12 | .89 | .58 | .86 | .69 | .45 |
| 18 | .87 | .54 | .89 | .71 | |
| 24 | .88 | .55 | .88 | .59 | .83 |
| 30 | .87 | .55 | .93 | .59 | |
| 3 yr | .89 | .52 | .93 | .59 | .89 |
| 4 | .85 | .50 | .94 | .59 | .94 |
| 5 | .86 | .54 | .94 | .51 | .95 |
| 6 | .87 | .57 | .94 | .56 | .97 |
| 7 | .88 | .54 | .94 | .51 | .98 |
| 8 | .88 | .54 | .95 | .49 | .99 |
| 9 | .88 | .62 | .93 | .49 | .99 |
| 15 | .88 | .51 | .94 | .45 | .98 |
| Adult | .95 | .56 | .98 | .46 | .95 |

*Note:* Sample size varies by age, but $N$ for MZ $> 70$ pairs, $N$ for DZ $> 60$ pairs for all but adult measures, when $N$ for MZ $= 24$ pairs, $N$ for DZ $= 19$ pairs.

and hospital records. Height was measured as recumbent length up to age 24 months and as standing height thereafter.

The determination of zygosity for same-sex (SS) pairs was based on bloodtyping when the twins were at least 3 years old. Twenty-two or more antigens were examined. If a twin pair was discordant for any of the antisera tests, the pair was classified as dizygotic (DZ). Pairs concordant for the antisera tests were classified as monozygotic (MZ) twins (Wilson, 1980).

Multiple comparisons between MZ and DZ pairs for weight and height revealed no systematic differences between the two groups at any age. By contrast, both MZ and DZ children had a substantial deficit in birth size compared with singletons. The deficit was offset over time, however, so that by age 7 or 8 years twins had reached the normal range of growth curves for weight and height (Wilson, 1979b).

## Zygosity and Pair Concordance

Measures of concordance for physical growth at each age were obtained from within-pair (intraclass) correlations computed for the MZ and DZ groups. The results for the MZ and same-sex DZ (SS-DZ)[1] pairs are presented in Table 2-1. For both weight and height, the MZ pairs were less concordant at birth than the SS-DZ pairs, but by age 3 months the MZ correlation exceeded the SS-DZ correlation. The MZ correlation increased into the adult years, the increase being sharper for height than for

---

[1] The comparisons being made in this paper will be between MZ and same-sex DZ twin pairs.

weight. By contrast, the magnitude of the SS-DZ correlations decreased over time, so that by the adult years the correlations fell within an intermediate range. The general trend over the years was for each zygosity group to move consistently toward a level of concordance commensurate with the genetic similarity for each type of twin. Height particularly appeared to illustrate the regulatory role of the genotype, perhaps because height is better buffered against within-pair variations in diet.

### Age-to-Age Stability

Because height appeared to be less susceptible to relatively transient features of diet, the stability of physical growth can be illustrated for height. The year-to-year correlations are provided in Table 2-1. The magnitude of the correlations indicated that the rank ordering of the children's height was markedly stable after they reached 3 years of age. Apparently, the accelerations and lags of growth in height produced the largest reordering of children's relative stature during their first one or two years. In effect, the directional trends of the correlations over age indicated that the gains in height were steadily proceeding toward an end point targeted by a plan intrinsically founded. When one takes into account the systematic increase in the MZ pair correlations, it seems highly plausible that the plan is dictated by the genotype.

## Mental Development

If discontinuities of physical development exemplify genetic influences over time, then discontinuities of mental development may be considered from the same interpretive framework. More so than physical measures, mental test scores vary considerably for young children tested several times, and the developmental paths tracked by the scores may be distinctive for each child. Bayley described the idiosyncratic features of the pathways by the following: "Growth curves will enable us to observe a child's periods of fast and slow progress, his spurts and plateaus, and even regressions, in relation to his own past and future. Each child appears to develop at a rate that is unique for him" (Bayley, 1955, p. 814).

Bayley examined mental test scores in children to see if large increases or decreases could be explained by contiguous environmental events. Although she provided plausible explanations for some discontinuities, she could not readily explain other discontinuities of mental development, especially among young children. Nevertheless, discontinuities have been depicted as the most telling argument against genetic influences on the emergent and persisting intelligent behaviors.

Wilson (1972) suggested that there could be genetic influences on mental development even in view of irregularities of mental growth over time. The changes in mental test scores obtained from a series of tests on the same person were an example of the genetically influenced discontinuities so evident in physical growth. Wilson saw that longitudinal assessments of twins' mental abilities were a powerful resource for demonstrating genetic influences, accordingly, in a series of studies (1972, 1974b, 1977, 1978, 1981a, 1983), Wilson reported the concordance of MZ and DZ twins in the longitudinal sample of the Louisville Twin Study for a series of different mental tests.

## Mental Tests

The tests used in the longitudinal program were selected from the best standardized tests for individual assessment of children's general mental abilities. The tests included the Bayley Scales of Mental Development (Bayley, 1969); the restandardized version of the Stanford-Binet (Terman & Merrill, 1973); the Wechsler Preschool and Primary Scale of Intelligence, or WPPSI (Wechsler, 1967); the Wechsler Intelligence Scale for Children, or WISC, and its revised version, WISC-R, (Wechsler, 1974), and the McCarthy Scales of Children's Abilities (McCarthy, 1972). Young adult twins recalled for testing were given the Wechsler Adult Intelligence Scale, or WAIS (Wechsler, 1955). All of these tests and their revised versions furnish a means of comparing individual children with a representative sample of age-matched peers. Each child's relative placement is expressed by a standard score format that remains constant across ages and tests. Therefore, changes—spurts and lags—in mental development become interpretable according to a fixed set of measuring scales.

The Bayley mental scale was administered at 3, 6, 9, 12, 18, and 24 months of age; the Stanford-Binet at 30 months and 3 years; the WPPSI at 4, 5, and 6 years; and the WISC or WISC-R at 7, 8, 9, and 15 years. In recent years the McCarthy scale was substituted for the WPPSI at age 4 years. Adult twins were tested by means of the WAIS when they were between 21 and 25 years of age. All twin pairs were tested by two separate examiners, and the scoring was verified by a third examiner.

## Zygosity and Pair Concordance

Measures of physical concordance for mental test scores were calculated for the MZ and DZ groups at each age. The results for the MZ and same-sex DZ (SS-DZ) pairs are presented in Table 2-2. The within-pair correlations during the first year were comparable for the two types of twins, but with increasing age the correlations of the MZ pairs increased and the correlations of the SS-DZ pairs steadily decreased. By age 15 the MZ correlation had reached $r = .88$, whereas the SS-DZ correlation had become $r = .54$.

The increase in MZ and decrease in SS-DZ correlations were noted for physical measures, and the question was raised that weight might be linked to the trend of the correlations for both groups. Correlations between weight and mental test scores were significant only during the first year ($rs = .50, .48,$ and $.30$ at 3, 6, and 12 months, respectively). Prematurity thus might account for the similarity of the MZ and SS-DZ correlations during the first year, but it could not account for the progressive separation of the correlations for the two groups thereafter. Moreover, the decline in the SS-DZ correlations took place during constant exposure to the same family environments. While one might expect that the accumulating experiences common to both SS-DZ twins would be accompanied by decreased variability within SS-DZ pairs, the opposite effect was found.

## Developmental Change

Table 2-2 also shows the stability correlations of mental test scores obtained year to year. These correlations indicate that there was considerable change in scores, and

**Table 2-2** Monozygotic (MZ) and dizygotic (DZ) twin pair correlations and year-to-year stability for IQ

|       | Correlations |       |                        |
| ----- | ------------ | ----- | ---------------------- |
| Age   | MZ           | SS-DZ | Year to-year stability |
| 3 mos | .66          | .67   |                        |
| 6     | .75          | .72   |                        |
| 9     | .67          | .51   |                        |
| 12    | .68          | .63   | .44                    |
| 18    | .82          | .65   |                        |
| 24    | .81          | .73   | .48                    |
| 30    | .85          | .65   |                        |
| 3 yr  | .88          | .79   | .74                    |
| 4     | .83          | .71   | .76                    |
| 5     | .85          | .66   | .80                    |
| 6     | .86          | .59   | .87                    |
| 7     | .84          | .59   | .86                    |
| 8     | .83          | .66   | .87                    |
| 9     | .83          | .65   | .90                    |
| 15    | .88          | .54   | .80                    |
| Adult | .85          | .56   | .89                    |

*Note:* Sample size varies by age, but $N$ for MZ > 70 pairs, $N$ for DZ > 60 pairs for all but adult scores, when $N$ for MZ = 24 pairs, $N$ for DZ = 19 pairs.

that the change was most pronounced in the first years of childhood. Not until ages 4 and 5 did the ordering of individual differences stabilize at $r = .80$ and higher. Beyond age 6 the predictive power of the correlations remained virtually the same except for the interval between 9 and 15 years. The reduction in the magnitude of the correlation ($r = .80$) between ages 9 and 15 years can be attributed to the long six-year interval between tests; yet an equally long interval between age 15 and adult provided a higher correlation ($r = .89$). Perhaps the change attributed to adolescence may account for the small downturn in correlation between 9 and 15 years of age. By and large, however, adolescence is not a period of pronounced rearrangement of mental test scores.

## Concordance for Change

To this point, two developmental trends are clear:

1. Mental test scores for each twin became more stable with increasing age.
2. The magnitude of the MZ pair correlations became more pronounced with increasing age.

If the MZ twins became increasingly more concordant with one another, even when scores changed from one age to the next, the twins within MZ pairs apparently were closely yoked to the same pattern of mental development, even during spurts and lags. That is, within MZ pairs a profile of mental scores plotted over several ages for one twin (Twin A) should closely match those for the other twin (Twin B). To illustrate this point, profiles for two MZ twin pairs are provided in Figure 2–1. (For comparison, Figure 2–2 provides profiles for two DZ twin pairs.)

In Figure 2–1a, the MZ twin pair's scores from infancy to mid teens show a men-

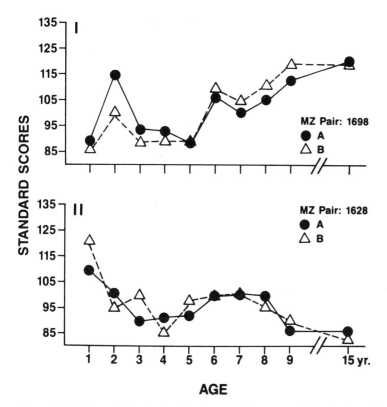

**Figure 2-1** Trends in standardized scores from mental tests given from 1 to 15 years of age to two monozygotic twin pairs.

tal growth spurt from Year 1 to Year 2, a lag from Year 2 to Year 5, and a steady rise from Year 5 until Year 15. Scores for the second MZ twin pair (Figure 2-1b) show a general lag after the first year. Undoubtedly each pair has a distinctive pattern, but within each pair there is a high degree of congruence for the profile of scores. The DZ pairs' scores (Figure 2-2) also show distinctive profiles, but the profile for one twin in each pair is not as closely matched by the profile for other twin as in the MZ pairs.

The concordance of twin pairs for the profile of scores can be distinguished in two ways: contour and overall trend. The contour profile represents the concordance of age-to-age changes in scores; the overall trend profile jointly appraises the concordance for both contour of scores and elevation of scores. In order to describe both aspects of the profiles, Wilson (1968, 1979a) derived means to express the concordance estimates of correlation and compared those means with analysis of variance procedures specifically adapted for twin pairs (see Haggard, 1958).

When scores fluctuate markedly from one age to the next, as do mental test scores during infancy, the profile correlations for MZ and DZ pairs indicate that the MZ correlations are significantly higher than the DZ correlations (Wilson, 1974b, 1978). For example, the MZ and DZ correlations for contour (age-to-age change) are, respectively, .40 and .15 during the first year and .67 and .42 during the second year.

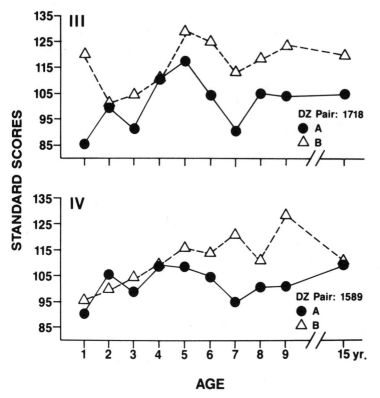

**Figure 2-2** Trends in standardized scores from mental tests given from 1 to 15 years of age to two dizygotic twin pairs.

As the scores stabilize, however, the contour correlations become less informative. The correlations for trend then become informative because they take into account the concordance for overall level of scores as well as change in scores. The trend correlations therefore represent a more consistent expression of MZ and DZ concordance for many age ranges than do the contour correlations.

Trend analyses have been applied to several different ranges of ages, as published elsewhere (Wilson, 1972, 1974b, 1978, 1981b). The entire age range represented by all of the assessments in the longitudinal study can be divided into six periods. These periods and the trend correlations for MZ and DZ pairs within each of them are shown in Table 2-3.

During the first age range—3, 6, 9, and 12 months—tests are presumably influenced, in part, by the effects of prematurity. The second period covers the transition from rudimentary development to beginning language skills. The third period—3 to 6 years—spans the preschool years, and the fourth period covers the early school years—6, 7, and 8 years. The bridge from childhood to adolescence is represented by the 8-to-15-year period, and the final period represents adolescence to young adulthood.

The first period did not show significant differences in concordant trends in mental development for MZ and DZ pairs. Beginning in the second period, the MZ pairs

# CONTRIBUTIONS FROM THE LOUISVILLE TWIN STUDY

**Table 2-3** Monozygotic (MZ) and dizygotic (DZ) twin pair correlations for developmental trend in IQ scores

| | Trend correlations | |
|---|---|---|
| Age periods | MZ | SS-DZ |
| 3, 6, 9, 12 mos | .69 | .63 |
| 12, 18, 24, 36 mos | .80 | .72 |
| 3, 4, 5, 6 yr | .87 | .65 |
| 6, 7, 8 yr | .81 | .66 |
| 8, 15 yr | .87 | .63 |
| 15, adult | .86 | .55 |

were significantly more concordant. Among the remaining periods, the trend correlations remained in the mid- .80s for the MZ pairs and in the mid- .60s for the DZ pairs.

## *Environment and Heritage*

Characteristics of the home setting, interactions between parents and children, and parental attainments—parental education and occupation—are known to be linked with mental development (e.g., Caldwell, 1978). Therefore, the twins' home environments were assessed for the physical attributes of the home and neighborhood, the social and intellectual influences in the home, and the opportunities for and quality of play and stimulation available to children (Matheny, Thoben, & Wilson, 1982).

The home visit protocol included measures from Caldwell's (1978) HOME inventory as well as a number of items ultimately condensed to a large general factor, "adequacy of home environment." Scores from the HOME inventory, factor scores from the general factor, years of parental education, and socioeconomic status (Reiss, 1961) became the composite profile of the home/family environment. These measures were subsequently analyzed (Wilson & Matheny, 1983a) for their relations with the twins' mental development scores. Some of the results are summarized in Table 2-4.

**Table 2-4** Correlations between home/family characteristics and mental test scores

| | Age at mental test | | | |
|---|---|---|---|---|
| Characteristics | 6 mos | 2 yr | 3 yr | 6 yr |
| Education | | | | |
|   Father | .28 | .40 | .51 | .55 |
|   Mother | .19 | .36 | .49 | .49 |
| Socioeconomic status | .16 | .42 | .52 | .51 |
| Adequacy of home environment | .22 | .47 | .55 | .55 |
| HOME[a]: Total | .17 | .39 | .44 | .42 |
| No. of twin sets | 205 | 207 | 220 | 204 |

*Note:* Full set of measures can be found in Wilson and Matheny (1983). All correlations shown are significant: $p \leq .05$.
[a] HOME: inventory developed by Caldwell (1978).

The attributes of the home and parents clearly became more highly correlated with the twins' mental development scores as the twins increased in age. The more robust correlations leveled off in the mid .50s. All characteristics were interdependent, however, and additional analyses indicated that parental education and socioeconomic status accounted for the biggest proportion of the explained variance at all four ages. Thus, as the measures of the twins' mental abilities progressively stabilized by school age (see Table 2-2), parental characteristics played an increasingly large role in the prediction of the twins' scores on the mental tests.

In view of the fact that the parental characteristics could serve as broad indicators of parents' realized mental abilities by the time of the twins' birth, the stabilized predictive correlations represented the possible genetic association between parents and offspring. The genetic association between parents and children is, of course, equivalent to the DZ pair correlations and to sibling correlations. Wilson (1983) argued that the concordances for parents–child, DZ twins, twin–sibling, and sibling–sibling pointed to the same general effect of genetically related pairs converging toward the intermediate level expected for genetic correlations.

## Temperament

The initial research on temperament was dependent on interviews in which the mother was asked about the twins' typical behavior at home. At each visit the mother reported whether the twins were alike or different for a wide variety of behaviors, many of which were linked with temperament as defined by Thomas, Chess, Birch, Hertzig, and Korn (1963). The classification of each twin pair for concordance or discordance for each behavior led to the identification of a nuclear temperament cluster made up of frequency and intensity of temperamental outbursts, irritability, crying, and demanding of attention (Wilson, Brown, & Matheny, 1971). This cluster, distinct from a sociability cluster, also provided evidence for genetic influences in that MZ pairs were clearly more concordant than DZ pairs (Matheny, Wilson, Dolan, & Krantz, 1981).

While the interview data provided informative distinctions for differences within twin pairs, the magnitude of the differences was not compared. Moreover, the method was dependent on parental reports. As an adjunct to the mental testing program, the Infant Behavior Record (IBR) (Bayley, 1969) provided an objective evaluation of each infant's behavior during mental testing, and the IBR became the vehicle for the initial studies of temperament (Matheny, Dolan, & Wilson, 1976).

A full-scale factor analysis of the IBR yielded three factors—task orientation, test affect-extraversion, and activity—that represented test-taking aspects of temperament (Matheny, 1980, 1983). Subsequent analyses suggested that concordance for the factors tended to be greater in MZ twins than in DZ twins, especially from the age of 12 months to 24 months. The analyses also revealed a genetic influence on the patterning of scores from the temperament profile. First, the organization of temperament was more tightly duplicated for MZ pairs than for DZ pairs. Also, there was a closer synchronization of the temperament scores over age for the MZ pairs. These patterns—organization of temperament scores at each age and synchronization of temperament scores across ages—are shown by the MZ and DZ correlations in Table 2-5.

**Table 2-5** Monozygotic (MZ) and dizygotic (DZ) twin-pair correlations for factor scores from Infant Behavior Record

| Age (mos) | Pair correlations for: | | | | | | Profile $r$ for pattern across factors | |
| --- | --- | --- | --- | --- | --- | --- | --- | --- |
| | Task orientation | | Test Affect-Extraversion | | Activity | | | |
| | MZ | SS-DZ | MZ | SS-DZ | MZ | SS-DZ | MZ | SS-DZ |
| 3 | .55[a] | .10 | .18 | .26 | .30 | .33 | .40[a] | .18 |
| 6 | .61[a] | .36 | .55[a] | .10 | .24 | .11 | .46[a] | .25 |
| 9 | .28 | .00 | .35 | .33 | .25 | .22 | .35[a] | .20 |
| 12 | .49[a] | .23 | .43[a] | .07 | .33 | .28 | .39[a] | .25 |
| 18 | .56[a] | .19 | .49 | .37 | .43[a] | .14 | .38[a] | .19 |
| 24 | .48[a] | .21 | .53[a] | .03 | .58[a] | .14 | .39[a] | .24 |
| *Trend correlations for age-to-age change* | | | | | | | | |
| 6, 12, and 18 | .53[a] | .18 | .35[a] | .06 | .27 | .19 | | |
| 12, 18, and 24 | .49[a] | .21 | .37[a] | .12 | .52[a] | .18 | | |

*Note:* N for MZ 60–91 pairs; N for DZ, 35–54 pairs.
[a] $p \leq .05$, one-tailed, MZ $r$ > SS-DZ $r$.

Parental reports of contrast and direct observations of infants were the background for the research program to direct attention to a more complete appraisal of temperament. The appraisal took two forms: laboratory observations of twin infants participating in standardized play routines, and parental ratings from temperament questionnaires.

For the lab observations, age-related routines were devised so that each twin or both twins were engaged with staff members, with and without a parent present. The routines were videotaped and the staff subsequently rated each infant's behavior from the videotapes. The primary rating scales were the following: (1) emotional tone; (2) activity; (3) attention span; and (4) social orientation to staff members. Ratings were made for each two-minute period of the videotape, and then the periods were combined to yield a composite score for each rating scale. A complete description of the lab routines and rating scales may be found elsewhere (Matheny & Wilson, 1981; Wilson & Matheny, 1983b).

Initial lab observations were done at age 12 months (Wilson & Matheny, 1983b) and were followed by longitudinal observations of the same twins at age 18 and 24 months (Matheny, Wilson & Nuss, 1984). Analyses indicated that a general factor, largely defined by ratings of emotional tone, remained moderately stable between 12 and 18 months ($r = .38$) and between 18 and 24 months ($r = .66$).

Ratings from the lab observations were supplemented by parental ratings from a temperament questionnaire for toddlers (Fullard, McDevitt, & Carey, 1984). The nine temperament characteristics scored from the questionnaire were those defined by Thomas, Chess, Birch, Hertzig, and Korn (1963). Factor analyses of the nine scores yielded a general factor pertaining to approach/withdrawal, adaptability, and mood that was replicated at ages 12, 18, and 24 months and was moderately stable from one age to the next: 12–18-month $r = .44$; 18–24-month $r = .65$.

Additional analyses of the lab and questionnaire measures of temperament for expanded samples of twins indicated that emotional tone from the lab data and approach/withdrawal from the temperament questionnaire were the temperament variables bridging the two sources of ratings and providing moderate stability from one age to the next (Wilson & Matheny, 1986). More recent unpublished data also indicate that emotional tone from the lab data and approach/withdrawal from the questionnaire are linked to infant fearfulness as rated on the IBR. Therefore, at each age, three different sources of ratings of the twin infants and toddlers point to a common behavioral dimension.

## Synchronization of Temperament Change

The sample of twin pairs in the longitudinal study of temperament only recently has increased to a size large enough to compare MZ and DZ concordance for lab, questionnaire, and IBR ratings at several ages: 12, 18, 24, and 30 months. Preliminary analyses have concentrated on the ratings of emotional tone, fearfulness, and approach/withdrawal for these four ages; the MZ and DZ correlations for each measure at each age are shown in Table 2–6.

In view of the fact that all three measures are moderately stable from one age to the next (average $r = .48$), it is apparent that there are developmental changes for the rated temperament of each twin. The question then turns to whether these changes are synchronized for twin pairs. Table 2–6 provides trend correlations for the MZ and DZ pairs. It is evident that the age-to-age changes in temperament were more closely synchronized for the MZ pairs. The more coherent pattern for the MZ pairs is a further illustration of the genetic influence on behavioral change during development.

## PROSPECTS FOR A GENETIC PERSPECTIVE ON BEHAVIORAL DEVELOPMENT

Theories of behavioral development have often strongly emphasized the role of environmental influences and have been unalloyed by a recognition of genetic contri-

**Table 2–6** Monozygotic (MZ) and dizygotic (SS-DZ) twin correlations for temperament measures at 12, 18, 24, and 30 months of age

| | Measure | | | | | |
|---|---|---|---|---|---|---|
| | Emotional tone | | Fearfulness | | Approach/withdrawal | |
| Ages (mos) | MZ | SS-DZ | MZ | SS-DZ | MZ | SS-DZ |
| 12 | .59 | .27 | .76[a] | .48 | .67[a] | −.21 |
| 18 | .83[a] | .28 | .77[a] | .02 | .83[a] | −.07 |
| 24 | .87[a] | .26 | .80[a] | .20 | .15 | −.16 |
| 30 | .79[a] | .25 | .63[a] | −.09 | .48[a] | −.18 |
| Trend correlations for change | | | | | | |
| 12–30 mos | .79[a] | .26 | .75[a] | .21 | .53[a] | −.15 |

*Note:* N for MZ = 33, N for SS-DZ = 19.
[a] $p \leq .05$, one-tailed, MZ $r$ > SS-DZ $r$.

butions continuing beyond the initiation of human developmental themes. The longitudinal study of twin children qualifies that strong emphasis and reiterates the theme that genetic influences guide development beyond the prenatal period or infancy. The essence of the theme, clearly stated by McClearn (1970), is that we must "appreciate that the influence of genes is not manifested only at conception or at birth or at any other single time in the individual's life history. Developmental processes are subject to continuing genetic influence, and different genes are effective at different times" (p. 61). The empirical results from the Louisville Twin Study are in harmony with this theme.

The results do not ignore environment, particularly as conceptualized for its interactions with the genotype, for its significant potentiating influences on development. Yet, it is important to note, the developmental genetic processes foster individual differences even within equivalent environments and, if anything, guide the emergence of individual differences when developmental changes occur (Plomin, 1986).

From this view, behavioral development for the human species and for individual differences among humans is guided by genetic plans that are analogous to those traced for physical development.

> The end product—the phenotypic behavior of the human, cradle to grave—is distilled from the constant interplay of genetic material and the environmental surroundings. But the message, the conserved microfilm of evolution's choices, is preserved in the genotype, and it is progressively actualized throughout the life span. Perhaps an appreciation of this fact can help anchor the concepts in developmental psychology and lead to a more comprehensive model for assaying the determinants of behavior. [Wilson, 1983, p. 315]

## ACKNOWLEDGMENTS

Preparation of this chapter drew largely from discussions with Ronald S. Wilson during the month preceding his death, and from two of his influential papers (Wilson, 1978, 1983). Prior study directors Drs. Frank Falkner and Steven Vandenberg also made extensive contributions. This chapter is based on data collected by the Louisville Twin Study during the past two decades. That research has been supported in part by grants from the National Institute of Child Health and Human Development; Grant Foundation, Office of Child Development; John D. and Catherine T. MacArthur Foundation; and the Graduate School of the University of Louisville.

## REFERENCES

Bayley, N. (1955). On the growth of intelligence. *American Psychologist, 10,* 805–818.
Bayley, N. (1969). *Bayley scales of infant development.* New York: Psychological Corporation.
Caldwell, B. M. (1978). *Home observation for measurement of the environment: HOME.* Little Rock, AR: University of Arkansas.
Falkner, F. (1957). An appraisal of the potential contribution of longitudinal twin studies. In *The Nature and Transmission of the Genetic and Cultural Characteristics of Human Populations.* Philadelphia: Milbank Memorial Fund.

Falkner, F., & Tanner, J. M. (1986). *Human Growth: A Comprehensive Treatise* (2nd ed.). New York: Plenum.
Fullard, W., McDevitt, S. C., & Carey, W. B. (1984). Assessing temperament in one to three-year-old children. *Journal of Pediatric Psychology, 9,* 205-217.
Goldsmith, H. H. (1983). Genetic influences on personality from infancy to adulthood. *Child Development, 54,* 331-355.
Haggard, E. A. (1958). *Intraclass Correlation and the Analysis of Variance.* New York: Dryden Press.
Matheny, A. P. Jr. (1980). Bayley's Infant Behavior Record: Behavioral components and twin analyses. *Child Development, 51,* 1157-1167.
Matheny, A. P. Jr. (1983). A longitudinal twin study of stability of components from Bayley's Infant Behavior Record. *Child Development, 54,* 356-360.
Matheny, A. P. Jr., Dolan, A. B., & Wilson, R. S. (1976). Within-pair similarity on Bayley's Infant Behavior Record. *Journal of Genetic Psychology, 128,* 263-270.
Matheny, A. P. Jr., Thoben, A. S., & Wilson, R. S. (1982). Appraisals of Basic Opportunities for Developmental Experiences (ABODE): Manual for home assessments of twin children. *Catalog of Selected Documents in Psychology, 12,* Ms. 2472.
Matheny, A. P., Jr., & Wilson, R. S. (1981). Developmental tasks and rating scales for the laboratory assessment of infant temperament. *Catalog of Selected Documents in Psychology, 11,* Ms. 2367.
Matheny, A. P. Jr., Wilson, R. S., Dolan, A. B., & Krantz, J. Z. (1981). Behavioral contrasts in twinships: Stability and patterns of differences in childhood. *Child Development, 52,* 579-588.
Matheny, A. P., Jr., Wilson, R. S., & Nuss, S. M. (1984). Toddler temperament: Stability across settings and over ages. *Child Development, 55,* 1200-1211.
McCarthy, D. (1972). *McCarthy Scales of Children's Abilities.* New York: Psychological Corporation.
McClearn, G. E. (1970). Genetic influences on behavior and development. In Ph. H. Mussen (Ed.), *Carmichael's Manual of Child Psychology* (Vol. 1). New York: John Wiley.
Plomin, R. (1986). *Development, Genetics, and Psychology.* Hillsdale, NJ: Lawrence Erlbaum.
Reiss, A. J. (1961). *Occupations and Social Status.* New York: Free Press of Glencoe.
Rowe, D. C. (1987). Resolving the person-situation debate: Invitation to an interdisciplinary dialogue. *American Psychologist, 42,* 218-227.
Terman, L. M., & Merrill, M. A. (1973). *Stanford-Binet Intelligence Scale: 1972 Norms Edition.* Boston: Houghton-Mifflin.
Thomas, A., Chess, S., Birch, H. G., Hertzig, M. E., & Korn, S. (1963). *Behavioral Individuality in Early Childhood.* New York: New York University Press.
Wechsler, D. (1955). *Wechsler Adult Intelligence Scale.* New York: Psychological Corporation.
Wechsler, D. (1967). *Wechsler Preschool and Primary Scale of Intelligence.* New York: Psychological Corporation.
Wechsler, D. (1974). *Wechsler Intelligence Scale for Children—Revised.* New York: Psychological Corporation.
Wilson, R. S. (1968). Autonomic research with twins: Methods of analysis. In S. G. Vandenberg (Ed.), *Progress in Human Behavior Genetics.* Baltimore: Johns Hopkins Press.
Wilson, R. S. (1972). Twins: Early mental development. *Science, 175,* 914-917.
Wilson, R. S. (1974a). Growth standards for twins from birth to four years. *Annals of Human Biology, 1,* 175-188.
Wilson, R. S. (1974b). Twins: Mental development in the preschool years. *Developmental Psychology, 10,* 580-588.
Wilson, R. S. (1977). Mental development in twins. In A. Oliverio (Ed.), *Genetics, Environment, and Intelligence.* Amsterdam: Elsevier.

Wilson, R. S. (1978). Synchronies in mental development: An epigenetic perspective. *Science, 202,* 939–948.
Wilson, R. S. (1979a). Analysis of longitudinal twin data: Basic model and applications to physical growth measures. *Acta Geneticae Medicae et Gemellologiae, 28,* 93–105.
Wilson, R. S. (1979b). Twin growth: Initial deficit, recovery, and trends in concordance from birth to nine years. *Annals of Human Biology, 6,* 205–220.
Wilson, R. S. (1980). Bloodtyping and twin zygosity: Reassessment and extension. *Acta Geneticae Medicae et Gemellologiae, 29,* 103–120.
Wilson, R. S. (1981a). Mental Development: Concordance for same-sex and opposite-sex dizygotic twins. *Developmental Psychology, 17,* 626–629.
Wilson, R. S. (1981b). Synchronized developmental pathways for infant twins. In L. Gedda, P. Parisi, & W. E. Nance (Eds.), *Twin Research. 3: Intelligence, Personality, and Development.* New York: Alan Liss.
Wilson, R. S. (1983). The Louisville Twin Study: Developmental synchronies in behavior. *Child Development, 54,* 298–316.
Wilson, R. S., Brown, A. M., & Matheny, A. P. Jr. (1971). Emergence and persistence of behavioral differences in twins. *Child Development, 42,* 1381–1398.
Wilson, R. S., & Matheny, A. P. Jr. (1983a). Mental development: Family environment and genetic influences. *Intelligence, 7,* 195–215.
Wilson, R. S., & Matheny, A. P., Jr. (1983b). Assessment of temperament in infant twins. *Developmental Psychology, 19,* 172–183.
Wilson, R. S., & Matheny, A. P. Jr. (1986). Behavior-genetics research in infant temperament: The Louisville Twin Study. In R. Plomin & J. Dunn (Eds.), *The Study of Temperament: Changes, Continuities, and Challenges.* Hillsdale, NJ: Lawrence Erlbaum.

# 3
# Dual Genesis and the Puzzle of Aggressive Mediation

ROBERT B. CAIRNS AND JEAN-LOUIS GARIÉPY

Over ten years ago William Paterson College hosted a symposium on communicative behavior and evolution (Hahn & Simmel, 1976). At that time one of us reported on the outcomes of studies on the ontogeny and phylogeny of social interactions (Cairns, 1976). In this chapter we report on the use of convergent microevolutionary, developmental, and interactional methods to address a specific issue of behavioral mediation. Our starting point is the state of our understanding as reflected in that earlier volume and we will bring that report up to date, summarizing what we have since learned about the linkages between the development and microevolution of aggressive behavior. We will then consider the implications of that work for a proposal on dual genesis and the ontogeny and phylogeny of social behaviors.

Since our research strategy differs in important ways from those often adopted in developmental and genetic behavioral investigations, a brief review of the concepts that provide the framework for this chapter is in order.

## BASIC CONCEPTS

### Developmental Genetic Integration in Social Behavior

A new synthesis on behavioral development has emerged in the behavioral and biological sciences over the past two decades. This modern developmental orientation has broad implications for biobehavioral research and the study of genetic-behavioral processes. The perspective was described by T. C. Schneirla (1966) in the *Quarterly Review of Biology,* along with his conclusion that "Behavioral ontogenesis is the backbone of comparative psychology. Shortcomings in its study inevitably handicap other lines of investigation from behavioral evolution and psychogenetics to the study of individual and group behavior" (p. 283).

The developmental perspective cuts across disciplines, methodologies, and applications. Its kernel assumption is that significant social adaptations reflect the fusion of factors over the life course and over evolution (Cairns, 1979; Scott, 1977). Within this perspective, behavior patterns represent the melding of neurobiological, social, and genetic processes in development. Because the relative weights of these components vary, longitudinal analyses are required to unravel the ways the components are coalesced over ontogeny. Developmental research begins where the nature–nurture debate ends.

Though the need for a developmental perspective has been clearly documented in behavioral biology and comparative psychology (Gottlieb, 1983), the empirical implications of such a perspective for genetic and microevolutionary behavioral processes have rarely been investigated (see Chapter 4 in this volume). The developmental concept presupposes attention to the emergence, change, and continuity of behavioral patterns. This developmental focus on behavioral dynamics and processes historically has been seen as antithetical to a genetic concern with heritable traits. A fundamental conflict on how to conceptualize behavior is perhaps one reason why joint developmental-genetic empirical analyses have been given short shrift.

In addition, existing methodologies and statistical models have their own shortcomings. The developmental perspective on social behavior requires research designs that address processes of genetic-ontogenetic integration over time (Cairns, 1979; Wohlwill, 1973). As Magnusson (1985) has cogently argued, longitudinal methods must include ways to capture the integration of biological, social, and learning processes. With this goal in mind, it has been necessary to generate new research strategies and to use traditional procedures in nontraditional ways. For example, methods of genetic analysis (i.e., selective breeding) may be adopted to address issues of developmental process. Similarly, interactional analyses of behavior are not limited to the study of learning but can be used to clarify neurobiological processes. Identification of effective mediators across levels of analysis and across time has the highest priority on the developmental research agenda.

## Dual Genesis

A central assumption of the concept of dual genesis is that factors in development and evolution typically collaborate, not compete, in behavioral determination. Hence there is rarely a nature–nurture contest because both sets of determinants reflect a similar design for adaptation. The idea is not new. At the turn of the century, J. M. Baldwin (1895) provided a succinct statement of this concept in his discussion of the futile "antithesis between 'heredity' and 'environment.'"

> It is clear that we are led to two relatively distinct questions: questions which are now familiar to us when put in the terms covered by the words, "phylogenesis" and "ontogenesis." First, how has the development of organic life proceeded, showing constantly, as it does, forms of greater complexity and higher adaptation? This is the phylogenetic question.... But the second question, the ontogenetic question, is of equal importance: the question, How does the individual organism manage to adjust itself better and better to its environment?... This later problem is the most urgent, difficult, and neglected question of the new genetic psychology. (pp. 180–181)

Given the modest advances on these issues in this century, the developmental question remains "most urgent, difficult, and neglected."

But the matter has not been wholly neglected. One of the more promising steps in contemporary approaches has been the empirical reexamination of the relations between ontogenetic and phylogenetic analyses of behavior (see Cairns, 1976; Gottlieb, 1983; Mason, 1979). In this regard, it has been argued recently that modifications in behavioral development stand at the leading edge of microevolutionary changes in adaptation (Cairns, 1986). The unique functional properties of behavior for organismic adaptation and evolution have been downplayed, which is a pity. In the present view, behaviors provide a mechanism for rapid, dynamic, and potentially reversible adaptations to changing circumstances and challenges to survival. This presupposes that the neurobiological processes that support the behavioral changes should themselves be susceptible to short-term and reversible changes. To the extent that behavioral accommodations work over successive generations in the short term, they may be consolidated and preserved for the long term by structural modifications. Such function-structure bidirectionality has been supposed to operate in individual ontogeny by Schneirla (1966), Kuo (1967), and Gottlieb (1983). We suggest that the same principles apply to adaptive changes in phylogeny as well.

This brings us to the issue central to the dual genesis proposal: how to explain the actual linkages between accommodations in development and adaptations in evolution. In this regard, Ernst Mayr (1974) has proposed that some behaviors are "open" and others "closed" to individual experience in development. He assigned social communicative actions to the closed category. To us this seems a mistake for two reasons. First, the categorization presupposes that social behaviors should be viewed as structures of the organism rather than as dynamic outcomes of structural processes. Second, social acts necessarily incorporate ongoing, contemporaneous information (from the self, from interactive partners, and from the physical and social milieu), and no social process can be entirely closed to experience and remain adaptive. It may thus be appropriate to view social actions as significantly constrained but to avoid reifying behaviors as structures.

A revision of Mayr's (1974) valuable contribution would emphasize that most but not all components of social action are closed to rapid change in development and microevolution. A good illustration of the high level of sterotypy that occurs in social behavior may be seen in the agonistic patterns of mice. The several lines of mice in the present selection studies are not distinguishable from each other in appearance and in the performance of most behavior patterns. In their attack behavior, mice show a species-typical stereotyped behavior regardless of selection line. Although attack form does not differ, attack frequency and latency do. We therefore speculate that most aspects of this social behavior system are closed to rapid modification, whether from development or from microevolution.

Despite a strong bias toward stability in development and microevolution, social behavior systems are not entirely resistant to change. Even a slight window that remains open for behavioral/neurobiological change appears to allow significant modifications in social behavior and organization. Progress in understanding how seemingly modest genetic effects are magnified in social behavior requires information about the nature of the proximal controls in both ontogeny and microevolution (e.g., King & Wilson, 1975). That has been the essential task of the present research program.

## THE PAST

Some findings on the interrelations between development and genetics in understanding aggressive behavior were presented by Cairns in 1976. Those findings may be summarized in three points:

- Selection for aggressive behavior in male mice proceeded very rapidly. The distributions were virtually nonoverlapping after the $S_1$ generation (i.e., the first generation of selective breeding).
- Differences in aggression between the selectively bred lines were age dependent and highly vulnerable to experiential influence. Line differences were nonexistent prior to puberty, were robust in early maturity, and were modest in late maturity.
- Selection effects on aggressive behavior appeared to be mediated by line differences in arousal-reactivity. We proposed that the initial attacks were supported by the higher reactivity levels of the high aggressive line and/or by the greater probability of freezing or immobility in the low aggressive line. When the high-intensity reactions occurred among previously isolated animals in the dyadic tests, they typically escalated to attacks.

Quoting from that article, "we had apparently varied—through selective breeding—the maturation rate of the physiological systems related to arousal-reactivity" (Cairns, 1976, p. 123). Modest changes in basic physiological systems seem to have been associated with marked differences in aggressive behavior. To interpret the findings on selection, deBeer's (1958) concepts of heterochrony were invoked along with the proposal that behavior may be at the leading edge of developmental and microevolutionary change.

As a confession of ignorance and a hedge against overinterpretation, Cairns added, "Our best guess is that it is not the whole story" (Cairns, 1976, p. 124). A major problem was the ambiguity about the precise nature of the mediational process. That is, the major physiological-behavioral pathway through which the effects of genetic selection were channeled was speculative. In the absence of between-line differences in physical features or morphology, discussions of behavioral mediation focused on bias in activity and reactivity (Cairns, 1976; Lagerspetz, 1964). Our proposal on the possible role of enhanced reactivity was based on (1) the salient effect of isolation rearing, which is a key manipulation for bringing out aggressive behavior in male mice, and (2) the direct observation of aggressive interchanges in previously isolated animals. It had been earlier reported that initial attacks occurred as an outcome of an escalation sequence (Cairns & Nakelski, 1971). Investigatory actions by nonisolated social partners typically gave rise to startle counterresponses in the isolated animal. These in turn led to enchainment of the two participants in a vigorous encounter where bites and attacks rapidly followed (Banks, 1962; Cairns, 1973). In accord with the escalation hypothesis, it was found that diminishing the counter-reactions of the nonisolated animal by drugs sharply diminished the probability of its being attacked (Cairns & Scholz, 1973).

From these observations it was expected that a selective breeding program that imposed a dual criterion (increased attack and heightened reactivity to stimulation) would lead to effective line differentiation. When the two criteria were jointly employed, behavioral differentiation occurred very rapidly. But there was a problem. Simultaneous selection for heightened attacks and heightened reactivity obscured the interpretation of mediation. It was unclear whether the same effects would have been obtained if only a single criterion had been employed—either aggression without attention to reactivity or reactivity without attention to aggression.

Moreover, a key feature of the evidence reported in 1976 was inconsistent with the idea that the effective mediational factor was heightened reactivity in the high aggressive line. If the effective factor were indeed heightened reactivity, one would expect increased aggressiveness in the descendent high line rather than decreased aggressiveness in the low line. The results were opposite to this expectation. Although the selection program was bidirectional, the primary outcome was production of a line of mice less aggressive than the foundational stock. In this regard, a correlated behavior emerged in the course of selection that distinguished the low aggressive lines from the high aggressive ones: The low aggressive animals developed a strong tendency to become immobile and nonreactive (i.e., behavioral "freezing") on being approached by a conspecific.

The last outcome—increased immobility in the low-aggression line—suggests an alternative interpretation of the results. Selection may change the low line rather than the high line. Indeed, the observation that the selection experiment produced a line that was lower in aggressive behavior than the founding generation indicates that the analysis should focus on the properties of the low line rather than the high line.

## THE PRESENT

One of the first things done following the work presented in 1976 was to extend the research and attempt to replicate it with a fresh selective breeding program. Now, after more than a decade, the independent replication is in its sixteenth generation. With regard to the three original points, the results indicate the following.

1. The rapidity of selection. The intial finding of very rapid line differentiation held up over the next five generations. Figure 3-1 shows the results of the generations beyond the first two of the original experiment. The abscissa shows the generations in the I (Indiana) breeding series with ICR mice begun in 1972, and the ordinate shows attack frequency. I-100 refers to the line selected for low aggression and I-900 to the line selected for high aggression.

    The differentiation of high- and low-aggression lines proceeded almost as fast as in the original series when the selection experiment was replicated in another site with ICR foundational animals from a different supplier (Figure 3-2). The line separation occurred at the third generation in the NC (North Carolina) lines begun in 1976. These results do not stand alone. Lagerspetz (1964) and van Oortmerssen and Bakker (1981) found virtually the same rapid selection effects.

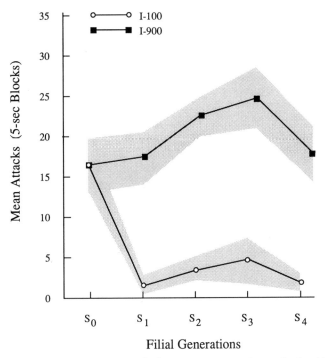

**Figure 3-1** Generational changes in rates of attacks for high-aggression (I-900) and low-aggression (I-100) aggressive lines (first selection series). Standard error of the mean shown in gray. (Adapted from Cairns, MacCombie, and Hood, 1983.)

2. Relativity of behavioral genetic outcomes to age and experience. The effects of developmental change and experience were harder to pin down. Although the tests were brief (ten minutes in duration) and the intervals between them were long (four tests over six months), there were salient carry-over effects from one test to the next. To assess simultaneously the effects of repeated testing, genetic differences, and their interaction, comparisons were made between the standard longitudinal research design and a new procedure that has been labeled the *cosibial longitudinal design* (Cairns, MacCombie, & Hood, 1983). In the cosibial design, members of the same litter (i.e., full siblings) are assigned to be tested at different ages. The litter members are tested just once except for one animal chosen at random to be repeatedly measured (i.e., the longitudinal subject). As in a standard longitudinal design, this individual is tested on all days. Taken together, the cosibial and standard animals provide a sensitive assessment of developmental changes, experiential effects, genetic differences, and their interaction.

The longitudinal/cosibial comparisons in $S_4$ demonstrate one technique by which these potential sources of variance may be disentangled. For the standard longitudinal animals, the lines converge in the number of attacks observed by late maturity. These outcomes, shown in Figure 3-3, replicate those in the original report (Cairns, 1976). However, a different picture on

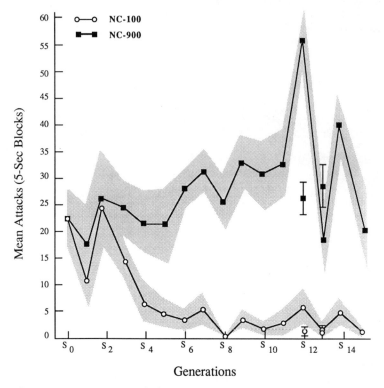

**Figure 3–2** Generational changes in rates of attacks for high-aggression (NC-900) and low-aggression (NC-100) aggressive lines (second selection series). Standard error of the mean shown in gray. (Adapted from Gariépy, Hood, and Cairns, 1988.)

convergence is obtained when the cosibial longitudinal results are analyzed (Figure 3–4). The complete convergence obtained among repeatedly tested animals is not evident in siblings tested just once but at different ages. Moreover, repeated testing was associated with an increase in attack frequency in the low aggressive line, which was the basis for convergence between lines in later maturity.

Striking evidence for the joint effects of experience and age also appear in attack latency. Figure 3–5 shows the combined effects of development and experience in the $S_4$ generation. Each curve represents a separate longitudinal study in that testing began at different ages. As the animals age and gain experience, the latency to attack becomes shorter. This effect holds as strongly for the low-aggression line (NC-100) as for the high aggression line (NC-900). The difference in latency to first attack shows the effect of development; the slope of the line shows the effect of experience. These results suggest that both repeated testing and age have a great deal to do with the magnitude of genetic effects.

To sum up, these data support our original two points: the rapidity of genetic selection and the developmental relativity of the genetic effects. So far, so good. But

# THE PUZZLE OF AGGRESSIVE MEDIATION

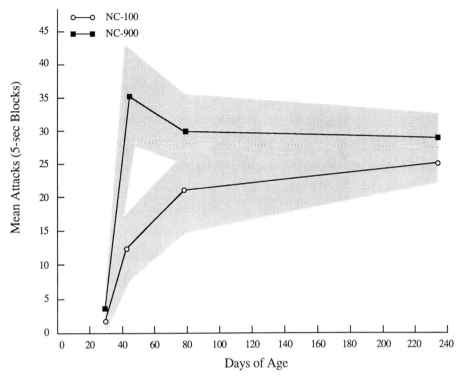

**Figure 3-3** Standard longitudinal investigation of attack frequency as a function of age in the high-aggression (NC-900) and low-aggression (NC-100) lines. Each animal was tested on four occasions ($S_4$ generation in second selection series). Standard error of the mean shown in gray. (Adapted from Cairns, MacCombie, and Hood, 1983.)

the third point—that heightened arousal mediates the selection effects in social interactions—required a major revision.

## MEDIATION OF AGGRESSIVE BEHAVIOR: HEIGHTENED REACTIVITY OR INCREASED INHIBITION?

In 1976 it was argued that the line differences reflected the higher reactivity and arousability of the high-aggression lines. As the picture was constructed, the higher reactivity in the high aggressive line was responsible for escalation in their species-typical exchanges and biased the mice toward aggression. However, the low aggressive lines were the ones found to depart from the foundational generation, not the high aggressive lines. This contradicted the notion that there should be generational increases in the reactivity of the high aggressive lines.

The problem was that we were focusing on changes in our high-aggression lines when we should have been paying attention to the low-aggression ones. Once we began to look at immobility in the low-aggression lines, the pieces began to fall into place. Three separate lines of evidence obtained over the past several years—micro-

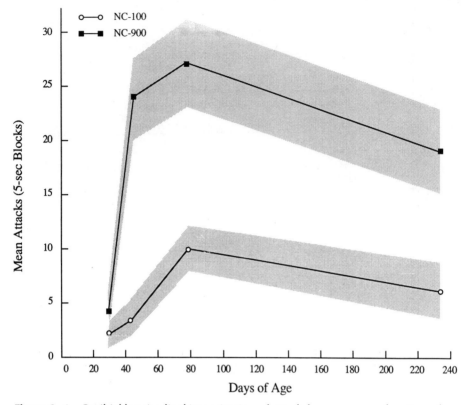

**Figure 3-4** Cosibial longitudinal investigation of attack frequency as a function of age in the high-aggression (NC-900) and low-aggression (NC-100) lines. One animal from each litter was tested on a single occasion, with the same litters represented on each test day ($S_4$ generation in second selection series). Standard error of the mean shown in gray. (Adapted from Cairns, MacCombie, and Hood, 1983.)

evolutionary, developmental, and interactional—point to the role of increased inhibition rather than heightened reactivity as the effective mediator.

### Microevolution, Aggression, and Inhibition

First we consider the evidence from selection. In the NC selection series, we found that the generational onset of line differences in behavioral immobility closely tracked the onset of line differences in aggressive behavior. The between-line difference first appeared in the $S_4$ generation, the same generation in which differences in rate of attack first appeared (Figure 3-6). After having discovered this strong effect in the NC series, we reanalyzed raw data from the original selection study. The question was whether a focus on heightened reactivity had overshadowed fundamental differences in behavioral immobility. Apparently it had. As shown in Figure 3-7, precisely the same outcomes with respect to immobility in a dyadic test had been

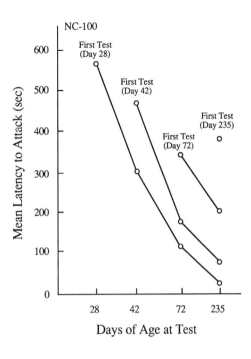
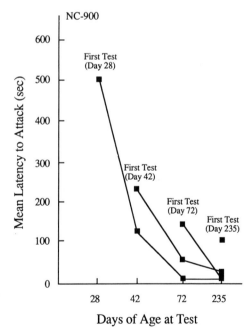

**Figure 3–5** Combined cosibial and standard longitudinal investigations of attack latency as a function of age in the high-aggression (NC-900) and low-aggression (NC-100) lines ($S_4$ generation in second selection series). (Adapted from Cairns, MacCombie, and Hood, 1983.)

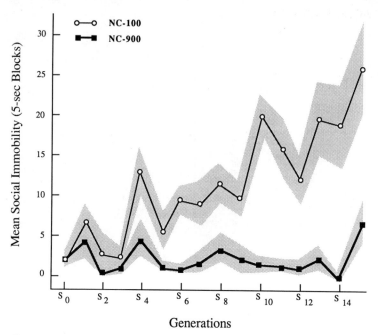

**Figure 3–6** Generational changes in social immobility for the high-aggression (NC-900) and low-aggression (NC-100) lines (second selection series). Standard error of the mean shown in gray. (Adapted from Gariépy, Hood, and Cairns, 1988.)

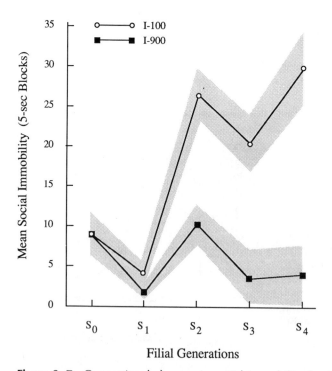

**Figure 3–7** Generational changes in social immobility for the high-aggression (I-900) and low-aggression (I-100) lines (first selection series). Standard error of the mean shown in gray.

obtained in the original selection study as in the subsequent work, but this disposition had been assigned only a supporting role at the time.[1]

Ironically, the disposition that had been our original focus—heightened reactivity—showed only modest differentiation of the lines in the subsequent selection study. For example, the primary measure of reactivity that had been described in initial publications on the isolation-aggression effect—reactivity to tactile stimulation (RTS) (e.g., Cairns, 1972)—has been deemed important enough to be a systematic part of the test protocol followed for all animals in all generations in both selection studies (see the appendix for a description of the procedure). In all generations of selection, the isolated animals differed markedly in RTS from the nonisolated or group-reared animals, as had been originally described (Cairns, 1973). The difference confirmed the sensitivity of the measure of reactivity employed.

Although isolated animals in both lines differed from their nonisolated counterparts, the high- and low-aggression lines did not differ from each other in the RTs measure (Figure 3-8). Convergent findings across generations were obtained in dyadic assessments of social reactivity (Gariépy, Hood, & Cairns, 1988). To sum up the microevolutionary evidence, behavioral immobility was strongly associated with line differentiation. Moreover, the directionality of the effects was such that the low-aggression line differed from the foundational generation in both rate of attack and behavioral immobility.

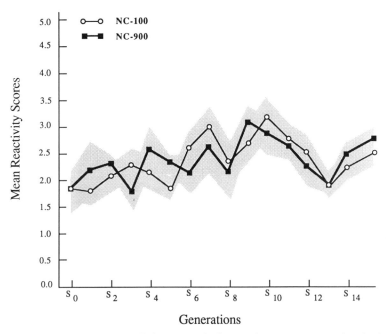

**Figure 3-8** Generational changes in behavioral reactivity (RTS) for the high-aggression (NC-900) and low-aggression (NC-100) lines (second selection series). Standard error of the mean shown in gray. (Adapted from Gariépy, Hood, and Cairns, 1988).

## Developmental Differentiation

Following Schneirla's (1966) dictum, a key avenue of evidence on the possible mediational role of behavioral inhibition should be found in the analysis of behavioral development. The evidence was clear with respect to developmental trends. In all generations, behavioral immobility diminished with age after day 30. This age effect was seen in the $S_1$ generation of the second selection study and was replicated in each subsequent generation.

What emerged over succeeding generations was a between-line difference in the developmental trajectory of freezing/immobility. Early in the selection series the two lines showed the same developmental trajectory. Line differences appeared by the fourth generation and continued through the fifteenth generation (Figure 3-9). When social immobility persisted into later development, it was associated with diminished levels of aggressive behavior. The developmental establishment of attack behavior directly paralleled the developmental diminution of the freezing response.

## Interactions, Inhibition, and Attacks

The third line of evidence on mediation is microbehavioral and interactional. A standard procedure, including test protocol and interactional coding, has been employed virtually unchanged for two decades in this laboratory (see Cairns & Nakelski, 1971). As a concrete picture of how interchanges were coded and analyzed, a brief account of how the tests were actually conducted is given in the appendix to this chapter.

Line differences in social interactions provided strong evidence for correlated genetic effects (Cairns et al., 1983; Gariépy et al., 1988). The biggest line difference was obtained for behavioral indices of inhibition. This result was consistent across both social and nonsocial testing conditions. To determine the importance of social immobility relative to other behavioral and physical factors, regression analyses were computed to determine the factors associated with attack latency at day 45 across generations, regardless of line. The multiple correlations were quite high, ranging from $r = .67$ to $r = .75$. The stepwise regression equations indicate that immobility was the first factor to emerge in all analyses, and it captured most of the predictive variance (Gariépy et al., 1988).

## RELATED INVESTIGATIONS

Significant variations in aggressive behavior in mice may be attributed to genetic factors (e.g., Fuller & Thompson, 1978; Ginsburg & Allee, 1942; Lagerspetz, 1964; Michard & Carlier, 1985; Scott, 1942; van Oortmerssen & Bakker, 1981). When behavioral performance is altered through selective breeding, the line differences obtained often extend beyond differences in the characteristics selected for. Selection for aggressive behavior in mice provides a case in point. Correlated differences have been demonstrated in patterns of lateralization (Scott, Bradt & Collins, 1986), in elimination behavior (Annen & Fujita, 1984; Sandnabba, 1985), and in emotionality (Annen & Fujita, 1984; Hall & Klein, 1942). The ubiquitous effects of selection for aggression have often been interpreted as evidence for the pleiotropic effects of the

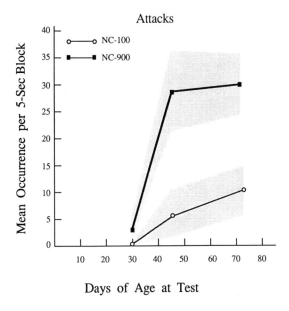

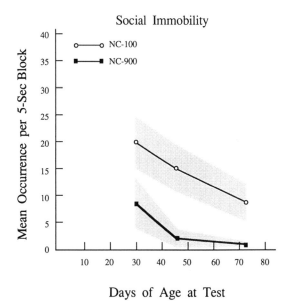

**Figure 3–9** Developmental changes in attacks and social immobility for the high-aggression (NC-900) and low-aggression (NC-100) lines as a function of age in cosibial longitudinal investigation ($S_{15}$ generation in second selection series). Standard error of the mean shown in gray. (Adapted form Gariépy, Hood, and Cairns, 1988).

genes involved. In our view, these correlated responses also provide clues regarding the physiological pathway through which genetic effects are mediated into behavioral actions.

It was initially proposed that modest shifts in arousal/reactivity threshold may provide an economical genetic route for bringing about manifold changes in social behavior and social organization (Cairns, 1976). Subsequent data permit that proposal to be amended and sharpened, by pointing to the threshold for behavioral inhibition as the likely route. It should be noted that the research on the developmental-genetic analysis of aggression in mice has been limited, save for some notable exceptions, to studies of males.[2] Might it be an entirely different story for females of the same species? To address this matter directly, Hood and Cairns (1988) conducted related work with females in the NC-100 (low-aggression) and NC-900 (high-aggression) lines. They observed that females, when tested in a sex-appropriate setting (i.e., postpartum), exhibited differences parallel to those obtained in the male lines. NC-900 females vigorously attacked "intruders" in the postpartum period whereas NC-100 females did not. Moreover, low-aggression line females were more likely to show inhibition than high-aggression line females. These outcomes are consistent with the Lagerspetz and Lagerspetz (1975) conclusion that line differences in aggressive behavior are transmitted by autosomal genetic mechanisms.

## CONCLUSIONS

Modest shifts in inhibition threshold provide a parsimonious route for bringing about manifold changes in aggression characteristics and social organization. Changes appeared in the investigations reported here in both individual development and in the microevolution of selected lines. Such flexibility in inhibition thresholds presumably is adaptive in natural settings. In this regard, it has been shown that behavioral immobility reduces the probability of continued attacks in various species (e.g., Annen & Fujita, 1984; Hennig, 1978). Freezing, whether supported by line differences in inhibition thresholds, by prior defeat, or by drugs, diminishes the probablity of further injury or death for the individual (Cairns & Scholz, 1973). In a view akin to this proposal, Kummer (1971) has argued that the same factors that account for variability within the individual over development may account for variability within species over evolution (see also deBeer, 1958; Gottlieb, 1983; Gould, 1977; and Piaget, 1978).

We propose that semireversible behavioral/neurobiological mechanisms exist that permit the fail-safe assessment of the success of adaptive changes prior to the emergence of durable physiological and anatomical scaffolds. The behavioral mechanisms should be sufficiently plastic to permit developmental and evolutionary risk taking. Moreover, the neurobiological processes that support these behavioral accommodations should themselves be susceptible to short-term and reversible change in development and microevolution. If these modifications prove repeatedly adaptive and reliable, they can be consolidated by more permanent and durable structures.

The issues of dual genesis and the mediational links between levels of analysis of social behavior remain as central for the discipline today as they were nearly a century ago (Baldwin, 1895). The contemporary evidence on the matter may be summarized in three points:

1. Developmental and genetic contributions to social behaviors typically collaborate, not compete, in achieving adaptation.
2. Social behavior patterns and the neurobiological processes that support them are mostly closed (i.e., fixed) in the course of development and across generations, but windows exist in both ontogeny and microevolution for dynamic, rapid, and reversible changes.
3. The same features of social behavior and neurobiological organization that are open to rapid ontogenetic change are typically open as well to rapid genetic change.

On the first point, Fuller and Thompson indicated in their classic volume, *Behavior Genetics* (1960), that there are virtually no competently done studies of selective breeding for specific social behaviors that have failed to achieve a separation of animals into distinct lines. Moreover, these separations are usually achieved rapidly, typically within two or three generations. These powerful findings have been downplayed, ignored, or obscured in psychology and behavioral biology. Even though these outcomes are inconsistent with some widely held beliefs on the conservative nature of gene effects, they nicely illustrate the sensitivity of complex social behaviors to genetic manipulation.

On the second point, (1) ontogenetic changes in behavior are more limited in scope and range than has been assumed by most contemporary developmental theories, and (2) microevolutionary changes are more open and rapid than assumed by most contemporary evolutionary theories. Although social behaviors and the neurobiological processes that support them are relatively more versatile and reversible than other organismic systems, they are nonetheless highly constrained in an absolute sense. A serious error of the nature–nurture dichotomy has been to equate plasticity with development and stability with evolution.

The third point implies a significant step toward the solution of the Lamarckian problem (i.e., how accommodations in development may be coordinated with and guide microevolutionary adaptations). Modern concepts that address this issue (e.g., phenocopies [Piaget, 1978]; genetic assimilation [Waddington, 1957]) can claim only modest empirical support. According to the present analysis, a key lies in the conceptualization of behavioral events as outcomes of processes instead of behavior as an entity. This step leads to a fresh look at behavioral-neurobiological systems, including analyses of how their several components are coalesced in development and in microevolution.

Given the nature of constraints on behavior and neurobiology, it should not be surprising to find that the windows for change that are open in development are the same ones that are open in evolution. An immediate goal for research is to identify the neurobiological substrata that promote the social behavior accommodations, whether in ontogeny or phylogeny. What is new is the idea that those behavioral-neurobiological components that are sensitive to rapid modification in individual ontogeny are also sensitive to rapid modification by genetic selection. A recognition of the dual genesis of social behaviors should sharpen both levels of analysis.

## APPENDIX: DYADIC TEST PROCEDURES

The first assessment day involved the reactivity to tactile stimulation (RTS) test, habituation in the dyadic test chamber, and weighing. In the RTS procedure, each animal was taken alone from the individual home compartment and placed in a Plexiglas cage with an ebonite floor. No bedding was supplied. After 30 seconds of habituation to the new surrounds, the subject was tapped lightly on its flank with a cotton swab. Six stimulations were delivered on a 15-second variable-interval schedule. After two minutes the animal was replaced in his cage. Reactivity was rated on a 5-point scale. A score of 1 was given if the animal exhibited no reaction to the stimulation, merely continuing his activity. Higher scores were given when the animal reacted with any combination of reflexive kicks, vocalizations, and jumps. Interobserver reliability for this measure was high, ranging from $r = .90$ to $r = .94$ throughout the testing periods.

Following the RTS test the animals were weighed and placed alone for ten minutes in the dyadic test chamber with the dividing panel in position. The animals were then returned to their home compartments and were undisturbed for 24 hours. Animals in the selected lines had been reared in individual compartments since weaning at day 21; they had had no previous experience in attacking or being attacked.

On the second day test, the subjects were observed in pairs in a dyadic test. The compartment used for this test was made of Plexiglas (20 cm × 21 cm × 31 cm) with a removable sheet-metal panel, which, when in place, divided the chamber into two equal parts. The subject (which had been isolated since weaning at 21 days of age) was placed on one side of the compartment and a same-age, group-reared male was placed on the other side. After five minutes of habituation, the panel was removed and all interactions between the subject and his partner were scored. The dyadic test terminated after ten minutes and both animals were removed and weighed.

The interactions observed in the dyadic test were scored following a dyadic syntax designed to identify both the initiator's actions and the other's response. The goal was to capture, in real-time coding, a continuous record of social events that occurred from the onset of the interchange to its termination. The record captured (1) the content of both individuals' behaviors toward the other, including the behavioral events associated with the initiation of the action and the counteractions that occurred, and (2) the sequence in which the social events occurred, including interactional sequences (between-animal, action-reaction patterns) and intraorganismic action successions (within-animal, autocorrelated behaviors and states). Social interchanges were recorded on specially designed sheets divided into 120 five-second blocks, but the coding was continuous. The method preserved the sequential dependencies of social interactions for subsequent coding. Thirty-three categories of social actions were used in the coding of the dyadic tests (see Cairns & Nakelski, 1971; Cairns & Scholz, 1973). A sample of the behavioral definitions is given in Table 3–1. In the several generations in which these categories have been used, interobserver agreement has always exceeded 90 percent. The interactional behavioral coding is versatile, reasonably simple to master, and powerful for the detection of subtle behavioral differences in sequence and content.

**Table 3-1** Selected categories of the dyadic behavior code

| Category | Behavioral criteria |
|---|---|
| Attack | Vigorous lunging toward the other animal, with intense biting/slashing directed at its back, head, belly, or rump. The entire body of the subject is coordinated in this highly vigorous and forceful species-typical action. |
| Withdrawal[a] | Sharp startle response and/or withdrawal when the animal is stimulated by a conspecific, including the mild stimulation involved in approaching, sniffing, and touching. It may also occur under more intense stimulation, including being bitten or attacked. |
| Kick[a] | Reflexive kick response, typically upon being touched, sniffed, or nosed by the other animal. |
| Immobility (freezing) | Remaining in rigid posture, nonmoving and seemingly frozen, upon and following stimulation by the other animal. The immobility may be brief ($\geq$ one s) or may extend for several seconds. |
| Tunnel | Species-typical, nose-first burrowing beneath the body of the other animal, sometimes accompanied by remaining still with head covered by the torso of the other animal. The approach may be from the front, side, or rear. Distinguished from "nosing" in that it does not involve active anogenital sniffing and investigation. |

[a] *Dyadic reactivity* $(W/K)$ is the sum of the two correlated categories, $W$ (withdrawal) and $K$ (kicking).

## ACKNOWLEDGMENTS

We thank students and colleagues who have collaborated in this work over the past 20 years, particularly Dennis J. MacCombie, Susan D. Scholz, Jane Neff Midlam, James A. Green, and Darlene DeSantis. Kathryn E. Hood has been a full collaborator in much of the research reported here. Work reported here has been supported by a National Institute of Child Health and Human Development grant to R. B,. Cairns (R01-14628) and a Natural Science and Engineering Research Council (Canada) postdoctoral fellowship to J. L. Gariépy.

## NOTES

1. But not entirely ignored. Although the lack of emphasis on social inhibition has been underscored in this chapter, it must be noted that the effect was not totally overlooked. For instance, in the original report it was observed that "when sniffed by another animal, those in the nonaggressive line tended to freeze or become rigidly immobile . . . [in contrast,] animals in the high aggressive line tended to respond to their test partners with vigorous counter-actions; they rarely froze" (Cairns, 1976, pp. 121–123).

2. The exceptions include the valuable work of Hyde and Ebert (1976), Ebert and Hyde (1976), Hyde and Sawyer (1979, 1980), Lagerspetz and Lagerspetz (1975), St. John and Corning (1973), Vale, Ray, and Vale (1972), and van Oortmerssen and Bakker (1981).

## REFERENCES

Annen, Y., & Fujita, O. (1984). Intermale aggression in rats selected for emotional reactivity and their reciprocal $F_1$ and $F_2$ hybrids. *Aggressive Behavior, 10,* 11–19.

Baldwin, J. M. (1895). *Mental Development in the Child and the Race: Methods and Processes.* New York: Macmillan.

Banks, E. M. (1962). A time and motion study of prefighting behavior in mice. *Journal of Genetic Psychology, 101,* 165–183.

Cairns, R. B. (1972). Fighting and punishment from a developmental perspective. In J. K. Cole & D. D. Jensen (Eds.), *Nebraska Symposium on Motivation* (Vol. 20, pp. 59–124). Lincoln, NE: University of Nebraska Press.

Cairns, R. B. (1976). The ontogeny and phylogeny of social behavior. In M. E. Hahn & E. C. Simmel (Eds.), *Evolution and Communicative Behavior.* New York: Academic Press (pp. 115–139).

Cairns, R. B. (1979). *Social Development: The Origins and Plasticity of Development.* San Francisco: Freeman.

Cairns, R. B. (1986). Development and evolution of aggressive behavior. In C. Zahn-Waxler, M. Cummings, & S. Inanotti (Eds.), *Altruism and Aggression: Social and Biological Origins.* New York: Cambridge University Press (pp. 58–86).

Cairns, R. B., MacCombie, D. J., & Hood, K. A. (1983). A developmental-genetic analysis of aggressive behavior in mice: I. Behavioral outcomes. *Journal of Comparative Psychology, 97,* 69–89.

Cairns, R. B., & Nakelski, J. S. (1971). On fighting in mice: Ontogenetic and experimental determinants. *Journal of Comparative and Physiological Psychology, 74,* 354–364.

Cairns, R. B., & Scholz, S. D. (1973). Fighting in mice: Dyadic escalation and what is learned. *Journal of Comparative and Physiological Psychology, 85,* 540–550.

deBeer, G. (1958). *Embryos and Ancestors* (3rd ed.). London: Oxford University Press.

Ebert, P. D., & Hyde, J. S. (1976). Selection for agonistic behavior in wild female *Mus musculus. Behavior Genetics, 6,* 291–304.

Fuller, J. L., & Thompson, W. R. (1960). *Behavior Genetics.* New York: John Wiley.

Fuller, J. L., & Thompson, W. R. (1978). *Foundations of Behavior Genetics.* St. Louis, Mo.: Mosby.

Gariépy, J.-L., Hood, K. E., & Cairns, R. B. (1988). A developmental-genetic analysis of aggressive behavior in mice: III. Behavioral mediation by heightened reactivity or increased immobility? *Journal of comparative Psychology, 102* (4), 392–399.

Ginsburg, B., & Allee, W. C. (1942). Some effects of conditioning on social dominance and subordination in inbred strains of mice. *Physiological Zoology, 15,* 485–506.

Gottlieb, G. (1983). The psychobiological approach to developmental issues. In P. H. Mussen (Ed.), *Handbook of Child Psychology* (Vol. 2, 4th ed.). New York: John Wiley (pp. 1–26).

Gould, S. J. *Ontogeny and Phylogeny.* Cambridge, MA: Harvard University Press.

Hahn, M. E., & Simmel, E. C. (Eds.). (1976). *Evolution and Communicative Behavior.* New York: Academic Press.

Hall, C. S., & Klein, S. J. (1942). Individual differences in aggressiveness in rats. *Journal of Comparative Psychology, 33,* 371–383.

Hennig, C. W. (1978). Tonic immobility in the squirrel monkey (*Saimiri sciureus*). *Primates, 19,* 281–299.

Hood, K. E., & Cairns, R. B. (1988). A developmental-genetic analysis of aggressive behavior in mice: II. Cross-sex inheritance. *Behavior Genetics, 18,* 605–619.

Hyde, J. S., & Ebert, P. D. (1976). Correlated response to selection for aggressiveness in female mice: I. Male aggressiveness. *Behavior Genetics, 6,* 421–428.

Hyde, J. S., & Sawyer, R. G. (1980). Selection for agonistic behavior in wild female mice. *Behavior Genetics, 10,* 349–360.

Hyde, J. S., & Sawyer, T. F. (1979). Correlated characters in selection for aggressiveness in female mice: II. Maternal aggressiveness. *Behavior Genetics, 9,* 571–578.

King, M. C., & Wilson, A. C. (1975). Evolution at two levels in humans and chimpanzees. *Science, 188,* 107–116.
Kummer, H. (1971). *Primate Societies: Group Techniques of Ecological Adaptation.* Chicago: Aldine Press.
Kuo, Z.-Y. (1967). *The Dynamics of Behavioral Development.* New York: Random House.
Lagerspetz, K. (1964). Studies on the aggressive behavior of mice. *Annales Acadamiae Scientiarum Fennicae,* Sarja-ser. B 131, 3, pp. 1–131.
Lagerspetz, K. M. J., & Lagerspetz, K. Y. H. (1975). The expression of the genes of aggressiveness in mice: The effect of androgen on aggression and sexual behavior in females. *Aggressive Behavior, 1,* 291–296.
Magnusson, D. (1985). Implications of an interactional paradigm for research on human development. *International Journal of Behavioral Development, 8,* 115–137.
Mason, W. A. (1979). Ontogeny of social behavior. In P. Marler & J. G. Vandenbergh (Eds.), *Handbook of Behavioral Neurobiology V.III: Social Behavior and Communication.* New York: Plenum Press (pp. 1–28).
Mayr, E. (1974). *American Scientist, 62,* 650.
Michard, C., & Carlier, M. (1985). Les conduites d'agression intraspecifique chez la souris domestique: Differences individuelles et analyses genetiques. [Intraspecific aggressive behaviors in the domestic mouse: Individual differences and genetic analyses.] *Biology of Behaviour, 10,* 123–146.
Piaget, J. (1978). *Behavior and Evolution.* New York: Pantheon.
Sandnabba, N. K. (1985). Differences in the capacity of male odours to affect investigatory behaviour and different urinary marking patterns in two strains of mice, selectively bred for high and low aggressiveness. *Behavioural Processes, 11,* 257–267.
Schneirla, T. C. (1966). Behavioral development and comparative psychology. *Quarterly Review of Biology, 41,* 283–302.
Scott, J. P. (1942). Genetic differences in the social behavior of inbred strains of mice. *Journal of Heredity, 33,* 11–15.
Scott, J. P. (1977). Social genetics. *Behavior Genetics, 7,* 327–346.
Scott, J. P., Bradt, D., & Collins, R. L. (1986). Fighting in female mice in lines selected for laterality. *Aggressive Behavior, 12,* 41–44.
St. John, R. D., & Corning, P. A. (1973). Maternal aggression in mice. *Behavioral Biology, 9,* 635–639.
Vale, J. R., Ray, D., & Vale, C. A. (1972). The interaction of genotype and exogenous neonatal androgen: Agonistic behavior in female mice. *Behavioral Biology, 7,* 321–334.
van Oortmerssen, G. A., & Bakker, Th. C. M. (1981). Artificial selection for short and long attack latencies in wild *Mus musculus* domesticus. *Behavior Genetics, 11,* 115–126.
Waddington, C. H. (1957). *The Strategy of the Genes.* London: Allen & Unwin.
Wohlwill, J. (1973). *The Study of Behavioral Development.* New York: Academic Press.

# 4
# Approaches to the Study of Genetic Influence on Developing Social Behavior

MARTIN E. HAHN

The relationships among three research areas—social behavior, development, and genetics—are at the heart of this chapter. Figure 4-1 illustrates these relationships. A careful survey of the literature reveals that the relationship between genes and development is under active investigation, the development of social behavior has been a vigorous research area for years, and the genetics of social behavior is a small but growing area of interest. At this time, however, with the exception of the pioneering dog research of Scott and Fuller (1965) and the work of a few other investigators, the interaction of genetics, development, and social behavior is virtually a null set. Computer searches of the relevant research literature list only a handful of studies. In spite of the lack of completed work, the study of genetic and developmental influences on social behavior is important from several perspectives.

First, studies of genetic influence on social behaviors and the development of such behaviors over the life span of animals will be of critical importance in evaluating theoretical statements in sociobiology (for example, Wilson, 1975; Dawkins, 1976) which states that many important social behaviors have responded to the pressures of natural selection and thus have a genetic basis. Or as Fuller (1983) has stated, "The fundamental theorem of sociobiology can be expressed as: genes influence important characteristics of social behavior that, as a result of natural selection, maximize the probability that copies of these genes will be represented in later generations" (p. 439).

An interesting and topical example of the potential contribution of behavior genetic study to sociobiology can be found in the phenomenon of infanticide. In several species of rodents, male intruders frequently kill any young they encounter, and it has been suggested that such killing (infanticide) is a reproductive strategy of those males in which the young of a female from a previous mating are destroyed and a subsequent mating with that female assures that a new pregnancy is rapidly begun (Brooks and Schwarzkopf, 1982). A male successfully carrying out such behaviors, that is who gains access to the reproductive capacity of a female, will have begun to send his genes into the next generation and simultaneously has removed some of the competition for animals carrying his genes. For infanticide to have a "payoff" for the intruding male, that is, for it to increase the probability that copies of the genes

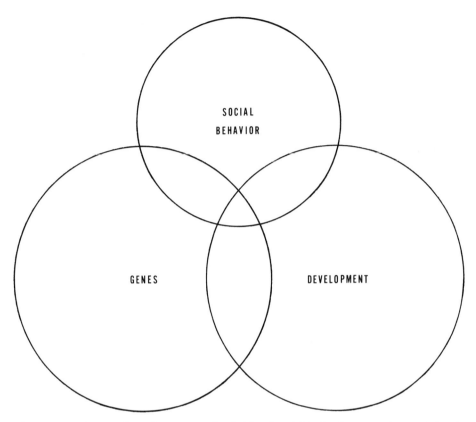

**Figure 4-1** Areas of overlap among the fields of social behavior, genetics, and development. The small triangle in the center illustrates the multidisciplinary field of developmental social behavior genetics.

of the intruder will be represented in later generations, infanticide must be heritable. To the best of my knowledge no studies have examined the inheritance of infanticide. Developmental studies of infanticide in male rodents could shed light on the origins of the behavior, eliciting stimuli and any experiences required to "enable" the behavior in adults.

Second, the developmental study of genetic influences on social behavior should be very helpful in attacking both of the major problems that Gottlieb (1983) set for studies of the development of behavior: (1) the normative description of the developing behaviors of the animal of interest and (2) the experimental analysis of those developing behaviors to establish linkages between earlier and later stages of development. Such study should obviate nature–nurture arguments about the origins of social behaviors by describing the ways in which genes and environment interact to produce social behaviors.

Third, such study will aid in understanding the relationship between the phylogeny and ontogeny of social behavior. Though an old view of this relationship suggested that developmental stages in an organism recapped the ancestral history of that organism's species, other approaches are possible. Two alternate versions of the relationship are available in the work of Cairns (1976; Chapter 3 in this volume) and Henderson (1981, 1986). Cairns is interested in changes in the rate of development, that is, heterochronies, of social behaviors in a species from a phyletic perspective. He asks, for example, whether the developmental features of social development in humans are accelerated or retarded with respect to the timing of the same features in closely related nonhuman species. Henderson, on the other hand, proposed a different view of the traditional life-span approach to the study of development, suggesting that behavior may be functional and adaptive at each stage in the developing organism. That is, rather than examining how the behavior of a young animal assists in the development of behaviors in the adult, Henderson examined the genetic architecture of behaviors expressed at all ages to determine their fit to existing environmental circumstances.

The chapter to follow is organized into two main sections. In the first I examine issues in methodology that are central to the study of developmental social behavior genetics. In the second I illustrate successful approaches to the field by describing four model studies.

## METHODOLOGICAL ISSUES

What methodological issues are involved in the study of the influences of genes and development on social behavior? A glance back at Figure 4-1 illustrates the disciplines of development, genetics, and social behavior; the interdisciplines of developmental genetics, the development of social behavior, and the genetics of social behaviors; and the multidiscipline of developmental social behavior genetics. Each of the single disciplines and each of the interdisciplines carries with it unique methodological issues. The multidisciplinary field of developmental social behavior genetics will contain issues of method derived from its constituent disciplines and new and unique issues as well. My approach here will be to outline important methodological issues in developmental social behavior genetics under the three questions, how to study, what to study, and when to study genetic influences on developing social behaviors.

### Question 1. How Should Genetic Influences on Developing Social Behavior Be Studied?

Lorenz (1981) described a strategy for researchers to make "an organic system understandable as a whole." The strategy involves beginning with as wide a view as possible—describing the ecological and evolutionary circumstances of the system of interest. Following this general description of the system, there should be statements identifying all of its parts. The final step is to understand the relationships among the parts of the system. Lorenz argued that this final step is meaningful only after the earlier steps have been completed. The overview of the system is indispensible in understanding the functioning of each part. As Lorenz stated:

> This progression in a direction from the entirety of the system to its parts is, in biology, obligatory. Naturally a researcher is free to make any part of an organic entirety the object

of his investigation; it is equally legitimate and equally accurate to examine the whole organic system within the context of its environmental niche, as the ecologist does, or to concentrate interest on molecular processes, as the biogeneticist does. But an insight into the network of the system must always be present so that the specialized researcher can be oriented with respect to where, in the total organic structure, the subsystem he is studying has its place. Only the sequence of the research steps is prescribed. (p. 40)

In applying the ideas of Lorenz, the developing organism should first be viewed in its ecological and evolutionary setting.

For example, Arnold (Chapter 9) discusses the interaction of ecological niche and ontogeny on eating behaviors in a species of water snake. Arnold's discussion of the literature indicates that young snakes feed primarily on aquatic insect larvae, while adults prey on crayfish. A behavioral shift accompanies the food source shift, because the insect larvae are swallowed head first and the crayfish is swallowed tail first. An investigator studying the snake would be hard pressed to understand its eating behavior without knowledge of the snake's prey selection in the natural setting.

Second, in the scheme of Lorenz, the parts of the system need to be described in detail. Applied to the investigation of a developing organism's behavior, this means that the naturally occurring or normally occurring behaviors of the organism at all ages should be described. This point was also made by Gottlieb (1983) as part one of a two-part study of any developmental phenomenon. A description of the behaviors of a developing organism and the normal time course of their appearance in an organism (Part I), is basic to any experimental analysis of the causation of a behavior (Part II). Gottlieb further pointed out that such basic analysis of the normal developmental events and their time course provides a starting point for observations of continuities and discontinuities in behavioral development. Such continuities and discontinuities would seem to be well suited to genetic analysis, particularly since starts and stops in development might be underlaid by the action of regulatory genetic systems.

Third in the prescription provided by Lorenz is study designed to allow statements describing the relationships among the parts. Applied to the development of a behavior, such experiments would be designed to search for causal linkages between early and later stages of development. Massive amounts of data have been collected and vigorous theoretical arguments have raged on the question of causal linkages between early and later events, since the outcome of such study has been viewed as being at the core of the nature–nurture question. Genetic analysis of development, as described by Eaves et al. (Chapter 13), Hewitt (Chapter 11), and Henderson (Chapter 10) should provide critical data for understanding the relationships among the parts in the developing organism.

## Question 2. What Phenomena Are to Be Studied in Developmental Social Behavior Genetics?

In his presidental address to the Behavior Genetics Association entitled "Social Genetics," J. P. Scott (1977) posed two problems for the genetic study of social behavior:

> ... almost all behavior that is exhibited by the members of a highly social species such as man is expressed within social relationships. What little solitary behavior remains is

expressed within social contexts derived from these relationships.... A member of a highly social species always acts as a part of a social relationship, which is the only unit of behavioral measurement that has a large degree of independence. (pp. 327–328)

First, just as Lorenz argued that an organism must be viewed in its ecological setting, Scott argued that an organism must be viewed within the context of its social relationships. The environment of social relationships as described by Scott is not fundamentally different from the ecological environment discussed by Lorenz, since both provide a rich, continuous, and varying source of cues for the behavior of an organism. While cues in the environment may come from air temperature, wind velocity, or animals of another species, cues in "Scott's environment" come primarily from animals of the same species engaged in relationships with an animal of interest. Making sense of the developing social behavior of an animal outside of social relationships or without knowledge of its social relationships would seem to be as problematic as doing so without knowledge of its fit to an ecological niche.

Second, by his statement on the nature of social behavior and relationships, Scott made the interaction between two or more organisms the unit of behavior to be measured in studies of genetic influence on social behavior. This is a critical issue, since it would appear that the phenotype being measured—the outcome of the interaction—cannot be clearly attributed to the genotype or to experiences of any single organism involved in an interaction. In contrast, in a behavior genetic study of rodent activity, the "number of squares entered" is readily attributed to a single animal, and comparisons between animals of different genotypes is straightforward. Methods have been devised to deal with this issue; Fuller and Hahn (1976) placed them into three categories: (1) a method that homogenizes genetic and experiential influences by pairing animals of similar genotype and/or experience—the homogeneous set design, (2) methods that use a "standard" partner for social interactions—the standard-tester design (the standard partners behave minimally or in a standardized fashion, allowing the experimenter to attribute the outcome of interactions to genetic and/or experiential influences of the nonstandard animal) and (3) methods that employ a panel of social partners—the panel-of-testers design. In this third case the investigator can sum the interactions of a test animal with the partners in its panel and calculate an average score and differences between the average scores of test animals with their panels can be attributed to genetic and/or experiential effects. Examples of the use of these designs are seen in the studies by Scott and Fuller (1965); by Cairns, MacCombie, and Hood (1983); and by Plomin and Rowe (1979). Naturally, there are advantages and disadvantages associated with each of these designs, which were detailed by Scott (1983) and by Fuller and Hahn (1976). For example, a disadvantage of the homogeneous set design is that the results may have little generalizability, since a social group composed of organisms of very similar genotype and/or experience may be quite rare in nature. Without these methods, however, genetic analysis of social behaviors would not be possible.

## *Definitions of Social Phenomena*

To effectively examine Scott's ideas on the behavioral context provided by social relationships and to integrate those ideas into methods for the genetic study of developing social behavior, it is important to have clear definitions of the social phenomena to be studied. Three basic social phenomena can be identified: social behavior, social interactions, and social relationships.

*Social Behavior.* Social behavior has been defined in a variety of ways. Some definitions have been exclusive—for example, social behavior is an exchange of behaviors between two individuals of the same species. Some definitions have been inclusive—for example, social behavior is a behavior of one individual that has an effect on another individual. These definitions as well as a range of possible definitions that fall between the extremes differ on two points: (1) the complexity of the interaction between the animals and (2) whether the animals engaged in the behavior are from the same species. I prefer an inclusive definition, since this type includes behaviors that show a range of interaction between individuals and that are engaged in by animals of different species:

> Social behavior is the behavior of one animal directed toward another animal or that has an effect on another animal.

Adopting an inclusive definition of social behavior allows study of a number of types of behavior of great interest. A survey of texts on animal behavior and animal social behavior finds the following social behaviors, all of which fit into the category of social behavior as I have defined it:

| | |
|---|---|
| Aggression | Dominance |
| Attachment | Maternal, paternal behavior |
| Attraction | Mating |
| Coaction | Recognition |
| Communication | Socialization |
| Courtship | Territoriality |

While most of these types of behavior would be included in a more restrictive definition of social behavior, because they involve members of the same species interacting, some of them, such as coaction or other types of noninteractive, primitive, social behaviors, would not fit. Predatory and antipredatory behaviors, which are in the category of aggression, would be excluded from study if "social behaviors" were restricted to members of the same species.

*Social Interactions.* Social interactions refer to exchanges of behaviors within any of the types of social behavior just listed. The intent of the phrase, *social interaction,* is to describe the interlocking of behavior between two or more organisms by virtue of the cue functions that each behavior or act has for the other organism. Such cue or stimulus functions serve to attract, direct, synchronize, facilitate, or inhibit the activities of another organism and thus shape the interaction. Classic examples of social interactions are found in the courtship and mating sequence of many birds and fish. The most familiar is illustrated in the mating pattern of three-spined stickleback fish, discussed by Tinbergen (1969) and illustrated in Figure 4-2. Tinbergen described this social behavior as a sequence of acts, each serving as a releaser for a succeeding act through attracting, directing, facilitating, or inhibiting.

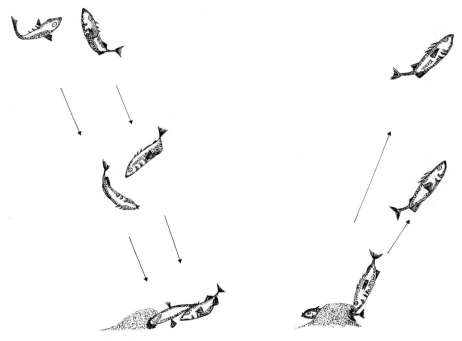

**Figure 4–2** Example of a social interaction: the courtship pattern of the three-spined stickleback as observed by Tinbergen. (Adapted from Tinbergen 1969).

*Social Relationships.* Social relationships are based on a series of interactions between two or more organisms over some time span. Since relationships are based on experientially acquired or in some cases genetically stored information, as the number of interactions between two organisms increases, usually the character of the interactions becomes more stable and more predictable. Scott (1977) listed several features of relationships that are of interest.

1. Relationships require feedback so that the behaviors of two organisms affect each other.
2. Relationships become more stable through adjustment and learning.
3. A relationship, such as dominance-subordinance, may be asymmetrical, with one organism exerting more influence on its partner than is reciprocated.
4. Aspects of relationships and the behavior of individuals within relationships change over time.

Developmental events such as weaning or the onset of sexual maturity will likely alter the nature of a relationship in a fundamental manner. A mother-pup relationship in dogs, with its rich exchange of behaviors, its initial lack of symmetry, and its changing character over time well illustrates the points made by Scott. Studies of genetic and environmental influences on stability and change in social relationships should prove to be rewarding; models that can be employed for such analyses are presented by Hewitt (Chapter 11) and by Eaves et al. (Chapter 13).

In addition to studying social relationships for their intrinsic interest, we may find that studying the behaviors of organisms within relationships may serve to clarify the literature on certain social behaviors. For example, Simon (1979) in reviewing genetic studies of intermale fighting in mice, discovered that an array of seemingly minor methodological differences between and even within studies led to very different results and conclusions about the role of genotype in mouse aggression. Such factors as the number of males interacting, the level of lighting, and the use of a home or neutral cage for testing altered the outcome of comparisons between males or different inbred strains. In virtually all studies of intermale mouse aggression, observations are made on mice that are strangers when placed together. In addition, observations of their interactions are characteristically of short duration. The disparate results Simon described may be the result of varying gene expression in different situations. However, it may be that such differences are due to instability associated with brief encounters and that they would disappear in mice that had a history of prior interactions-relationships. Studies comparing the behaviors of sets of animals with a history of interactions versus sets of animals without such a history would provide useful information on the predicted stabilizing nature of relationships. An example of such a study is that of Plomin and Rowe (1979), who compared the social behaviors of human babies toward their mothers and toward a stranger (see the second part of this chapter for a more detailed description of this study).

## *Units of Behavior*

Precise definition is an essential first step in the analysis of factors that influence social behavior. However, there are ways of classifying social behaviors that go beyond simply generating a set of defined categories. For example, Scott (1958, 1972) proposed a set of categories for animal behavior, as follows:

Investigatory
Shelter seeking, shelter building
Ingestive
Sexual
Epimeletic (care-giving)
Etepimeletic (care soliciting)
Agonistic
Allelomimetic
Eliminative

Scott went on to carefully name and define each category in a way designed to isolate the behavior from confusion that had surrounded research on it in the past. Confusion had occurred either because the behavior had been loosely defined or because a behavior name had been chosen that was loaded with connotations from nontechnical use. For example, the term *aggressive behavior* or *aggression* has been widely used to mean such divergent acts as a bee sting or a hostile verbal exchange. Scott renamed aggression *agonistic behavior* and defined it as "any adaptation which is connected with a contest or conflict between two animals of the same species, whether fighting, escaping or freezing, may be included under this term" (Scott, 1958). Careful naming of behaviors and precise definitions of behavioral categories will assist future investigators in avoiding misunderstandings about the behaviors under study.

Fuller and Hahn (1976) took a different approach to classification. They placed social behaviors in categories on the basis of the behaviors' probable genetic consequences.

Class 1. Some social behaviors increase or decrease the probability of transmitting the genes of all participants. Sexual behaviors, caretaking, and care-soliciting behaviors, if successful, increase the probability of transmitting the genes of all participants. If these behaviors are unsuccessful, the probability that the genes will be transmitted to all participants is reduced.

Class 2. Some behaviors alter the probability of transmitting the genes of some of the participants. Agonistic behaviors (e.g., fighting and threatening), dominance, territoriality, and infanticide alter the probability of gene transmission of the participants. Infanticide, for instance, may eliminate offspring carrying the genes of a certain animal, while allowing the offspring of the infanticidal animal a greater chance for survival.

A third approach is represented in the work of Tinbergen (1969). Tinbergen's system places behaviors into a hierarchy, an "if, then" sequence of behaviors. Figure 4–3 illustrates such a sequence for the behaviors involved in stickleback reproduction. With this system, Tinbergen has not only named the behaviors of interest but outlined relationships among those behaviors. The system allows an investigator to examine sequences of behaviors and begin to document causal links in the establishment of social behaviors.

A fourth approach, a multivariate developmental approach, is exemplified in the work of Henderson (1979, 1986). Multivariate approaches are achieving great popularity, especially in the analysis of genetic and environmental influences on developing behaviors. (See the entire issue of *Behavior Genetics* devoted to multivariate techniques, with an introduction by DeFries and Fulker, 1986). Until now, most genetic analysis of animal behavior has been carried out on individual behaviors (e.g., seconds of fighting or trials to the acquisition of a learned response) measured in a

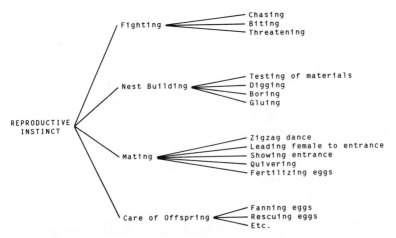

**Figure 4–3** Tinbergen's (1969) hierarchy of reproductive behaviors: one method of organizing behaviors into units.

highly specific and controlled situation using animals of one age. Henderson (1986) presented a detailed argument that genetic analyses based on such narrow slices of behavior do not allow a general understanding of the workings of the genotype or of the relationship between behavior and evolution. Applying his argument to a data set he had collected over the years, he organized the behaviors into a cluster, "activity." His measures (Henderson, 1986, Table IV) were all of activity though obtained in different situations and at different ages. Among the behaviors were open field activity at 4 days of age, nest return at 11 days of age, cage jump at 14 days of age, and rod and pole activity in adults. Henderson concluded that a multivariate developmental approach (examining clusters of individual behaviors across situations) provided a more coherent picture of the relationship between behavior and evolution than did a single-variable approach.

Social behaviors can be organized as Henderson organized activity. For example, dominance in a litter of dogs can be observed through studies on nipple competition at 8 days of age, rough play and mild fighting at 60 days, and adult male dominance at 400 days. However, to successfully integrate multivariate approaches and the development of social behavior, investigators will need to give careful consideration to the behavioral units that are organized within clusters. Social behaviors usually involve interactions, and thus sequences of behaviors have to be recorded rather than a single activity like the number of squares entered in an open field. The results of detailed genetic analyses on social behaviors studied in a developmental multivariate manner will allow a much greater understanding of the influence of genotype and environment on continuities and changes in social behavior during development.

## Question 3. When Should Genetic Influences on Developing Social Behaviors Be Studied?

One serious problem that will need to be addressed if a literature on developmental social behavior genetics is to be developed is the time and resources required to complete the appropriate longitudinal studies. For example, in a genetic analysis of the development of dominance relationships within litters of mice, the focus of the investigation would be the set of behaviors and their time course, which lead to defining a dominance structure. The study would span the period birth through 40 days of age (just after the onset of sexual maturity). The genetic method would involve a diallel cross breeding design—a complete intercrossing of a set of inbred strains. The diallel design (which is referred to by several other authors in this volume), allows highly sophisticated genetic analysis. With use of a spartan methodology to conserve resources, four inbred strains would be completely intercrossed by the diallel method, to produce 16 genetic groups. Ten litters of each genetic group would be observed for one hour each day from birth through 40 days of age. Such a procedure would require 6,560 hours of observation, or 820 eight-hour days! Many additional hours would be required for collapsing the observations and completing an analysis of the data. Though such a study would be extremely valuable in assessing the constant and changing interactions of heredity and environment on dominance behavior in developing mice, the cost in time and other resources would be prohibitive. To my knowledge, no such study exists in the literature. Clearly, an important methodological issue for developmental social behavior genetics will be to find ways to economize in methodologies while not abandoning those methodological principles delineated by Lorenz (1981), Gottlieb (1983), and Scott (1977).

One way to achieve such economy would be a two-stage procedure, to (1) describe the entire developmental sequence of a particular social behavior and (2) employ probes to isolate particularly fertile periods for detailed genetic investigation. For example, in the hypothetical study just described, instead of a survey being made of the behaviors of litters of mice of all the groups generated by the breeding design, the behaviors of two inbred strains or of two hybrids might be examined for 40 days and then particular periods chosen for observation on the 16 genetic groups generated by the diallel cross. Such a two-stage strategy should satisfy the methodological prescriptions of Lorenz, Gottlieb, and Scott as discussed earlier and provide an economy in the use of resources.

Guidance on the optimal timing for probes could come from knowledge of significant physiological events that take place during the development of the organism. Such events could be as molar as the onset of sexual maturity or the day the eyes of an animal open, or as specific as the maturation of specific neurotransmitter systems (see Chapter 7 for more detail).

Another source of information for obtaining probes is theory. One example of a potentially applicable theory is the widely known critical period theory of behavioral development detailed by Scott (1978, 1968, 1962). A second example is the recent theory of genotype–environment correlation proposed by Plomin, DeFries, and Loehlin (1977) and elaborated by Scarr and McCartney (1983). These theories should be useful because each contains statements about the relationship of time to development. Briefly, Scott (1962) cited embryological evidence indicating that rapidly growing tissue is most susceptible to changes in the environment. Using this finding, Scott showed that there are periods during which certain behaviors are most susceptible to environmental stimulation. Among these behaviors are species preference and attachment (e.g., imprinting). Using the dog as an example, Scott showed that at about 3 to 4 weeks of age, there is a strong attraction for other organisms. At about 12 to 14 weeks that attraction is replaced by fearfulness. The social object that is present during the critical or sensitive period (about 6 to 12 weeks of age) is the object toward which the dog will behave socially as an adult. Thus, over a span of six to eight weeks the dog is attracted to others, is maximally susceptible to the cues they present, and then avoids organisms that do not carry those cues. This has been described as a period of primary socialization. It occurs on different time lines in such diverse species as dogs, ducks, and humans.

Knowledge of this theory and of others like it allows investigators to concentrate their study of genetics and social behavior on specific, manageable periods of time. Plomin, DeFries, and Loehlin (1977) and Scarr and McCartney (1983) have proposed a theory that genotype drives the selection of aspects of the environment. Further, the theory as described by Scarr and McCartney has a developmental component, because across age, selection of aspects of the environment is alternately "passive, evocative or active." A very young organism has little control over its environment, but since it shares genes with its parents, their contribution to its environment will not be random but related to its genotype. In the second phase, the more developed organism's mode of behaving will begin to elicit certain aspects of the behavior of its parents more than other aspects. In the third phase, the older organism will actively move to control its environment. This theory could result in economies in methodology because, for instance, it proposes that the development of social relationships is not uniform across age. An investigator interested in individual differences in continuities and changes in the development of interpersonal relationships could concentrate on one of the periods described as passive, evocative, or active or could attempt to describe the events that characterize transitions from passive to evocative or evocative to active.

## AN AGENDA FOR RESEARCH IN DEVELOPMENTAL SOCIAL BEHAVIOR GENETICS

To this point I have discussed several issues relating to the methodology of the study of developmental social behavior genetics. Further, I have presented definitions of the social phenomena to be studied. I would like to build on this discussion by suggesting questions that could serve as the primary agenda for study. These five questions are an update and extension of a list proposed by Cairns (1979) for the study of social development.

1. How are social behaviors, social interactions, and social relationships initially established?
2. How do social behaviors, social interactions, and social relationships remain stable over time?
3. How are social behaviors, social interactions, and social relationships modified over time?
4. What influences are instrumental in producing individual differences in the establishment, maintenance, and modification of social behaviors, interactions, and relationships?
5. How do individual differences in the establishment, maintenance, and modification of social behaviors, interactions, and relationships affect individual fitness?

## MODEL STUDIES

In this section I will describe aspects of model research studies that illustrate approaches to the genetic study of developing social behaviors as outlined in the agenda proposed in the previous section. The four studies examine critical periods in the development of social behavior in dogs (Scott & Fuller, 1965), the evolution of ontogenies of fighting behavior in mice (Cairns, MacCombie, & Hood, 1983), features of ultrasounds in developing rodents (Sales & Smith, 1978) and social behaviors in young human twins (Plomin & Rowe, 1979).

## Critical Periods in the Behavioral Development of the Dog

Scott and Fuller's 1965 text is over 20 years old. Yet their comprehensive approach to the study of behavioral development in the dog makes their methods and data a unique set that retains its value today. A careful reading of their monograph and the minutes of the planning conference that preceded the project (Scott & Fuller, 1965; Scott & Beach, 1947) indicates that, with the exception of the detailed developmental genetic analysis being developed today, their methods provide benchmarks for current and future studies of developmental behavior genetics.

The work that is of particular interest to my chapter on social behavior is found in their Chapter 5, "The Critical Period," at the beginning of which they stated:

> Our studies of the development of behavior show that a puppy enters into a period of great change and sensitivity with regard to social relationships at approximately 3 weeks of age, and that his experiences at this time determine which animals and humans will become his closest social relatives. ... The period is therefore a critical one. (Scott & Fuller, 1965, p. 117)

For their work they employed five pure breeds of dogs—cocker spaniel, wire-haired fox terrier, African basenji, Shetland sheep dog, and beagle—and various cross-bred generation animals. Most of their subject dogs were reared in the laboratory, but some were reared with a minimum of human contact in large fenced fields and some with a maximum of human contact in human families. They prepared for studying the critical period of socialization by making detailed observations of several developmental events in the sensory, locomotive, and learning capacities of their dogs. They observed that at about 2 to 3 weeks of age, the eyes opened. At about that same time the dogs showed an auditory startle response, their locomotion began to mature from crawling to walking, and they began to eat hard food. Perhaps most important, the puppies were beginning to discriminate among their littermates. Thus, during this sensitive period beginning at about 3 weeks, the puppies could see and hear, could approach or avoid each other by walking, and were capable of remembering the consequences of interactions with other dogs or humans.

An important component of socialization during the critical period is approach to another organism. This behavior was observed during the "handling test," in which a puppy was allowed to interact with a human who presented a standard set of stimuli such as standing 2 feet from the puppy, walking several feet away, squatting down and holding out a hand to the puppy, and stroking the puppy. Figure 4-4. shows the attraction and following responses exhibited by dogs of the five breeds at 5 through 15 weeks and later at 52 weeks of age. There are clear overall breed differences, with the cocker spaniel showing the most attraction and following behavior and the basenji showing the least. Over the ten-week test period, the cocker remained constant in its behaviors, while the basenji changed dramatically, becoming more attracted to humans with increasing age. When tested for following at 52 weeks of age, the Shetland sheep dog followed humans the least, while the cocker spaniel followed the most.

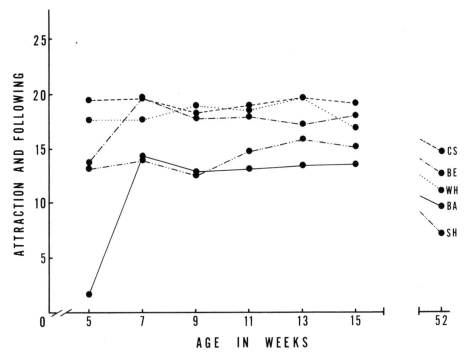

**Figure 4-4** Data from the studies of Scott and Fuller illustrating changes in a social behavior in dogs as a function of breed and age. (Adapted from Scott and Fuller, 1965.)

With this study Scott and Fuller provided data on the dog for several of the questions proposed previously in this chapter for the study of developmental social behavior genetics:

1. They described several of the physiological and behavioral antecedents to the establishment of social behaviors and relationships.
2. They showed that breed differences contribute to the stability and change of social behaviors over developmental time.
3. They showed that individual variation in the establishment of social behaviors is to some extent inherited.

Furthermore, they conducted their research in a manner consistent with the comments of Lorenz on treating a biological system as an entirety. That is, they placed their data in the context of the overall social structure and social organization of the dog, noting the differences in social organization that occur due to artificial selection by humans. Finally, they employed theory to suggest times of particular importance in the development of the dog, carefully studied one of those periods, and then used their data to answer questions about the causes of continuity and change in social behavior during ontogeny.

## Evolution and the Ontogenies of Fighting in Mice

Anchored in their general examination of the possible relationships between ontogeny and phylogeny (Cairns, 1976; Chapter 3 in this volume), the research program reported in Cairns et al. (1983) was to examine the ways in which evolution and ontogeny might interact in the expression of a social behavior in the house mouse. More specifically, these investigators argued that "modifications in the rate of development might mediate the generational differences obtained by selective breeding differences in attack occurrence" (p. 70). They reported the results of two separate selection experiments of five generations each on intraspecific attack in male mice. High-scoring and low-scoring males were selected for breeding with an index of attack (based on attack frequency/attack latency) in a ten-minute interaction with another male of the same age. Both selection programs achieved a substantial separation of high-aggression and low-aggression lines within four generations. Both selections were asymmetrical, as the high-attack line increased only slightly over the mean attack score of the base population, while the mean attack score of the low-attack lines dropped dramatically over generations. Cross-fostering failed to reliably alter the attack behavior of animals of either line in either selection.

Having shown that attack behavior in the male mouse has a genetic basis and responds to selective breeding, Cairns et al. switched the focus of their investigation to a search for the mediator of the selection. They first studied attack behavior in the high-attack and low-attack lines at several ages. The results showed that attacks varied with age in both lines. There was no difference at 28 days of age, but reliable differences occurred at 42, 72, and 235 days of age. Thus, developmental age plays a role in the expression of attack, and the maximum differences between lines occurs around the age when the selection index was applied. A further finding was that when animals are tested in more than one social encounter, their latency to attack drops dramatically. This finding serves as a reminder of the classic problem of the effects of order of treatment on experimental results. The problem of potential bias introduced by repeated measurements on the same organism or social unit will need to be carefully addressed in any developmental behavior genetic research.

To test the hypothesis that selection may operate on differences in developmental rates in mice (heterochronies), Cairns et al. (1983) compared attack rates in each line at different ages in the first and fourth generations of selection. In the first generation, a comparison of animals tested for attack rates for the first time at 28, 42, 72, or 235 days of age showed that the lines did not differ substantially at any age. By the fourth generation, however, the low-attack line was depressed in attack rate at ages 42, 72, and 235 days of age. Though the animals differed at the three ages, the two lines showed signs of convergence of attack rates with increasing age. The difference in attack score at 235 days was less than the difference at 72 days. The investigators interpreted these results to suggest that line differences after several generations of selection were associated with a delay in the development of attack behavior in the low-attack line.

As did the work of Scott and Fuller, this work by Cairns and colleagues contributes to a better understanding of the influences of development and genes on the expression of a social behavior:

1. With their demonstration of successful genetic selection for attack behavior, they showed that genes contribute to individual differences in social behavior in the mouse.
2. They demonstrated that the genetic influence on attack rate was not uniform over age. Rather it was greatest around the age at which the selection criterion was applied and it declined somewhat thereafter.
3. With the use of cross-sectional and longitudinal designs, they were able to suggest strongly that the line difference that resulted from genetic selection was mediated by a developmental retardation of attack in the low-attack line.

## Comparative Studies of the Ultrasonic Calls of Infant Rodents

A survey of the literature indicates that ultrasonic calls have been detected from the young of many species of murid (family of mice and rats) rodents. The calls have been elicited by several types of stimuli, but cold and handling are especially effective. Ultrasonic calls enhance the retrieval of pups emitting them by adult females and males and also enhance more general parental behaviors, such as nest building. Since young mice and rats have immature abilities to move and thermoregulate until at least 2 weeks of age, ultrasonic calling seems to be social signaling of great evolutionary importance. (For a biometrical genetic analysis that supports this statement, see Hahn, Hewitt, Adams, & Tully, 1987.) Sales and Smith (1978) described a study involving the young of ten species of murid rodents and three strains of one species, *Mus musculus.* They studied these animals for varying numbers of days between 1 and 35 days of age. They employed sophisticated measurements of ultrasonic calls and described such parameters as amplitude, length, rate, maximum, and minimum frequencies and measures of shape characteristics such as frequency drift, frequency steps, and frequency sweeps.

Their results revealed a rich complex of similarities and differences among the groups they examined. For example, the rate and intensity of calling decreased with age in all species. The young of all species emitted calls of shorter duration and higher pitch as they aged. Unlike what might have been expected, no species-typical frequency patterns emerged from their data. The results produced interesting species × age interactions in both the rate of calling and the peak intensities of calls. Sales and Smith interpreted these results to indicate that different rates of physiological maturation, which are present in the various species, influence the ability of pups to produce calls and alter the utility of the calls.

With their work, Sales and Smith have provided important information on the development and evolutionary significance of ultrasonic calls produced by infant murid rodents:

1. They showed that a social behavior—infant mouse ultrasonic calling—changes with development and that species differences contribute to the changes. This portion of their data fits nicely with Cairns' notions (1976;

Chapter 3 in this volume) about one relationship between ontogeny and phylogeny, "dual genesis," in the construction of behaviors in general and of social behaviors in particular.

2. As we saw in the work of Scott and Fuller discussed previously, Sales and Smith placed the behavior they studied within the context of social relationships (i.e., pup-parent relationships) and in an ecological setting (the nest and burrow habitat occupied by naturally living murid rodents). The interpretations they made of their own data and of the data of other investigators was facilitated by an understanding of the biological system in which infant rodent ultrasounds are heard.

## Individual Differences in Infant Human Social Behavior

Rounding out the set of studies that I wish to discuss is a study on basic human infant social behavior. In terms of the behaviors examined and the methods used, this study is reminiscent of that of Scott and Fuller (1965) already discussed. Plomin and Rowe (1979) set out to examine the etiology of individual differences in the reactions of infants to brief experiences with strangers and to a brief separation from their mother. Genetic studies of human social behavior are rare, yet when properly done they provide important information on the ways in which genotypes and environments influence behavior. Comparisons of twin sets are particularly useful, since they can yield estimates of the amounts of variability that can be attributed to genetic, within-family, and between-family sources.

Plomin and Rowe observed 42 male and 50 female twins who ranged in age from 13 to 37 months. The twins were observed in their own home in seven different standardized situations. These included stranger approach, play with stranger, cuddle with mother, and separation from mother. Among the behaviors measured were latency of approach to stranger, positive vocalizations, smiles, squirming, and negative facial expressions.

The results of this study can be summarized in five points: (1) The children varied greatly in their social responding. (2) The children's responses tended to be situation specific. A factor analysis of the results yielded five separate dimensions of responding. (3) The factors of age and sex were overcome by individual differences in the behavior of infants at a given age and sex. (4) Using intraclass correlations to compare monozygotic and dizygotic twins, Plomin and Rowe found that social behaviors toward strangers were heritable across several situations. The same analyses revealed that only one behavior between mother and child (touching) was heritable. (5) While within-family variables influenced a number of behaviors, between-family factors influenced only three behaviors.

This study is an especially important addition to the study of genetic influences on developing social behavior, since it illustrates the commonalities in approach between "human" and "animal" studies. Plomin and Rowe's "mother" and "stranger" paradigms are very much like the "homogeneous set" and "standard tester" paradigms used in genetic analyses of animal social behavior described by Fuller and Hahn (1976) and discussed above. Increasing the amount of interaction between those researchers studying humans and those studying other animals would be of considerable benefit to both.

In addition, Plomin and Rowe's study addresses two questions of the agenda for research in developmental social behavior genetics:

1. Plomin and Rowe partitioned influences on human social behavior into genetic, within-family, and between-family sources. Their finding that within-family factors are more prominent than between-family factors is surprising but fits with other recent data on personality development. They suggested the possibility that the important formative influences on personality may be genetically influenced, individual reactions to specific events during development rather than broad, general, and pervasive influences.
2. With a spirit that is in line with the issue that began this chapter, Plomin and Rowe related their findings on human infant social behavior to larger questions about the influences of natural selection on human behavior. Specifically, they discussed the selective pressures that may be applied to social gregariousness and attachment in humans. Their discussion, within the context of work on attachment and critical periods, makes clear that any study of the adaptive features of gregariousness and attachment must be developmental in nature. Just as a behavioral shift occurs in the eating style of the water snake (discussed by Arnold in Chapter 9 of this volume), a behavioral shift occurs in gregariousness in humans as a function of age. An understanding of these behavioral shifts requires in each case an understanding of the ecological and relational environments in which the organism exists.

## Comments on the Model Studies

The four studies just described are pioneering efforts to study genetic influences on the development of social behavior. As a group, the studies provide data on all of the questions that compose the agenda presented earlier for developmental social behavior genetics. The investigators chose a variety of ways to manipulate genotype in the search for genetic effects on behavior. Each of the methods—breed comparisons and Mendelian crosses, artificial selection, species and strain comparisons, and human twin comparisons—has potential for future studies, each having particular stengths and limitations (Fuller & Thompson, 1978; Plomin, DeFries, & McClearn, 1980).

The setting in which social behaviors were observed also varied in each study. Probably the most useful approach was that of Plomin and Rowe (1979) since they analyzed human infant behaviors that were directed to their mother and to a stranger. This manipulation allowed a contrast of social behaviors expressed inside and outside a relationship.

While there were differences in approach among studies, there were important commonalities. First, each investigator used the ecological or evolutionary niche of the organism of interest as a touchstone in the investigation. Particularly noteworthy in this regard was the attempt by Cairns et al. (1983) to quantify a hypothesized relationship between ontogeny and phylogeny. Second, despite the variety of methods and species employed, in three of the four studies the results demonstrate genetic influences operating on stability and change in social behavior over time. The results of these studies alone validate the field of developmental social behavior genetics and preview findings of the near future.

## CONCLUSIONS

Studies of genetic influence on the development of social behavior are rare. This is the case despite the importance the results of such studies could have for a range of issues, from testing of theoretical statements in sociobiology to evaluation of child-rearing practices. The purpose of this chapter is to facilitate study on developing social behaviors by discussing relevant methodological issues.

The first issue raised here is: How does one study a biological system? Lorenz showed the way by advocating a three-part strategy. First, the general characteristics of the system should be described, which places the system in an ecological and/or environmental setting. Second, the parts of the system should be identified and described. Third, the relationships among the parts of the system should be described. For Lorenz, this order of operations is critical. He argued that an investigator cannot list the parts of a system or understand the relationships among those parts without understanding the fit of the biological system into its niche.

The second issue in methodology is: What are the social phenomena to be studied? Various approaches to defining and categorizing social behaviors are discussed, and it becomes clear that investigators should pay careful attention to defining and organizing the social behaviors they study. The unit of behavior becomes especially important in genetic studies of social behavior, since, " . . . social behavior, by definition, involves the interaction of at least two individuals. The unit observed is the group" (Fuller & Hahn, 1976, p. 392).

An intriguing potential approach to genetic analysis of behavior, a multivariate developmental approach, was found in the work of Henderson (1986). Applying this approach to social behavior will be difficult but rewarding.

The third issue raised here is timing: When during development should social behavior be studied? Given that studying genetic influences on social behavior over the entire life span of all the species of interest is not possible, what times or periods should be chosen? I think the challenge here is to build theory that will suggest the best periods for detailed study. Scott's theory of critical periods (Scott, 1978, 1968, 1962) is an example of such a theory.

The discussion of issues in methodology led to the centerpiece of this chapter, a proposed agenda for future studies of developmental social behavior genetics. Inspired by the work of Cairns (1979), the agenda covers the establishment of social phenomena, continuity, and change in social phenomena over time; the nature of individual differences in the timings of change; and the relationship between developmental patterns and genetic fitness. This agenda is admittedly ambitious.

To show that interesting data can be gathered on the questions of the agenda, I presented a selected set of four model studies that examine the intersection of genes, development, and social behavior. The four studies vary in organisms studied and in methods employed, but all are worthy of imitation in building a science of developmental social behavior genetics.

## REFERENCES

Brooks, R. J., & Schwarzkopf, L. (1982). Factors affecting incidence of infanticide and discrimination of related and unrelated neonates in male *Mus musculus. Behavioral and Neural Biology, 37,* 149–161.

Cairns, R. B. (1976). The ontogeny and phylogeny of social interactions. In M. E. Hahn & E. C. Simmel (Eds.), *Communicative Behavior and Evolution.* New York: Academic Press.

Cairns, R. B. (1979). *Social Development: The Origins and Plasticity of Interchanges.* San Francisco: W. H. Freeman.

Cairns, R. B., MacCombie, D. J., & Hood, K. E. (1983). A developmental-genetic analysis of aggressive behavior in mice. I. Behavioral outcomes. *Journal of Comparative Psychology, 97* (1), 69–89.

Dawkins, R. (1976). *The Selfish Gene.* New York: Oxford University Press.

DeFries, J. C., & Fulker, D. W. (1986). Multivariate behavioral genetics and development: An overview. *Behavior Genetics, 16,* 1–10.

Fuller, J. L. (1983). Sociobiology and behavior genetics. In J. L. Fuller & E. C. Simmel (Eds.), *Behavior Genetics: Principles and Applications.* Hillsdale, NJ: Lawrence Erlbaum.

Fuller, J. L., & Hahn, M. E. (1976). Issues in the genetics of social behavior. *Behavior Genetics, 6,* 391–406.

Fuller, J. L., & Thompson, W. R. (1978). *Foundations of Behavior Genetics.* St. Louis, MO: C. V. Mosby.

Gottlieb, G. (1983). The psychobiological approach to developmental issues. In P. H. Mussen (Ed.), *Handbook of Child Psychology* (4th Ed.), *Vol. 2. Infancy and Developmental Psychobiology.* New York: John Wiley, pp. 1–26.

Hahn, M. E., Hewitt, J. K., Adams, M., & Tully, T. (1987). Genetic influences on ultrasonic vocalizations in young mice. *Behavior Genetics, 17,* 155–166.

Henderson, N. D. (1979). Adaptive significance of animal behavior: The role of gene-environment interaction. In J. R. Royce & L. P. Mos (Eds.), *Theoretical Advances in Behavior Genetics.* Alphen aan den Rijn, Netherlands: Sijthoff and Noordhoff.

Henderson, N. D. (1981). A fit mouse is a hoppy mouse: Jumping behavior in 15-day-old *Mus musculus. Developmental Psychobiology, 14* (5), 459–472.

Henderson, N. D. (1986). Predicting relationships between psychological constructs and genetic characters: An analysis of changing genetic influences on activity in mice. *Behavior Genetics, 16,* 201–220.

Lorenz, K. (1981). *The Foundations of Ethology.* New York: Simon and Schuster.

Plomin, R., DeFries, J. C., & Loehlin, J. (1977). Genotype-environment interaction and correlation in the analysis of human behavior. *Psychological Bulletin, 84,* 309–322.

Plomin, R., DeFries, J. C., & McClearn, G. E. (1980). *Behavioral Genetics: A Primer.* San Francisco: W. H. Freeman.

Plomin, R., & Rowe, D. C. (1979). Genetic and environmental etiology of social behavior in infancy. *Developmental Psychology, 15,* 62–72.

Sales, G. D., & Smith, J. C. (1978). Comparative studies of the ultrasonic calls of infant murid rodents. *Developmental Psychobiology, 11,* 595–619.

Scarr, S., & McCartney, K. (1983). How people make their own environments: A theory of genotype–environment effects. *Child Development, 54,* 424–435.

Scott, J. P. (1958). *Animal Behavior* (1st ed). Chicago: University of Chicago Press.
Scott, J. P. (1962). Critical periods in behavioral development. *Science, 138,* 949–958.
Scott, J. P. (1968). *Early Experience and the Organization of Behavior.* Belmont, CA: Brooks/Cole.
Scott, J. P. (1972). *Animal Behavior* (2nd ed.). Chicago: University of Chicago Press.
Scott, J. P. (1977). Social genetics. *Behavior Genetics, 7* (4), 327–346.
Scott, J. P. (Ed.) (1978). *Critical Periods.* Stroudsburg, PA: Dowden, Hutchinson and Ross.
Scott, J. P. (1983). Genetics of social behavior in non-human animals. In J. L. Fuller & E. C. Simmel (Eds.), *Behavior Genetics: Principles and Applications.* Hillsdale, NJ: Lawrence Erlbaum.
Scott, J. P., & Beach, F. (Eds.) (1947). *Minutes of the Conference on Genetics and Social Behavior.* Bar Harbor, ME: Roscoe B. Jackson Memorial Laboratory.
Scott, J. P., & Fuller, J. L. (1965). *Genetics and the Social Behavior of the Dog.* Chicago: University of Chicago Press.
Simon, N. G. (1979). The genetics of intermale aggressive behavior in mice: Recent research and alternative strategies. *Neuroscience and Biobehavioral Reviews, 3,* 97–106.
Tinbergen, N. (1969). *The Study of Instinct.* New York: Oxford University Press.
Wilson, E. O. (1975). *Sociobiology: The New Synthesis.* Cambridge, MA: Belknap-Harvard University Press.

# 11
# GENETIC APPROACHES TO THE DEVELOPING NERVOUS SYSTEM

ROBERT H. BENNO

The modern neurobiologist believes that all behavior is a reflection of brain function. The neurobiologist is thus faced with the formidable task of explaining how neural circuits are organized, in order to control these behaviors. This organization is defined during development largely under genetic constraints, although the environment undoubtedly plays a major role in the process.

Genes may affect neurodevelopment at several different organizational levels throughout the life of the organism. Thus there are several possible genetic approaches to the study of neuronal development. The particular orientation of the researcher will determine the research focus and the questions that can be posed and answered. Stent (1981) defined two related but distinct conceptual aspects of the genetic approach to developmental neurobiology: the "ideological" and the "instrumental."

In the ideological approach, individuals of genetically homogeneous populations are screened for differences in neurological phenotype. According to Stent, the goal of this approach is to determine the preciseness with which the developmental processes give rise to the structure of the nervous system. Stent suggested that phenotypic variance in a genetically homogeneous population may be due to what Waddington (1957) referred to as "developmental noise." Unfortunately, the term *noise* has a somewhat negative connotation, suggesting deviation from an ideal situation. However, we may consider developmental noise to be simply any variation in environment with which a member of a homogeneous population comes in contact, without implying an ideal environment or resultant phenotype. In this sense, developmental noise would refer to the individual environment to which a particular genotype is exposed. The interaction between the genotype and the environment, and the resultant influence that they have on one another, may be referred to as *epigenesis* (see Chapter 7 in this book for further discussion). Defining the exact nature of this interaction and the extent to which it might affect the development of a specific part of the nervous system is important to our understanding of developmental biology. A clear understanding of how epigenesis affects the developmental program is obviously clouded by the fact that genes influence the environment and the developmental environment influences the genes (Purves & Lichtman, 1985). Unraveling of this genetic/epigenetic interaction during development, especially in longitudinal developmental studies, can be best accomplished by an "ideological" approach using genetically homogeneous (isogenic) animals. Furthermore, analysis of phenotypic variance in isogenic animals using biometrical genetic procedures (detailed in Part III of this book) may provide knowledge about the evolutionary history of a particular trait.

The instrumental approach to genetic analysis of neuronal development attempts to establish if there is a causal link between a known difference in genotype and an observed difference in neuronal phenotype. Instrumental studies are performed on individuals of genetically heterogeneous populations, often comparing normal animals and conspecific genetic mutants that exhibit abnormal neurological phenotypes. Since straightforward interpretations of developmental processes in mutant animals in which all tissues are of mutant origin are difficult to interpret, it has been necessary to develop techniques whereby some cells of a developing organism are of mutant genotype and others are of normal genotype. The resultant *genetic mosaics* or *chimeras* may be used in developmental neurobiology to ascertain if a neurological abnormality of structure or function X is a direct result of the mutation or merely a secondary, noncausally related result of the abnormal phenotype of Y. Thus far, the use of genetic mosaics in developmental biology is limited mainly to studies using *Drosophila* (see Chapter 6) and the mouse (see Chapter 5). These species have proved to be valuable for mosaic studies because of the availability of a wide diversity of mutant neurological phenotypes and a sophisticated technology necessary for the production of the mosaics in both cases.

Using the ideological or instrumental aspects of the genetic approach, the geneticist may study the development of the nervous system from the molecular, cellular, or organismal viewpoint. Within this context the researcher may also choose a biochemical, morphological, behavioral, or physiological orientation. These orientations are not mutually exclusive, as demonstrated by the chapters that follow. The authors of these chapters represent a diverse group of neuroscientists using a wide variety of genetic approaches to study of the nervous system. Each author has attempted to relate his or her particular orientation to the interdisciplinary construct of developmental behavior genetics.

Cynthia Wimer (Chapter 5) discusses contributions made from both ideological and instrumental studies. She outlines three types of strategies the neurobiologist might utilize. These three—gene action, developmental process, and adult variation—represent studies directed respectively at the molecular, cellular, and organismal levels. At each of these levels, specific genetic tools may be used to answer the specified questions. For instance, Wimer describes how mutants and chimeras are used in an instrumental manner to understand mechanisms of gene action and developmental process. She then turns to an ideological approach, discusssing the importance of strain comparisons, breeding studies, recombinant inbred studies, and genetic selection as tools for describing variances in adult phenotype such as brain weight or hippocampal connectivity.

Jeffrey Hall (Chapter 6) uses a molecular genetic instrumental approach to understand the process of neuronal development and its subsequent role in the shaping of behavior. Hall's tool is the fruit fly, *Drosophila,* which is perhaps the best catergorized of all species from a genetic viewpoint and has the advantage of offering the researcher a wide diversity of mutants with altered phenotype. Hall concentrates on single-gene mutations that result in defective courtship behavior in *Drosophila*. He describes studies that demonstrate how courting abnormalities are related to aberrant formation of the *Drosophila's* nervous system at different stages of neuronal assembly (e.g., embryonic, pupal). He also presents evidence that there is plasticity of neuronal structure and resultant behavior in the fruit fly after metamorphosis. Finally, Hall describes studies using a *D. melanogaster per* mutant that lacks circadian rhythms and exhibits a courtship song in which the pulse rate is arbitrary. He concludes by attempting to show how isolating behaviorally interesting mutants such as *per* in the laboratory is of considerable importance for studies of population biology and evolution.

Robert Benno (Chapter 7) presents an overview of the processes of neuronal development operating at the cellular and neuronal circuit levels. He attempts to show the importance of genetics, epigenetics, and phylogenetics in the processes of neuronal determination and differentiation. His contention is that the relative roles of genetics and epigenetics may vary, depending on the specific process under investigation and the phylogenetic level of the organism. He concludes by illustrating how a genetic approach to the process of neuronal development may help elucidate some of the behavioral differences observed during development in Fuller brain-weight-selected (BWS) mice. Using this ideological BWS model, he attempts to show how genetic selection for brain weight may help elucidate underlying mechanisms that are important for development of the nervous system and subsequent differences between animals in behavioral repertoires.

Larry Leamy (Chapter 8) concludes this section on genetic approaches to the nervous system by describing how genes may be involved in the evolution of brain and body size. The ratio of brain to body size has been argued to be of important consequence for behavior, as detailed in classic work of Jerrison (1973). Leamy's approach is directed toward obtaining an understanding of how genes may be involved in determining brain and body size and the ratio of brain to body size. To fulfill this goal, Leamy details quantitative genetic models that permit an assessment of the relative importance of genetic and maternal factors on brain size and its scaling with body size. He presents empirical data from random-bred and inbred/hybrid populations of mice, which he demonstrates are basically compatible with genetic models that attempt to explain the evolution of brain and body size. Leamy concludes by stating that a large portion of the heritability for brain size stems from largely prenatal maternal influences and he suggests several ways by which one could elucidate these factors and their relative importance.

## REFERENCES

Jerrison, H. J. (1973). *Evolution of the Brain and Intelligence.* New York: Academic Press.
Purves, D., & Lichtman, J. W. (1985). *Principles of Neural Development.* Sunderland, MA: Sinauer Associates.
Stent, G. S. (1981). Strength and weakness of the genetic approach to the development of the nervous system. *Annual Review of Neuroscience, 4,* 163–194.
Waddington, C. H. (1957). *The Strategy of the Genes.* London: Allen andUnwin.

# 5
# Genetic Studies of Brain Development

CYNTHIA WIMER

The genetic approach to the study of development of the brain has led to a diversity of research contributions from a variety of disciplines. The goal of a particular study, and the direction the research takes, will depend on the orientation of the researcher. Thus, for example, the geneticist may focus on the site of gene action; the embryologist can use genetic preparations as a tool for the elucidation of developmental mechanisms; and the anatomist is often interested in genetically associated developmental differences that lead to variation in adult brain morphology. To have a conceptual framework within which to evaluate the contributions of the variety of studies, it is convenient to classify them according to their focus. Table 5-1 categorizes three types of studies that will be discussed in this chapter—their focus, some of the genetic tools that can be used, and specific examples that will be discussed here. These tools are not the only ones available, but are perhaps the most used. The examples will be taken almost exclusively from morphological studies in the mouse, although there are equally cogent examples from other orientations (see Chapter 6 in this volume for studies of the *Drosophila* nervous system).

## GENE ACTION

Studies that focus on gene action or expression ask such where, when, and how questions as: In what kind of cell or cells is the gene (or genes) acting? What exactly is it doing? At what point in development do the crucial events occur? Mullen and Herrup (1979) have discussed the advantages of this type of study and have emphasized the sorting out of primary, secondary, and pleiotropic effects. In the present classification, the category is similar to but more general than Stent's (1981) "ideological" approach, which he defined as the attempt to determine how genes specify the components of the nervous system and their assembly.

Researchers focusing on gene action work primarily with established single-gene differences in neurological mutants; most work at the level of cells and interactions among them. There is a long list of cerebellar mutants in the mouse, in which the most obvious defect is a loss of either the granule cells or their targets, the Purkinje cells. To review some aspects of cerebellar development, as discussed in greater detail by Robert Benno in Chapter 7 of this volume: The Purkinje cells are born (in the

**Table 5-1** Classification of genetic studies of brain development

| Focus | Gene action | Developmental process | Adult variation |
|---|---|---|---|
| Tools | Mutants | Mutants | Strain comparisons |
| | Chimeras | Chimeras | Breeding studies |
| | | | Recombinant inbred strains |
| | | | Genetic selection |
| Examples | Staggerer | Neuronal migration | Brain weight |
| | Weaver | Cell lineages | Corpus callosum |
| | | Cell death | Hippocampal connectivities |

mouse) between embryonic days 11 and 13 (E11 and E13) in the embryonic ventricular zone, and thence migrate outward to form several layers of cells below the molecular layer. Postnatally they spread out to form a single layer. The granule cells originate a bit later, from days E12 to E15, in a subventricular zone, and migrate across the surface of the cerebellum to form an *external* granule cell layer. Postnatally they emigrate along Bergmann glial fibers, through the molecular layer, and past the Purkinje cells to form a new granule cell layer, leaving behind parallel fibers that synapse on Purkinje cell dendrites. The relative positions of the two types of cells prenatally and postnatally are schematically illustrated in Figure 5-1.

In the mutant staggerer, the granule cells degenerate about three weeks postnatally, shortly after their final migration. There are fewer Purkinje cells, and they are smaller and ectopic (Mullen & Herrup, 1979). Qualitative studies have concentrated on where, when, and how the staggerer gene is expressed. Roffler-Tarlov and Herrup (1981) suggested that there is an early effect in the ventricular zone on cells that are destined to become Purkinje cells. Apparently the neural cell adhesion molecule fails to convert from the embryonic to the adult form, and thus its binding properties are altered at a critical period of development (Edelman & Chuong, 1982). Calcium channels either are missing or cannot be activated in the Purkinje cells (Crepel,

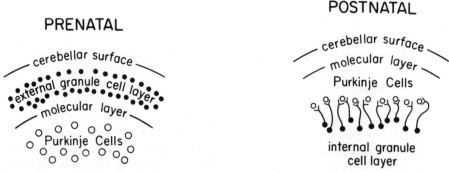

**Figure 5-1** Schematic diagram showing relative positions of Purkinje (o) and granule (●) cells prenatally and postnatally. See text for description.

# GENETIC STUDIES OF BRAIN DEVELOPMENT

Dupont, & Gardette 1984). Herrup and Mullen (1981) approached the question quantitatively, reasoning that qualitative differences are ultimately the result of the number of cells in a given place at a given time. They used the elegant technique of producing chimeric mice, in which cells from both the normal and the abnormal genotype are combined.

Since chimeras were used in many of the studies to be discussed in this chapter, a brief review of the technique is in order here. Embryos are removed from pregnant females at the eight-cell stage of development, and two embryos are combined in culture, where they aggregate to form a double-size morula and then a blastocyst. The blastocysts are transplanted into the uterus of pseudopregnant females, where they develop normally except that some of their cells are from one set of the original parents and the rest are from the other. The relative proportions vary from individual to individual. The cells themselves do not fuse, but rather are mixed together, so that each cell has a normal genetic complement from one or the other set of parents. If one set of parents carries the mutant gene and the other set carries the normal allele at that locus, both normal and abnormal cells will be found at the site of action of the gene. We can thus study the effects of a normal environment on an abnormal cell, and vice versa. For further details on the technique and its uses, see Mullen and Herrup (1979).

To study the genotypically normal and abnormal cells it is necessary to distinguish between them phenotypically. This is usually accomplished with a genetic marker, a gene for which the two sets of parents differ, and which affects the appearance of the cells being studied while not interfering with the primary effect of the mutant gene. For the study of staggerer (Herrup & Mullen, 1981), the marker used distinguished between genotypically normal and mutant Purkinje cells. All cells that were normal in size and location were genotypically normal, while all staggerer cells were both small and ectopic. These results supported the conclusion from qualitative studies of the mutant that the staggerer gene defect is expressed in the Purkinje cells rather than in the granule cells, and that granule cell degeneration is a secondary effect. Further support was found in a later study (Herrup, 1983), in which a marker for granule cells was employed and staggerer granule cells were found in the chimeras, that is, they were "rescued" by being placed in an environment with some normal Purkinje cells.

The weaver mutant is also characterized by granule cell degeneration, but unlike staggerer, the cells degenerate in the external layer before their final migration. Bergmann glial fibers are fewer in number and are abnormal in appearance, cerebellar size is decreased, and both types of cells are abnormally positioned as well as reduced in number (Mullen & Herrup, 1979). Rakic and Sidman (1973) attributed the deficit to a disorder of the Bergmann glia, resulting in faulty granule cell migration. This cannot be the whole story, however, because there is granule cell degeneration in the weaver/normal heterozygote, in which the Bergmann glia appear to be normal. Goldowitz and Mullen (1982) examined cell positioning in chimeras in which both Purkinje and granule cells were marked, and found no relationship between genotype and ectopic position for Purkinje cells. All ectopic granule cells, however, were of the weaver genotype, while there were normally positioned cells of both genotypes. Thus they concluded that, unlike staggerer, the defect was definitely extrinsic to the Purkinje cells and *probably* intrinsic to the granule cells (though Bergmann glia were not ruled out). The granule cells may exert an early influence on Purkinje cell survival and on gliogenesis. A recent study of cerebellar cultures taken from week-old weaver mutants (Willinger & Margolis, 1985) demonstrated abnormal neurite growth. These investigators concluded that the granule cell defect affects migration, with a secondary effect on viability. The site of gene action appears to be at the cell surface, resulting in a failure of the granule cells to make appropriate contacts during their migration through the Purkinje cell layer.

What can we learn from this kind of study? Stent (1981) has criticized the approach as being too narrow a context for understanding genetic processes. The specific role of genes, that is, specifying the structure of protein molecules, is too far from the developmental processes that we are trying to understand. He suggested that the solution will be at the cellular and intercellular levels, though genetic studies will of necessity contribute to that solution. Goldowitz and Mullen (1982), however, regarded determination of the cellular site of gene expression as a major step toward understanding the process by which genome is translated to gene product, and by which gene products act and interact to determine the structural organization of the nervous system. Basically, developmental studies of the type just described contribute to an understanding of the last half of this process, whereas the first half is the province of the molecular geneticist.

## DEVELOPMENTAL PROCESSES

In studies that focus on developmental processes or mechanisms, genetic tools are often no more than just that: convenient preparations in which to study normal ontogeny by looking at developmental differences in genetically disparate organisms. A genetically associated abnormality in development—in a neurological mutant mouse, for example—can often shed light on the events that are necessary for the normal processes to occur.

One of the cerebellar mutants that has been used to study cell migration in the central nervous system is reeler. This mutant, which has deficient foliation in the cerebellum, has very few granule cells, and the Purkinje cells are abnormally positioned, mostly below the granule cell layer but separate from the deep cerebellar nuclei. In reeler/normal chimeras, both reeler and normal Purkinje cells are found in both normal and abnormal positions, suggesting that a factor extrinsic to the cells themselves is responsible for the defect (Mullen & Herrup, 1979). Goffinet (1983) suggested that the critical factor involves interactions between neurons or between neurons and glial fibers at the end of migration. In spite of the architectonic disorganization, generation, early migration, axon growth, differentiation, and synaptogenesis appear normal. Goffinet hypothesized a loss of adhesivity between Purkinje cells and radial glial fibers. It has been argued, however, that in normal mice Purkinje cells are close to the glial fibers throughout migration, but subsequently they differentiate and lose their attachment. In reeler the attachment is *not* lost, so that early formed cells block later arriving cells and normal migration is thwarted (Caviness, 1982; Pinto-Lord, Evrard, & Caviness, 1982). Whichever hypothesis is correct, studies of reeler demonstrate the critical role of cell-cell interactions, and especially of glial guides, in the positioning of cells during migration.

A number of studies with a more intrinsically genetic orientation have focused on the function of cell lineages in the nervous system. One of the questions asked concerns the significance of clones in the developing brain: To what extent does the fate of a cell depend on its lineage, as opposed to its interactions with other cells? The mammalian central nervous system, unlike other parts of the body, does not appear to have a "clonal development" in a spatial sense, that is, large groups of cells all descended from the same progenitor (Mullen, 1978). Oster-Granite and Gearhart (1981) demonstrated the particular advantage of chimeras in the study of cell lineages. In chimeras between two inbred mouse strains, they showed that most of the brain (with the possible exception of the hippocampus) is a fine-grained mosaic, though not a completely random array. They postulated that, in the cerebellum, precursor cells might, after migration, undergo extensive intermingling and subsequently differentiate on the basis of their position rather than their lineage.

A series of studies by Herrup and his colleagues has taken a quantitative approach to this problem. Defining a clone as a group of cells descended from a single cell, they proceeded from the assumption that the smallest unit of cells of one genotype seen in a set of chimeras represents one clone. In chimeras between normals and lurcher mutants, they looked at Purkinje cells and found the "smallest number" to be 10,200. Counting the number of cells of this type in the remaining chimeras in the set, they discovered that, in each cerebellum, the number of Purkinje cells was some multiple of that base number. By dividing by 10,200 they could estimate the porportions of lurcher versus normal precursor cells, and this agreed with the proportions found in other organs of the body (Wetts & Herrup, 1982a). This is a sort of clonal development, then, but in a quantal rather than a spatial sense. It was subsequently determined (Herrup, 1986; Herrup & Sunter, 1986) that the Purkinje cells are derived from different numbers of clones with different clonal "sizes" in various inbred strains of mice. To this extent there are intrinsic mechanisms regulating cell number; the size of the clone is an autonomous property of the lineage and is thus intrinsic to the progenitor cell. The number of cells is apparently relatively insensitive to the surrounding neural environment. These investigators concluded that "cell lineage relationships play a significant role in control of cell number during mammalian development," a fact that is well established in invertebrates but was unexpected in the mouse.

Herrup and Bower (1986) attempted to determine the correspondence between a map of cerebellar granule cells based on cell lineage and a physiological map based on projections from the somatosensory area of the cerebral cortex. To a large extent the two maps are superimposable, suggesting that lineage relationships set up cues in the target area for later establishment of functional relationships.

Another phenomenon that has been studied in genetic preparations is that of natural cell death during development. Cell death is of particular interest in the context of this symposium: Katz and Lasek (1978) have proposed it as a mechanism for maintaining functional matches between connected neuron populations and have emphasized its importance in evolution. If there is a change in the number of cells in one of two matching populations, the change can be *heritable* only if functional concordance is maintained. Cell death is one possible mechanism for maintaining concordance: If, for example, the number of cells is decreased in a particular target population, some of the cells in the source population must die for the match to be preserved. Most of the evidence for cell death in a source population subsequent to a reduction in the target population comes from experimental embryology, but some recent studies have been done using chimeras between the cerebellar mutant lurcher and normal mice. Lurcher loses all of its target Purkinje cells and most of its source granule cells. Since the defect is intrinsic to the Purkinje cells (Wetts & Herrup, 1982b,c), the chimeras have been used as a model system to investigate the relationship between source and target, where the target size can be varied genetically (Wetts & Herrup, 1983). Since lurcher has its normal complement of Purkinje cells at the beginning of synaptogenesis, the granule cells may need only *transient* contact with Purkinje cells to survive. Support for this hypothesis comes from the fact that in the mutant staggerer, in which the Purkinje cells degenerate earlier in development, *all* granule cells die; in the mutant Purkinje cell degeneration (pcd), the Purkinje cells degenerate later, and *few* granule cells die. These results suggest that there is a critical period during which the source and target populations must interact. It was hypothesized, then, that earlier born granule cells in lurcher would be more likely to survive. It has recently been shown, however, that survival is not related to generation time (Vogel & Herrup, 1986), so the story is still incomplete.

In lurcher chimeras, more granule cells survive than expected. Caddy, Fieldman, and Herrup (1986) have addressed this phenomenon of the apparent rescue of granule cells. They hypothesized that more granule cells might synapse on each Purkinje cell, but found that the dendritic trees of Purkinje cells had *fewer* branch points, and thus fewer connections. It may be that granule cells in lurcher chimeras have abnormal targets, and this possibility is now under investigation. This potential "rerouting" of connections from the source population (compensatory innervation, Katz & Lasek, 1978) is another way in which the functional match between source and target populations could be preserved.

The value of these kinds of studies for understanding developmental processes must depend at least partly on the kinds of information they provide that cannot be obtained by more traditional methods. An obvious advantage of genetic preparations is that developmental variations are produced by nature rather than by experimental intervention and are thus likely to provide more direct clues concerning the mechanisms involved. A particular advantage of mutants and chimeras is that in many cases the cellular site of gene action can be precisely defined. In chimeras, a quantifiable range of effects, not available in any other preparation, can provide a broader basis for understanding than can "all-or-none" mutant preparations (Herrup & Sunter, 1987). As just one example, the weaver mutant mouse and associated chimeras have provided information on influences on dendritic structure, neuronal interactions in the formation of synaptic contacts, and the role of glial cells in migration (Goldowitz & Mullen, 1982). The approach has its critics, however. Stent (1981), who called it the instrumental approach, believed that instead of resulting in major breakthroughs it primarily provided support for established developmental concepts, though studies of reeler chimeras revealed a previously unknown feature of neuronal migration that could not be discovered in any other way. Most contributions have been made in systems in which the level of understanding is already high. Quinn and

Gould (1979) claimed that mouse mutants raise interesting questions about neurogenesis but are *useless* for understanding the subtleties of neural development. They further claimed that the studies ask one set of questions and answer another; some of the most central questions of neurobiology—the chemical basis of ion gating and synaptic transmission, the changes in neurons and circuits that underlie learning—remain untouched. But the studies described here (most of which, to be fair, have been published since Quinn and Gould's criticism) are *not* asking these questions. They are, however, addressing another issue that Quinn and Gould also felt to be central: that of the mechanisms involved in specifying a neuron's shape and determining its interaction with other cells. And the answers they have provided have surely added to our understanding of the developmental process.

## ADULT VARIATION

The approach that begins with the identification of individual variation in brain and/or behavior involves an integrative process, the combining of studies often from a variety of disciplines. The goal is to understand how genetic differences can lead to variations in developmental processes that, interacting with environmental effects, ultimately result in variability in adult phenotype. Figure 5-2 identifies the major aspects of this progression and illustrates the various levels at which investigations can contribute to the understanding of the phenotype.

An obvious example of a neural phenotype that has been studied in this way is brain weight. The allometric approach to the study of brain size is discussed by Leamy in Chapter 8 in this volume; the experiments he describes constitute an integral part of the body of research on the topic. The interest of behavioral and neural geneticists is often stimulated by the report of strain differences in a particular phenotype, and the recent interest in brain weight perhaps began with Storer's (1967) report of extensive variability in brain weight in inbred strains of mice. Both Atchley,

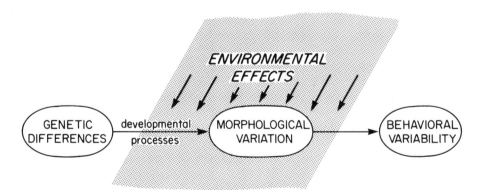

**Figure 5-2** Illustration of the general process studied by the "interactive approach," identifying the major aspects that can be investigated.

Riska, Kohn, Plummer, and Rutledge (1984) and Leamy (1985 and Chapter 8 in this volume) have carried out genetic analyses of brain and body weight in mice at several ages and found developmental changes in the genetic correlation between the two phenotypes. Brain and body size are more highly correlated during early growth, with the correlation decreasing later and perhaps even becoming negative. These results suggest that brain size is strongly associated with the cell multiplication phase of body size growth (as opposed to the later cell enlargement phase), with both responding simultaneously to systematic stimuli such as mitogens or hormones (Riska & Atchley, 1985). Genetic selection studies (Fuller, 1979; Roderick, Wimer, & Wimer, 1976) provided estimates of the heritability of brain weight that agreed with those of Leamy and Atchley (about .6–.7), and subsequent genetic analyses (Hahn & Haber, 1978, 1979; Henderson, 1979) estimated both the magnitude of genetic variance and the degree of directional dominance. The selection experiments, resulting in high- and low-brain-weight lines, furnished a rich resource for both behavioral and developmental studies. The behavioral studies, which have been summarized elsewhere (Fuller, 1979; Jensen, 1979), had somewhat equivocal results, and no firm conclusions could be drawn about the relationship between brain size and behavior.

Gene-environment interactions have also been analyzed, and it has been found, for example, that there is an interaction with maternal environment (Hahn, 1979; Wahlsten, 1975) and that aggressive behavior is an interactive function of brain weight and social experience (Hahn, Haber, & Fuller, 1973). Studies of development found different rates of brain growth (Fuller & Geils, 1972; Hahn, Walters, Lavooy, & Deluca, 1983) and of sensory and motor development (Fuller & Herman, 1974) in high- and low-brain-weight lines. A recent biometric analysis of early postnatal changes in brain weight (Hewitt, Hahn, & Karkowski 1987) suggested that selection for extreme values can disrupt developmental stability. Since it has also been shown that such selection can alter the genetic correlation between relative sizes of various brain regions (Wimer, 1979), it is clear that considerable caution must be exercised in generalizing from the results of experiments on selected lines. In spite of the equivocal results, this series of studies is of value as probably the first of its kind: an integrative approach to the study of phenotypic variability in a neural trait through examination of its genetic and developmental precursors and its behavioral associations.

This early approach to a rather gross and complex (i.e., polygenic) morphological phenotype has been followed by studies of specific regions of the brain, in fact of very precise aspects of these regions. An example is provided by a series of studies of the corpus callosum, a major commissure for connections between the two sides of the brain. The corpus callosum is of particular evolutionary interest: Unlike most other axon tracts, which have evolved by modified use of preexisting substrates, it has evolved through the emergence of a new substrate pathway, suggesting a more fundamental evolutionary change (Katz, Lasek, & Silver, 1983). This pathway, the "glial sling," connects the left and right forebrain only in placental mammals. It was of interest, then, to find one or two inbred strains of mice that had no crossing fibers in the corpus callosum (Wahlsten, 1974). This is a hereditary defect with incomplete penetrance or variable expressivity (i.e., it is found in less than 100 percent of the animals in the affected strains), although we have recently found that in strain I/Ln *all* animals are affected (unpublished results, this laboratory). The relatively mild

# GENETIC STUDIES OF BRAIN DEVELOPMENT

effects are within the normal range of variation and perhaps less related to fitness than are more severe defects. These characteristics suggest that the acallosal mouse should provide a good model for the study of developmental plasticity (Wahlsten, 1982a). The defect has a recessive mode of inheritance and is not attributable to a single Mendelian locus (Wahlsten, 1982b). In the nonplacental opossum, axons destined to provide interhemispheric connections cross through the anterior commissure (Silver et al., 1982), but they cannot do so in the acallosal mouse because the anterior commissure is mature before the callosal fibers normally cross the midplane (Wahlsten, 1981). If the glial sling is cut at precallosal stages in the normal mouse, the axons are displaced laterally but retain their potential to regrow if an artificial glia-coated sling is grafted in (Silver & Ogawa, 1983). In the genetically acallosal mouse, the fibers behave like those in the surgically acallosal mouse rather than like those in the opossum (Silver et al., 1982).

Wahlsten and his colleagues have carried out a long series of studies primarily concerned with the identification of environmental factors that affect the degree of penetrance of the defect in the affected strains. They have found that the percentage of affected animals varies depending on the source supplying the strains (Wahlsten, 1982c), that there is a higher incidence in overlapping litters (Wahlsten, 1982d), and that undernutrition of the dam during gestation has little or no effect (Wainwright & Gagnon, 1984). Studies employing ovarian grafts have shown no specific effect of maternal environment (Bulman-Fleming & Wahlsten, 1986). The developmental pattern is variable, sometimes associated with retarded growth of the brain as a whole, sometimes not. At embryonic day 17, almost all fetuses of the most affected strain have separated hemispheres; by the next day, some have developed crossing fibers; and by day E19 most have connecting fibers (Wahlsten, 1984). Thus, a critical developmental period has been pinpointed. Recent studies have shown that, even with an abnormal corpus callosum, contralateral visual connections are topographically correct (Serra-Oller, Olavarria, & VanSluyters, 1986). Evidence is accumulating at the various levels of analysis, all contributing to the understanding of the adult morphological variation. To complete the story, it would be gratifying to be able to report an associated behavioral variation (see Figure 5-2). Unfortunately, however, none has been found to date.

The remainder of this chapter will be devoted to a system that is currently being studied at several levels—the mossy fiber system in the hippocampus. This system is the connecting link between the dentate gyrus and CA3 (regio inferior) of the hippocampus. Efferent fibers from the granule cells of the dentate course through the hilus and synapse on apical and basal dendrites of CA3. As can be seen in Figure 5-3, the fibers pass both above (supra) and below (infra) the pyramidal cells, and in some cases through the cell layer itself (intra). The impetus for this series of studies came from the finding that inbred strains of mice differ in the relative proportions of mossy fibers in the three layers (Barber et al., 1974).

Thus, the "typical" strain has a large suprapyramidal input, a somewhat less extensive infrapyramidal input, and a rather small intrapyramidal layer. It is primarily in the extent of this intrapyramidal layer that the strains differ, and the relatively minor (though significant) differences suggest a polygenic effect. One strain, however, has no infrapyramidal layer at all, but rather a substantial intrapyramidal input (BALB/c in Figure 5-3), and this suggests a possible single gene effect. The

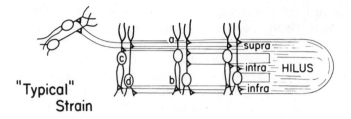

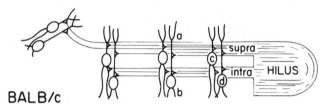

**Figure 5-3** Schematic diagram illustrating mossy fiber input to the pyramidal cells of CA3. Fibers synapse on both apical (a) and basal (b) dendrites of superficial (c) and deep (d) pyramids. See text for comparison of the upper and lower diagrams.

variation is of interest because it implies different patterns of connectivity within the hippocampus, with potential consequences for functional variation. Both the polygenic effect and the putative single gene effect have subsequently been examined.

Two biometric studies, a classic cross (Heimrich, Schwegler, & Crusio, 1985) and a diallel cross (Crusio, Genthner-Grimm, & Schwegler, 1986) have supported the suggestion of polygenic inheritance for the minor effect, and have found substantial additive genetic effects and only low levels of dominance, from which these investigators inferred an evolutionary history of stabilizing selection. Nowakowski (1984) has studied the major effect in BALB/c in classic crosses and recombinant inbred strains, obtaining results consistent with inheritance via a single autosomal dominant (or semidominant) gene, which he has named *Hld* (hippocampal lamination defect).

What specific developmental events could produce the connectivity differences? Barber, Vaughn, Wimer, and Wimer (1974) postulated a reversal of the normal "inside out" developmental sequence of hippocampal pyramidal cells, that is, early-generated cells located superficially in the pyramidal layer and late-generated cells located more deeply. A subsequent autoradiographic study of embryonic development (Vaughn et al., 1977) found the normal inside-out sequence except in strain BALB/c in that portion of the pyramidal layer closest to the hilus, where the infrapyramidal layer is usually prominent. If the earliest formed neurons in BALB/c occupy the most superficial positions, and if they synapse on the apical and basal dendrites of the superficial pyramids before the deep pyramids are formed, later arriving fibers may synapse on the apical dendrites of deep cells, precluding infrapyramidal connections with their later differentiating basal dendrites. Nowakowski and Davis (1985) confirmed this hypothesis, and analyzed dendritic arborization in the *Hld* mouse (BALB/c) and the wild-type mouse (C57BL/6). They found that in the wild type, late-generated (superficial) cells have dendritic excrescences on both apical and basal dendrites, whereas in *Hld,* the late (deep) cells have two sets on the apical dendrites and none on the basal. The genetically associated difference in mossy fiber pattern would lead to the prediction of variation in control of pyramidal neurons, and thus potentially to behavioral differences. A series of behavioral studies has been carried out by Lipp, Schwegler, and colleagues, who use as a morphological phenotype the ratio of intrapyramidal plus infrapyramidal mossy fibers to total (iip). They have been concerned with the polygenic effect, and have primarily compared inbred strains and lines selected for behavioral characteristics. In particular, they have found a strong *negative* association between iip and shuttle box performance in both mouse and rat (Lipp & Schwegler, 1982; Schwegler, Heimrich, Crusio, & Lipp, 1985). They have hypothesized that the pyramidal cells are important for avoidance learning and that the mossy fiber variation represents a shift in balance of the dentate input to the pyramids. This critical fiber system has a genetically associated plasticity at a critical period of development, and the pattern determined at that time is apparent as an adult morphological variation associated with behavioral differences. Other behaviors are currently being examined, with somewhat ambiguous results.

The above studies looked at the polygenic variation; behavioral associations with the *Hld* locus have been examined as well. Since major gene control of shuttle avoidance has also been proposed (Oliverio, Eleftheriou, & Bailey, 1973), Peeler (1987) compared shuttle box performance in recombinant inbred (RI) strains descended from two progenitor strains that differ in both the putative *Hld* and avoidance genes. The RI strains did not dichotomize on behavior (which fails to support the single avoidance gene hypothesis) and there was no relation between *Hld* and avoidance learning. Thus it was concluded that *Hld* is not determined by the same genes as avoidance learning.

The research on the mossy fiber variation is perhaps the best example of the integrative approach to the study of genetically associated variations in brain morphology, and in fact both Barber et al. (1974) and Lipp and Schwegler (1982) have presented convincing arguments for the approach. It depends on input from a number of disciplines: The genetic techniques that are being employed are elegant; the morphological measurements are for the most part quite precise; and new information about development and techniques for studying developmental events are constantly becoming available. If there is a weak link in the research, it appears to be in the study of behavior. Too often, behavioral results are ambiguous or contradictory. Furthermore, although we emphasize the "naturalistic" aspect of the morphological variations, we do not, for the most part, examine their consequences for naturalistic behavior.

## CONCLUDING REMARKS

It would not be appropriate here to attempt a critical comparison of the three genetic approaches to the study of brain development that have been discussed in this chapter. In their own way they are equally valid, and the approach an individual researcher chooses to adopt is a function of his or her interest, temperament, and background. It should be obvious from the relative amount of space devoted to each that, to me at least, the last approach, with its emphasis on adult variation in brain and behavior, is the most satisfying. And I suspect that this will be true for most behavior geneticists. But for the integrative approach to be a success, it will behoove us to improve the behavioral aspects of our research. After all, understanding of behavior is the goal that all of us are pursuing.

## ACKNOWLEDGMENT

This work was supported in part by Grant NS18860 from the National Institute of Neurological and Communicative Disorders and Stroke.

## REFERENCES

Atchley, W. R., Riska, B., Kohn, L. A., Plummer, A. A., & Rutledge, J. J. (1984). A quantitative genetic analysis of brain and body size associations, their origins and ontogeny: Data from mice. *Evolution, 38,* 1165–1179.

Barber, R. P., Vaughn, J. E., Wimer, R. E., & Wimer, C. C. (1974). Genetically associated variations in the distribution of dentate granule cell synapses upon the pyramidal cell dendrites in mouse hippocampus. *Journal of Comparative Neurology, 156,* 417–434.

Bulman-Fleming, B., & Wahlsten, D. (1986). Corpus callosum defects and maternal environment in the BALB/c mouse: Effects of ovarian grafting. *Neuroscience Abstracts, 12,* 1558.

Caddy, K. W. T., Fieldman, B., & Herrup, K. (1986). Quantitative analysis of Purkinje cell dendrites in normal and lurcher chimeric mice. *Neuroscience Abstracts, 12,* 1583.

Caviness, V. S., Jr. (1982). Neocortical histogenesis in normal and reeler mice: A developmental study based upon [3H]thymidine autoradiography. *Brain Research, 256,* 293–302.

Crepel, F., Dupont, J. L., & Gardette, R. (1984). Selective absence of calcium spikes in Purkinje cells of staggerer mutant mice in cerebellar slices maintained in vitro. *Journal of Physiology (London), 346,* 111–125.

Crusio, W. E., Genthner-Grimm, G., & Schwegler, H. (1986). A quantitative-genetic analysis of hippocampal variation in the mouse. *Journal of Neurogenetics, 3,* 203–214.

Edelman, G. M., & Chuong, C. M. (1982). Embryonic to adult conversion of neural cell adhesion molecule in normal and staggerer mice. *Proceedings of the National Academy of Science, 79,* 7036–7040.

Fuller, J. L. (1979). Fuller BWS lines: History and results. In M. E. Hahn, C. Jensen, & B. E. Dudek (Eds.), *Development and Evolution of Brain Size: Behavioral Implications.* New York: Academic Press, pp. 187–204.

Fuller, J. L., & Geils, H. (1972). Brain growth in mice selected for high and low brain weight. *Developmental Psychobiology, 5,* 307–318.

Fuller, J. L., & Herman, B. (1974). Effect of genotype and practice upon behavioral development in mice. *Developmental Psychobiology, 7,* 21–30.

Goffinet, A. M. (1983). The embryonic development of the cerebellum in normal and reeler mutant mice. *Anatomy and Embryology (Berlin), 168,* 73–86.

Goldowitz, D., & Mullen, R. J. (1982). Granule cell as a site of gene action in the weaver mouse cerebellum: Evidence from heterozygous mutant chimeras. *Journal of Neuroscience, 2,* 1474–1485.

Hahn, M. E. (1979). Fuller BWS lines: Parental influences on brain size and behavioral development. In M. E. Hahn, C. Jensen, & B. E. Dudek (Eds.), *Development and Evolution of Brain Size: Behavioral Implications.* New York: Academic Press, pp. 239–261.

Hahn, M. E., & Haber, S. B. (1978). A diallel analysis of brain and body weight in male inbred laboratory mice *(Mus musculus). Behavior Genetics, 8,* 251–260.

Hahn, M. E., & Haber, S. B. (1979). Dominance for large brains in laboratory mice: Reply to Henderson. *Behavior Genetics, 9,* 243–244.

Hahn, M. E., Haber, S. B., & Fuller, J. L. (1973). Differential agonistic behavior in mice selected for brain weight. *Physiology and Behavior, 10,* 759–762.

Hahn, M. E., Walters, J. K., Lavooy, J., & Deluca, J. (1983). Brain growth in young mice: Evidence on the theory of phrenoblysis. *Developmental Psychobiology, 16,* 377–383.

Heimrich, B., Schwegler, H., & Crusio, W. E. (1985). Hippocampal variation between the inbred mouse strains C3H/HeJ and DBA/2: A quantitative-genetic analysis. *Journal of Neurogenetics, 2,* 389–401.

Henderson, N. E. (1979). Dominance for large brains in laboratory mice. *Behavior Genetics, 9,* 45–50.

Herrup, K. (1983). Role of staggerer gene in determining cell number in cerebellar cortex. I. Granule cell death is an indirect consequence of staggerer gene action. *Developmental Brain Research, 11,* 267–274.

Herrup, K. (1986). Cell lineage relationships in the development of the mammalian CNS: Role of cell lineage in control of cerebellar Purkinje cell number. *Developmental Biology, 115,* 148–154.

Herrup, K., & Bower, J. (1986). A combined genetic and physiological study of neuronal specificity in mouse: Does Purkinje cell lineage play a role in establishing projection patterns in cerebellar cortex? *Neuroscience Abstracts, 12,* 769.

Herrup, K., & Mullen, R. J. (1981). Role of the staggerer gene in determining Purkinje cell number in the cerebellar cortex of mouse chimeras. *Developmental Brain Research, 1,* 475–485.

Herrup, K., & Sunter, K. (1986). Cell lineage dependent and independent control of Purkinje cell number in the mammalian CNS: Further quantitative studies of lurcher chimeric mice. *Developmental Biology, 117,* 417–427.

Herrup, K., & Sunter, K. (1987). Numerical matching during cerebellar development: Quantitative analysis of granule cell death in staggerer mouse chimeras. *Journal of Neuroscience, 7,* 829–836.

Hewitt, J. K., Hahn, M., & Karkowski, L. M. (1987). Genetic selection disrupts stability of mouse brain weight development. *Brian Research, 417,* 225–231.

Jensen, C. (1979). Learning performances in mice genetically selected for brain weight: Problems of generality. In M. E. Hahn, C. Jensen, B. E. Dudek (eds.), *Development and Evolution of Brain Size: Behavioral Implications.* New York: Academic Press, pp. 205–220.

Katz, M. J., & Lasek, R. J. (1978). Evolution of the nervous system: Role of ontogenetic mechanisms in the evolution of matching populations. *Proceedings of the National Academy of Sciences, 75,* 1349–1352.

Katz, M. J., Lasek, R. J., & Silver, J. (1983). Ontophyletics of the nervous system: Development of the corpus callosum and evolution of axon tracts. *Proceedings of the National Academy of Sciences, 80,* 5936–5940.

Leamy, L. (1985). Morphometric studies in inbred and hybrid house mice. VI. A genetical analysis of brain and body size. *Behavior Genetics, 15,* 251–263.

Lipp, H. P., & Schwegler, H. (1982). Hippocampal mossy fibers and avoidance learning. In I. Lieblich (Ed.), *Genetics of the Brain.* Amsterdam: Elsevier, pp. 325–364.

Mullen, R. J. (1978). Mosaicism in the central nervous system of mouse chimeras. In S. Subtelny & I. M. Sussex (Eds.), *The Clonal Basis of Development.* New York: Academic Press.
Mullen, R. J., & Herrup, K. (1979). Chimeric analysis of mouse cerebellar mutants. In X. O. Breakefield (Ed.), *Neurogenetics: Genetic Approaches to the Study of the Nervous System.* New York: Elsevier, pp. 173–196.
Nowakowski, R. S. (1984). The mode of inheritance of a defect in lamination in the hippocampus of BALB/c mice. *Journal of Neurogenetics, 1,* 249–258.
Nowakowski, R. S., & Davis, T. L. (1985). Dendritic arbors and dendritic excrescences of abnormally positioned neurons in area CA3c of mice carrying the mutation "hippocampal lamination defect." *Journal of Comparative Neurology, 239,* 267–275.
Oliverio, A., Eleftheriou, B. E., & Bailey, D. W. (1973). A gene influencing active avoidance performance in mice. *Physiology and Behavior, 11,* 497–501.
Oster-Granite, M. L., & Gearhart, J. (1981). Call lineage analysis of cerebellar Purkinje cells in mouse chimeras. *Developmental Biology, 85,* 199–208.
Peeler, D. F. (1987). Active avoidance performance in genetically defined mice. *Behavioral and Neural Biology, 48,* 83–89.
Pinto-Lord, M. C., Evrard, P., & Caviness, V. S., Jr. (1982). Obstructed neuronal migration along radial glial fibers in the neocortex of the reeler mouse: A Golgi-EM analysis. *Brain Research, 256,* 379–393.
Quinn, W. G., & Gould, J. L. (1979). Nerves and genes. *Nature, 278,* 19–23.
Rakic, P., & Sidman, R. L. (1973). Sequence of developmental abnormalities leading to granule cell deficit in cerebellar cortex of weaver mutant mice. *Journal of Comparative Neurology, 152,* 103–132.
Riska, B., & Atchley, W. R. (1985). Genetics of growth predict patterns of brain-size evolution. *Science, 229,* 668–671.
Roderick, T. H., Wimer, R. E., & Wimer, C. C. (1976). Genetic manipulations of neuroanatomical traits. In L. Petrinovich and J. McGaugh (Eds.), *Knowing, Thinking, and Believing.* New York: Pergamon, pp. 143–178.
Roffler-Tarlov, S., & Herrup, K. (1981). Quantitative examination of the deep cerebellar nuclei in the staggerer mutant mouse. *Brain Research, 215,* 49–59.
Schwegler, H., Heimrich, B., Crusio, W. E., & Lipp, H.-P. (1985). Hippocampal mossy fiber distribution and two-way avoidance learning in rats and mice. In B. E. Will, P. Schmitt & J. C. Dalrymple-Alford (Eds.), *Brain Plasticity, Learning, and Memory.* New York: Plenum, pp. 127–138.
Serra-Oller, M. M., Olavarria, J., & Van Sluyters, R. C. (1986). Pattern of interhemispheric connections in neocortex of mice with congenital deficiencies of the corpus callosum. *Neuroscience Abstracts, 12,* 1370.
Silver, J., Lorenz, S. E., Wahlsten, D., & Coughlin, J. (1982). Axonal guidance during development of the great cerebral commissures: Descriptive and experimental studies, in vivo, on the role of the preformed glial pathways. *Journal of Comparative Neurology, 210,* 10–29.
Silver, J., & Ogawa, M. Y. (1983). Postnatally induced formation of the corpus callosum in acallosal mice on glia-coated cellulose bridges. *Science, 220,* 1067–1069.
Stent, G. S. (1981). Strength and weakness of the genetic approach to the development of the nervous system. In W. M. Cowan (Ed.), *Studies in Developmental Neurobiology.* New York: Oxford, pp. 288–321.
Storer, J. B. (1967). Relation of lifespan to brain weight, body weight and metabolic rate among inbred mouse strains. *Experimental Gerontology, 2,* 173–182.
Vaughn, J. E., Matthews, D. A., Barber, R. P., Wimer, C. C., & Wimer, R. E. (1977). Genetically associated variations in the development of hippocampal pyramidal neurons may

produce differences in mossy fiber connectivity. *Journal of Comparative Neurology, 173,* 41–52.

Vogel, M. W., & Herrup, K. (1986). Target related cell death in cerebellar granule cells: Are early generated cells at a competitive advantage? *Neuroscience Abstracts, 12,* 869.

Wahlsten, D. (1974). Heritable aspects of anomalous myelinated fibre tracts in the forebrain of the laboratory mouse. *Brain Research, 68,* 1–18.

Wahlsten, D. (1975). Genetic variation in the development of mouse brain and behavior: Evidence from the middle postnatal period. *Developmental Psychobiology, 8,* 371–380.

Wahlsten, D. (1981). Prenatal schedule of appearance of mouse brain commissures. *Brain Research, 227,* 461–473.

Wahlsten, D. (1982a). Genes with incomplete penetrance and the analysis of brain development. In I. Lieblich (Ed.), *Genetics of the Brain.* Amsterdam: Elsevier, pp. 367–391.

Wahlsten, D. (1982b). Mode of inheritance of deficient corpus callosum in mice. *Journal of Heredity, 73,* 281–285.

Wahlsten, D. (1982c). Deficiency of corpus callosum varies with strain and supplier of the mice. *Brain Research, 239,* 329–347.

Wahlsten, D. (1982d). Mice in utero while their mother is lactating suffer higher frequency of deficient corpus callosum. *Brain Research, 281,* 354–357.

Wahlsten, D. (1984). Growth of the mouse corpus callosum. *Brain Research, 317,* 59–67.

Wainwright, P., & Gagnon, M. (1984). Effects of fasting during gestation on brain development in BALB/c mice. *Experimental Neurology, 85,* 223–228.

Wetts, R., & Herrup, K. (1982a). Cerebellar Purkinje cells are descended from a small number of progenitors committed during early development: Quantitative anaylsis of lurcher chimeric mice. *Journal of Neuroscience, 2,* 1494–1498.

Wetts, R., & Herrup, K. (1982b). Interaction of granule, Purkinje and inferior olivary neurons in lurcher chimeric mice. I. Qualitative studies. *Journal of Embryology and Experimental Morphology, 68,* 87–98.

Wetts, R., & Herrup, K. (1982c). Interaction of granule, Purkinje and inferior olivary neurons in lurcher chimeric mice. II. Granule cell death. *Brain Research, 250,* 358–362.

Wetts, R., & Herrup, K. (1983). Direct correlation between Purkinje and granule cell number in the cerebella of lurcher chimeras and wild-type mice. *Developmental Brain Research, 10,* 41–47.

Willinger, M., & Margolis, D. M. (1985). Effect of the weaver (wv) mutation on cerebellar neuron differentiation. I. Qualitative observations of neuron behavior in culture. *Developmental Biology, 107,* 156–172.

Wimer, C. (1979). Correlates of mouse brain weight; a search for component morphological traits. In M. E. Hahn, C. Jensen, and B. C. Dudek (Eds.), *Development and Evolution of Brain Size: Behavioral Implications.* New York: Academic Press, pp. 147–162.

# 6

# Genetic and Molecular Analysis of Neural Development and Behavior in *Drosophila*

JEFFREY C. HALL, SHANKAR J. KULKARNI,
CHARALAMBOS P. KYRIACOU, QIANG YU,
AND MICHAEL ROSBASH

Research in *Drosophila* behavioral genetics is often concerned with the development of the fruit fly's nervous system. A number of single-gene mutants, isolated as behaviorally defective adults, are caused by well-defined genetic changes that disturb particular features of cell differentiation and pattern formation in the organism's central nervous system (for review see Hall, 1985). These morphological defects can result in aberrant behavior and thus can indicate which neural systems are important in controlling these sensorimotor activities.

Anatomical brain mutants have also been isolated in "brute force" screens (for review see Fischbach & Heisenberg, 1984). At the outset these are interesting if only from the standpoint of the actual survival of the genetic variants, some of which seem severely brain damaged. However, the behavioral abnormalities (Heisenberg et al., 1985) and in some cases normalities (Heisenberg, 1980) of these mutants have also generated important clues about the neural substrates underlying the fly's actions.

This chapter will concentrate on *Drosophila* mutants that are defective in courtship. Some of these mutants are known to be, or intriguingly seem to be, aberrant in their neural development. This usually means embryonic, larval, or pupal assembly of the CNS. We also raise an issue that involves a newly discovered and surprising feature of brain development that occurs after *Drosophila* metamorphosis. In addition we provide information on the adaptive significance of normal versus mutant behaviors. For one gene, the detailed description of the courtship behavioral defects in the relevant mutants, and subsequent analysis of the wild-type gene defined by such variants, has evolutionary implications.

## MUTANTS AND COURTSHIP SONGS

A very conspicuous feature of *Drosophila* reproductive behavior is the song produced by male wing vibrations (Figure 6–1). These sounds influence female receptivity to mating attempts (for review see Hall, 1984).

**NORMAL**

35 ms

*cac*

Figure 6-1 Normal and mutant courtship songs. The two traces were produced from recordings of wing vibrations that males of *Drosophila melanogaster* direct at females. Such signals occur in trains, as shown here. These monocyclic to tricyclic "pulses" are produced at the rate of about 30 per second, reflected in the duration of a typical "interpulse interval" (i.e., 35 ms, as shown). The top trace is a pulse train from a wild-type male's song; below it is the abnormal song—with its polycyclic pulses—of a male expressing the *cacophony* mutation.

It was thought, therefore, that song mutants could be isolated by looking among the descendants of mutagenized flies for males that take "too long" to initiate mating with normal females. Indeed, the first song mutant, *cacophony,* was found in this manner (Schilcher, 1976a, 1977). Its singing abnormality, monitored in our recent recordings, is shown in Figure 6-1. The loud, polycyclic "tone pulses" caused by this X-chromosomal mutation (standard genetic abbreviation: *cac*) are so overtly abnormal that one might guess the female would be poorly stimulated by such cacophonous sounds.

A puzzling finding was that wingless *cac* males are even slower in mating latencies than are wild-type males from which the wings were removed (Schilcher, 1977; confirmed by Kulkarni & Hall, 1987). Wingless *Drosophila* males have long been known to exhibit lengthy mating latencies (review: Hall, 1984). The obvious prediction was that removing the wings from both *cac* and normal males would equalize their mating success. That the prediction failed indicates that an additional defect apart from the courtship song abnormality is present in this mutant.

This matter has recently been examined further, by precisely defining the *cac* locus genetically (Kulkarni & Hall, 1987). Included in the analysis was the creation of *cac*-bearing X chromosomes that had undergone recombination with marker-bearing X's. Certain classes of the recombinant males (involving crossovers that, by convention, are specified as having occurred to the left of the *cac* locus) still sang with the usual mutant singing defect. These males, however, exhibited normal, robust mating-initiation kinetics (Kulkarni & Hall, 1987). In other words, it is likely that another mutation, located to the left of *cac,* independently caused the abnormally long mating latencies associated with this strain.

Still, the action of the gene defined by the *cac* mutation is, it seems, interesting with respect to how the male becomes "programmed" to produce his normal song pulses. One way of looking into this further is to ask what part of the fly is influenced by *cac*. For this, genetic mosaics have been constructed that are composed of part *cac* mutant and part normal *(cac⁺/cac)* tissues (Figure 6–2). Examination of the songs produced by these mosaics, indicates that the *thorax* is the body region for this gene's site of action (Figure 6–2).

It will be interesting to determine whether a hypothetical "song circuit" in the thoracic nervous system has its development deranged in a specific manner by the *cac* mutation. Such a hypothesis implies that the normal allele is indeed active in developmental stages, during which portions of the imaginal CNS are forming. Alternatively the gene might meaningfully act only in adults to alter the physiology of a putative pattern generator, that is, against a background of what could be thoroughly normal neuroanatomy in the mutant.

These possibilities are raised because it should eventually be possible to ask *when*, as well as *where*, $cac^+$ is expressed. This is because the gene has been cloned molecularly, or should be readily clonable (T. Goralski, K. Konrad, & A. P. Mahowald,

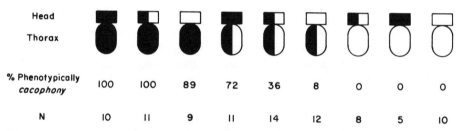

**Figure 6–2** Mosaic analysis of defective courtship song caused by the *cacophony* mutation. Each genetically mixed individual was part *cac* mutant (i.e., hemizygous for this X chromosomal factor) and part cac/+ (heterozygous for the mutation, which is recessive). In addition, these mosaics included the necessary embellishment of being thoroughly male (as opposed to sexually mixed, X//XX gynandromorphs); thus they were homozygous for the third chromosomal sex-transforming mutation, *tra* (see Hall, 1984; Kulkarni & Hall, 1987). Each mosaic was tested for singing in the presence of a virgin female, and the sounds produced from a given wing vibration (either left wing or right wing) were scored as to whether they were cac-like or normal (see Figure 6–1). All mosaics turned out to sing in the same manner when using either wing. Thus a given mosaic was "all-mutant" or "all-normal" in its behavior. Distributions of mutant versus normal tissues for the mosaic were scored externally (using a cac-tagging marker; see Hall, 1984), and the various mosaics were thus separated into their major categories. For example, the 11 individuals here with all or mostly mutant (black) heads had thoraces split (more or less) down the middle, such that one side was mutant (black), the other normal (white). Correlating the patterns of major body region genotypes with the associated singing phenotypes led to the strong suggestion that the *cac* mutation influences the development or function of thoracic tissues to cause a polycyclic song. Furthermore, since the "thorax-split" cases did not, overall, have a particularly high chance of singing abnormally, it could be that mutant tissue in both the left and right halves of this body region is required for the behavioral defect to be manifested.

unpublished research; see discussion in Kulkarni & Hall, 1987). If the precise portion of the cloned sequences that are necessary and sufficient for *cac* expression can be identified with precision (see following text for the pertinent method, described for another "behavioral gene"), one should be able to determine the stages during which the primary and secondary gene products are expressed.

Another component of the fruit fly's song, which is more subtle than the overt nature of the pulse trains, appears to be adaptively significant. The rate of pulse production fluctuates *rhythmically* in the songs of wild-type males, with the periodicities of such rhythms being 50 to 60 seconds in *D. melanogaster* (Kyriacou & Hall, 1980). Closely related *D. simulans* males also sing rhythmically, but with 30 to 40 second periods (Kyriacou & Hall, 1980,1986)—against a background rate of pulse production that is slower than in *D. melanogaster* (e.g., Cowling & Burnet, 1981).

Rhythm mutations in *D. melanogaster* affect the song rhythm in a manner that parallels their effects on circadian rhythms (Kyriacou & Hall, 1980,1986). These mutations, at an X-chromosomal locus called *period (per),* affect such daily cycles by definition, in that this is how these genetic changes were identified originally (Konopka & Benzer, 1971).

Thus, there is a way to change the song rhythmicity within a given species. Males carrying *per* mutations, whose songs have altered periodicities or, depending on the mutant allele, no regular patterns of pulse-rate fluctuations, are slower to initiate mating with normal females (Hall, 1981).

The function of the song rhythms has also been examined in "song simulator" experiments. These included assessing the effects of *pre*stimulating *D. melanogaster* females with rhythmic versus nonrhythmic pulse songs (Kyriacou & Hall, 1984). Playing the latter song to naive virgin females had no apparent effect on mating latencies (cf. Schilcher, 1976b) when naive males were introduced to the females after a given prestimulation. Yet programming the simulator to produce rhythmic songs with proper 55-second periodicities led to significant enhancements of mating-initiation kinetics; artificial song with 35-second, that is, *simulans*-like, cycle durations were ineffective (Kyriacou & Hall, 1984).

## EXPERIENCE-DEPENDENT COURTSHIP

The effect of rhythmic song prestimulation resembles a rudimentary form of memory, called sensitization. Application of such a term may or may not be valid—but is it interesting? The answer seems to be yes, because "generalized" learning and memory mutations in *Drosophila melanogaster* (for review see Tully, 1987) reduce the effectiveness of the song prestimulation (Kyriacou & Hall, 1984). These are the *dunce, rutabaga,* and *amnesiac* mutants, originally isolated with respect to abnormal avoidance conditioning (Tully, 1987). Such mutations were subsequently shown to disturb "simple" learning in noncourtship contexts, that is, to cause mediocre sensitization or habituation to sugar stimuli (Duerr & Quinn, 1982).

Aftereffects on *male* behaviors are also caused by certain courtship experiences to which these flies are exposed: (1) Courtship directed at a female is weak for a male who *previously* courted a mated female (review: Siegel et al., 1984). (2) The curious courtship that a mature male is stimulated to perform in the presence of an immature male (1 day old or less) leads to minimal courtship behavior with a *subsequent* immature male (Siegel et al., 1984).

Whereas the first aftereffect appears to be a case of associative conditioning (Ackerman & Siegel, 1986; Tompkins et al., 1983), the second seems to be a form of nonassociative habituation. This has recently been deduced from bioassay experiments (Gailey, Lacaillade, & Hall, 1986) involving manipulation of the chemosensory cues known to be largely responsible for an immature male eliciting high levels of courtship in the first place (Tompkins, Hall, & Hall, 1980).

Both kinds of behavioral aftereffects are abnormal to one degree or another when the courting males express a given learning/memory mutation (Hall, 1986; Siegel et al., 1984). Once more, the action of these genes seems significant adaptively (not concludable, perhaps, from the fact that *dunce* mutants and the like are aberrant in their ability to remember that electric shocks were coupled with a given artificial odorant; see Tully, 1987).

Do these experience-dependent male behaviors have "fitness values" in experimental situations? For aftereffects caused by courtship of fertilized females, the answer appears to be negative (Zawistowski & Richmond, 1985). Yet analogous experiments involving "training" of mature males by having them court immature ones showed that they subsequently performed better than did naive males in finding and mating with a virgin female among small populations that included potentially "distracting" immature males (Gailey, Hall, & Siegel, 1985; Zawistowski & Richmond, 1985). As was predicted, *dunce* males—even when trained in this manner—did relatively poorly in terms of ignoring the immature males and so were quite ineffective in locating the female (Gailey et al., 1985).

These experience-dependent phenomena also raise the question of whether there is functional significance to a young male's stimulating effects on an older male's courtship. The following points may be pertinent: (1) A portion of the fruit fly brain, the dorsally located and sexually dimorphic mushroom bodies, keeps developing even after the fly has gone through metamorphosis (Technau, 1984). (2) This weeklong increase in fiber number in the relevant mushroom body stalk does not occur when the flies are environmentally deprived after eclosion, that is, when they are kept in separate containers as opposed to being mixed with several other flies, odor sources, and so on (Balling, Technau, & Heisenberg, 1987; Technau, 1984). (3) Such postpupal CNS development could be connected to information storage and retrieval, because fiber numbers do not increase in *dunce* and *rutabaga* imagos, even when they are in an "enriched" environment (Balling et al., 1987). Alternatively, since *dunce* and *rutabaga* have increased and decreased levels of cyclic adenosine monophosphate (cAMP), respectively (review: Tully, 1984), it is possible that such biochemical abnormalities lead directly to the anatomical differences in the mutants, rather than indirectly affecting mushroom body development by "feeding back" via their learning impairments. (4) Young males, by stimulating older males to court them, are eliciting a good deal of *auditory* input, albeit for a limited period of about 24 hours (McRobert & Tompkins, 1983; Tompkins et al., 1980). If acoustic stimuli are relevant to the triggering of posteclosion mushroom body growth, perhaps the immature male "stores" some auditory information, which affects his subsequent singing ability in a positive manner.* (5) Whereas it is known (C. P. Kyriacou & J. C. Hall, unpublished data) that deprivation-reared *D. melanogaster* males will court and sing to females in a grossly normal manner, how effective this courtship is in securing a mate and how robust the more subtle features of the song are have not yet been investigated.

## DEVELOPMENTAL AND MOLECULAR BIOLOGY OF RHYTHMIC SINGING BEHAVIOR

The subtlety in question would be the aforementioned singing rhythm. Males produce song cycles by rhythmically changing the intervals between pulses of tone (see Figure 6-4). In addition to the alterations of song rhythm periods caused by certain of the *per* mutations of *D. melanogaster* (see preceding text), there are "arrhythmic" *(per$^o$)* mutant alleles that seem to eliminate circadian rhythms and cause the song's pulse rate variations to be apparently arbitrary (Hamblen et al., 1986; Kyriacou & Hall, 1980, 1986; Zehring et al., 1984).

Many of the data concerning *per*'s effects on circadian rhythms have been collected in locomotor activity tests of mutant versus normal flies (though the original *per* mutants were isolated with respect to changes in, or elimination of, pupal → adult eclosion rhythms; Konopka & Benzer, 1971). Whereas *per$^o$* adults are ostensibly arrhythmic in such tests, relatively high-frequency periodicities can be revealed (Dowse, Hall & Ringo, 1987). These cryptic rhythms have periods in the range of 4 to 22 hours, with the mean values being ultradian (in these cases, on the order of one-half day).

Such residual rhythmicity does not occur because there is a hypothetically low but meaningful level of activity remaining in the *per$^o$* mutants. This is because there are genotypes in which the *per* locus is deleted, not merely mutated (Bargiello & Young, 1984; Reddy et al., 1984; Smith & Konopka, 1981). These *per$^-$* flies are apparently arrhythmic (Hamblen et al., 1986; Smith & Konopka, 1981) but again exhibit cryptically periodic locomotor activity (Dowse, Hall, & Ringo, 1987).

Another obvious implication of this *per$^-$* genotype is that not only are the fly's overt rhythms dispensible, but the *per* gene is as well. One might have predicted that circadian rhythms would be vital and that a gene controlling them could have such widespread influence on CNS structure or function that mutating it to the null state, in a viable genetic variant, would be impossible. It is also notable that another behaviorally important gene mentioned in this chapter, *dunce,* has similarly been proved to be inessential, that is, by the creation of a viable *dunce$^-$* deletion (review: Davis & Kauvar, 1984). It must be considered, however, that a hypothetical gene (not *per*) whose "null allele" truly eliminates the possibility of any rhythmicity (cf. Dowse et al., 1987) would be a lethal mutation.

The proof that the so-called *per$^-$* type is indeed at the null state was not a simple

---

*Courtship-eliciting immature males do not themselves court females; the latter ability matures at about the same time that the young male ceases to elicit courtship from older males (Jallon & Hotta, 1979; Tompkins et al., 1980).

genetic exercise. It demanded a molecular analysis of DNA in and around the locus. These investigations have proceeded in several directions, aside from the application of *per* clones to analyze the chromosome aberrations that result in elimination of this clock gene (Bargiello & Young, 1984; Hamblen et al., 1986; Reddy et al., 1984).

"Northern blotting" studies and the generation of germ-line transformants have identified a 4.5-kb (kilobase) transcript from the *per* locus as the key RNA expressed by it (Bargiello & Young, 1984; Hamblen et al., 1986; Zehring et al., 1984). Expression of this transcript has neurodevelopmental significance: It is present in the CNS during the embryonic stage (James et al., 1986; Young et al., 1985). The transcript then becomes undetectable during larval stages, reappearing during metamorphosis and continuing in adults—where it is "head enriched" (James et al., 1986) and thus possibly concentrated in the brain. The roles that the early and/or late expression of this RNA species play in the control of various rhythmic phenotypes (review: Hall, 1985; Konopka, 1987) is under investigation. Studies include inquiries into whether cell differentiation and pattern formation in the fly's CNS could be influenced by expression at different developmental stages (cf. Konopka & Wells, 1980).

The informational content of *per*'s 4.5-kb transcript has been deduced from sequencing the corresponding genomic DNA (Jackson et al., 1986; Yu et al., 1987b) and also "cDNA" clones produced from this specific RNA (Citri et al., 1987). A salient feature of the inferred amino acid sequence is an alternating series of threonine-glycine pairs, rather centrally located in the conceptual *per* protein (Figure 6–3). This unusual domain of the gene has proved to be interesting from several perspectives, including matters of protein chemistry (Reddy et al., 1986). The current discussion, however, will focus on intraspecific and interspecific variability of this Thr-Gly repeat and the possible behavioral significance thereof.

The number of Thr-Gly pairs varies among *D. melanogaster* strains (Yu et al., 1987a), different ones being maintained in laboratories or collected from natural populations. For most strains there are 17 or 20 such pairs (in perfectly repeated sequence); a small number of 23-pair cases have also been detected.

Comparing the *D. melanogaster* data with data that established a Thr-Gly repeat configuration in a *D. simulans* strain, one finds a substantial number of differences, not so much in terms of the total number of Thr-Gly pairs, but in regard to the specific nucleotide sequence in the corresponding regions of the *per* gene cloned from these two species (D. A. Wheeler, A. C. Jacquier, and Q. Yu, unpublished data). Other portions of the *D. simulans* form of this gene that have been sequenced so far, encompassing about 60 percent of the total, are essentially identical to the *D. melanogaster* nucleotide sequence.

Does this variability within or between species have any relevance to behavioral variations? No data are yet available on rhythmic phenotypes among, say, the Thr-Gly repeat variable strains within *D. melanogaster*. However, recall that this species and *D. simulans* differ in their song rhythm characteristics (Kyriacou & Hall, 1980). Furthermore, it is intriguing that, in interspecific hybrids, the genetic etiology of this difference maps to the same chromosome (X) where *per* is located (Kyriacou & Hall, 1986).

A somewhat gross experimental manipulation of the Thr-Gly domain of *per* has been performed and assessed for its behavioral consequences (Yu et al., 1987b). The perfect repeat region was deleted by in vitro manipulation of cloned *D. melanogaster*

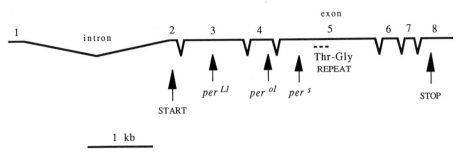

**Figure 6-3** Primary expression of the *period* gene in *Drosophila melanogaster*. This RNA, about 4500 nucleotides long, is the key transcript from this clock gene, as determined from Northern blotting and germ-line transformation experiments (Bargiello & Young, 1984; Hamblen et al., 1986; Zehring et al., 1984). The transcript is in reality a family of three closely related RNA species (Citri et al., 1987), each about 4.5 kb long and differing from each other by alternative splicing in the regions of exon 5 and further toward the right (3') end of the gene. The structure of the most abundant of these three species is shown here. The exons (horizontal lines) are numbered from the 5' to 3' direction. The first exon does not code for amino acids (hence the ATG translation "start" site in exon 2, nor do nucleotides downstream of the TAG "stop" in exon 8. Introns are designated by the angled lines between any two adjacent exons. Exon 4 is the site of the original arrhythmic *per* mutation, which is a translation-terminating nonsense codon; and the site of the change responsible for $per^s$, a serine-to-asparagine missense mutation, is in exon 5 (Yu et al., 1987b). The latter exon also contains the threonine-glycine amino acid repeat discussed in the text and Figure 6-4 (see Citri et al., 1987; Reddy et al., 1986; Yu et al., 1987a,b).

DNA, then reintroduced by transformation methods into arrhythmic $per^o$ hosts (in the same kinds of experiments performed earlier to identify the "core" of the *per* locus, e.g., Hamblen et al., 1986). This rather severe damage to the gene had essentially no effect on circadian rhythmicity, but the song rhythm cycles became 20 to 25 seconds faster than in the corresponding control transformants (Figure 6-4). Therefore the severity of this lesion is evidently of considerable significance for the rhythmic behavior that is in fact variable evolutionarily.

One might worry, though, about the meaning of such an artificial and rather substantial molecular modification of the *per* gene: What, if anything, can this kind of laboratory construct say about natural phenotypic variability? By the same token, what can severe mutations like the $per^o$ cases, chemically induced by behavioral geneticists, tell us about the adaptive significance of the *per* gene's function in "real" settings? One can speculate that $per^o$ alleles are outside the natural range of variation, even though such null alleles (Yu et al., 1987b) might, in this case, allow good survival in the wild (since such genotypes cause no overtly deleterious effects on the flies; see above). Also, *per* mutations do not, at least in the laboratory, impinge on the fly's normal lifespan (Konopka, 1987). These issues are taken up in the following summary section.

**A** $per^{ol}$; P[$per^+$] 8.0

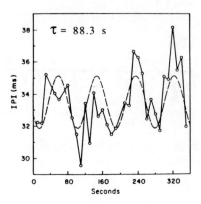
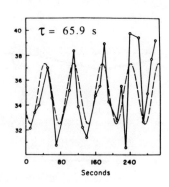

**B** $per^{ol}$; P[$per^+$] 8.0 ΔTG

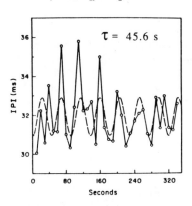
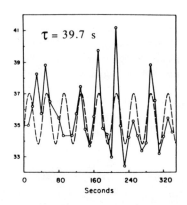

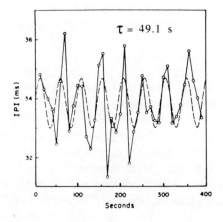
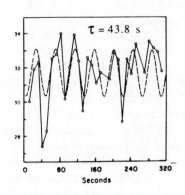

## IMPLICATIONS OF "SINGLE LOCUS" BEHAVIORAL AND MOLECULAR GENETICS

We submit that—whereas nonfunctioning forms of genes that influence behavior, or deletions from within a gene like *per,* might be irrelevant to matters of selective forces acting on genetic variants—mutations such as *per$^o$* have been of enormous heuristic value in allowing investigators to tap into bona fide natural variation. This is because, for the case of *per* and other genetic loci of neural and behavioral interest in this organism, the "artifical" severely mutated alleles have been crucial in the molecular identification of the genetic loci. Thus, for example, *per$^o$* or *per$^-$* types were the host genotypes that could be most meaningfully rescued by the relevant pieces of DNA in transformation experiments (see above).

One consequence of this molecular gene identification was to allow probing of flies from natural populations with sequences from the *per* locus. Results included the discovery of polymorphisms for the specific domain of the gene discussed in the preceding section. This Thr-Gly repeat region was then shown to be behaviorally important, albeit in the context of what is once more an artificial genotype created by further molecular manipulation of *per* sequences (Yu et al., 1987a). Such results perhaps involve behavioral alterations beyond what can (or do) occur in the wild. But what if the different polymorphic forms of the Thr-Gly repeat region turn out also to be associated with variation in singing periodicities? It is possible that such behavioral differences, however slight, would be detectable by females (see Kyriacou & Hall, 1982, 1986) and could lead to changes in mating dynamics. Such changes, of course, can be crucial in reproductive fitness and species isolation.

Therefore, all of this molecular work on *per,* including that revolving around the artificial forms of the gene, seems to have something to say about behavioral and

◄─────────────────────────────────────────────

**Figure 6-4** Courtship song rhythms of transformants involving in vitro mutagenesis of the *period* gene. An 8.0-kb DNA segment that includes all the amino acid coding information at this clock locus (see Figure 6-3 and Hamblen et al., 1986; Zehring et al., 1984) was transduced into the genome of arrhythmic *per$^o$* "hosts" (according to the strategies in the papers just cited). **A.** When this *per*-derived DNA fragment was from wild type, it, as usual, readily rescued the effects of this *per* mutation on both circadian and song rhythms (Yu et al., 1987a). For the latter phenotype, the periods ranged from wild-type taus (about one minute) to longer cycle durations, as exemplified by the songs of these two control transformant males. The plots involve average interpulse intervals (IPIs, see Figure 6-1), computed for a series of ten-second time frames for the several minutes of singing performed by these transformed flies in the presence of normal females (see Kyriacou & Hall, 1980, 1986). The patterns of IPI variations here could be fit to sine functions (dashed lines), from which one extracts the period estimate for a given song (see Kyriacou & Hall, 1980, for the method). **B.** When the 8.0-kb fragment was constructed deliberately to be deleted of the Thr-Gly repeat region (see Figure 6-3), IPI oscillations were routinely defined by shorter periods than in the control transformants. The rhythms from four of the experimental ΔTG males are shown. (See Yu et al., 1987a, for methods of deletion construction and summary of all the relevant behavioral results.)

evolutionary issues. The latter have come up, as well, in a host of other studies of this general type—for example, the molecular analysis of, and polymorphic variations among, naturally occurring forms of enzyme-encoding loci in various *Drosophila* species (e.g., Bodmer & Ashburner, 1984; Kreitman, 1983). These genes were "clonable," such that the population/molecular studies could be performed, without recourse to severe (and unnatural?) mutations induced at such loci, simply because the cloning strategies here could rely on a priori knowledge of the gene product. Yet this kind of molecular start for digging into a *behaviorally* interesting genetic factor is usually, perhaps almost always, not possible. So one must begin by isolating behavioral mutants that could be regarded as interesting only in the lab. Yet this allows inquiries into the concrete nature of the underlying genes. Some of the questions that stem from such studies, asked about the different *normal* forms of the genes, are potentially of considerable significance for population biology. Thus, the oft-cited criticism that laboratory-induced mutants are of little use in the study of evolution can nicely be laid to rest.

## ACKNOWLEDGMENTS

We thank our colleagues, especially Richard Siegel, Donald Gailey, Hildur Colot, Ronald Konopka, Harold Dowse, and John Ringo, for their many contributions to the work summarized here. Support was provided by the National Institutes of Health (grants GM-21473 and GM-33205) in the United States and by a Science & Engineering Research Council grant (GRD-18370) in the United Kingdom.

## REFERENCES

Ackerman, S. L., & Siegel, R. W. (1986). Chemically reinforced conditioned courtship in *Drosophila:* Responses of wild-type and the *dunce, amnesiac* and *don giovanni* mutants. *Journal of Neurogenetics, 3,* 111–123.

Balling, A., Technau, G., & Heisenberg, M. (1987). Are the structural changes in adult *Drosophila* mushroom bodies memory traces? Studies on biochemical learning mutants. *Journal of Neurogenetics, 4,* 65–73.

Bargiello, T. A., & Young, M. W. (1984). Molecular genetics of a biological clock in *Drosophila. Proceedings of the National Academy of Sciences USA, 81,* 2142–2146.

Bodmer, M., & Ashburner, M. (1984). Conservation and change in the DNA sequences coding for alcohol dehydrogenase in sibling species of *Drosophila. Nature, 309,* 425–430.

Citri, Y., Colot, H. V., Jacquier, A. C., Yu, Q., Hall, J. C., Baltimore, D., & Rosbash, M. (1987). A family of unusually spliced biologically active transcripts encoded by a *Drosophila* clock gene. *Nature, 326,* 42–47.

Cowling, D. E., & Burnet, B. (1981). Courtship songs and genetic control of their acoustic characteristics in sibling species of the *Drosophila melanogaster* subgroup. *Animal Behaviour, 29,* 924–935.

Davis, R. L., & Kauvar, L. M. (1984). *Drosophila* cyclic nucleotide phosphodiesterases. *Advances in Cyclic Nucleotide Protein Phosphorylation Research, 16,* 393–402.

Dowse, H. B., Hall, J. C., & Ringo, J. M. (1987). Circadian and ultradian rhythms in *period* mutants of *Drosophila melanogaster. Behavior Genetics, 17,* 19–35.

Dueer, J. S., & Quinn, W. G. (1982). Three *Drosophila* mutations that block associative learning also effect habituation and sensitization. *Proceedings of the National Academy of Sciences USA, 79*, 3646–3650.

Fischbach, K. F., & Heisenberg, M. (1984). Neurogenetics and behaviour in insects. *Journal of Experimental Biology, 112*, 65–93.

Gailey, D. A., Hall, J. C., & Siegel, R. W. (1985). Reduced reproductive success for a conditioning mutant in experimental populations of *Drosophila melanogaster. Genetics, 111*, 795–804.

Gailey, D. A., Lacaillade, R. C., & Hall, J. C. (1986). Chemosensory elements of courtship in normal and mutant, olfaction-deficient *Drosophila melanogaster. Behavior Genetics, 16*, 375–405.

Hall, J. C. (1981). Genetic connections between courtship and other complex behaviors of *Drosophila*. In Y. Hotta (Ed.), *Genetic Dissection of Drosophila Behavior.* Tokyo: Taniguchi Foundation, pp. 145–172.

Hall, J. C. (1984). Complex brain and behavioral functions disrupted by mutations in *Drosophila. Developmental Genetics, 4*, 355–378.

Hall, J. C. (1985). Genetic analysis of behavior in insects. In G. A. Kerkut & L. I. Gilbert (Eds.), *Comprehensive Insect Physiology, Biochemistry, and Pharmacology,* Vol. 9. Oxford: Pergamon, pp. 287–373.

Hall, J. C. (1986). Learning and rhythms in courting, mutant *Drosophila. Trends in Neuroscience, 9*, 414–418.

Hamblen, M., Zehring, W. A., Kyriacou, C. P., Reddy, P., Yu, Q., Wheeler, D. A., Zwiebel, L. J., Konopka, R. J., Rosbash, M., and Hall, J. C. (1986). Germ-line transformation involving DNA from the *period* locus in *Drosophila melanogaster:* Overlapping genomic fragments that restore circadian and ultradian rhythmicity to $per^o$ and $per^-$ mutants. *Journal of Neurogenetics, 3*, 249–291.

Heisenberg, M. (1980). Mutants of brain structure and function. In O. Siddigi, P. Babu, L. M. Hall, & J. C. Hall (Eds.), *Development and Neurobiology of Drosophila.* New York: Plenum Press, pp. 373–390.

Heisenberg, M., Borst, A., Wagner, S., & Byers, D. (1985). *Drosophila* mushroom body mutants are deficient in olfactory learning. *Journal of Neurogenetics, 2*, 1–30.

Jackson, F. R., Bargiello, T. A., Yun, S.-H., & Young, M. W. (1986). Product of *per* of *Drosophila* shares homology with proteoglycans. *Nature, 320*, 185–188.

Jallon, J.-M., & Hotta, Y. (1979). Genetic and behavioral studies of female sex appeal in *Drosophila. Behavior Genetics, 9*, 257–275.

James, A. A., Ewer, J., Reddy, P., Hall, J. C., and Rosbash, M. (1986). Embryonic expression of the *period* clock gene of *Drosophila melanogaster. European Molecular Biology Organization Journal, 5*, 2313–2320.

Konopka, R. J. (1987). Genetic dissection of the *Drosophila* circadian system. *Federation Proceedings, 38*, 2602–2605.

Konopka, R. J., & Benzer, S. (1971). Clock mutants of *Drosophila melanogaster. Proceedings of the National Academy of Sciences USA, 68*, 2112–2116.

Konopka, R. J., and Wells, S. (1980). *Drosophila* clock mutations affect the morphology of a brain neurosecretory cell group. *Journal of Neurobiology, 11*, 411–415.

Kreitman, M. (1983). Nucleotide polymorphism at the alcohol dehydrogenase locus of *Drosophila melanogaster, Nature, 304*, 412–417.

Kulkarni, S. J., and Hall, J. C. (1987). Behavioral and cytogenetic analysis of the *cacophony* courtship song mutant and interacting genetic variants in *Drosophila melanogaster. Genetics, 115*, 461–475.

Kyriacou, C. P., & Hall, J. C. (1980). Circadian rhythm mutations in *Drosophila* affect short-term fluctuations in the male's courtship song. *Proceedings of the National Academy of Sciences USA, 77,* 6929–6933.

Kyriacou, C. P., & Hall, J. C. (1982). The function of courtship song rhythms in *Drosophila*. *Animal Behaviour, 30,* 794–801.

Kyriacou, C. P., & Hall, J. C. (1984). Learning and memory mutations impair acoustic priming of mating behaviour in *Drosophila*. *Nature, 308,* 62–65.

Kyriacou, C. P., & Hall, J. C. (1986). Interspecific genetic control of courtship song production and reception in *Drosophila*. *Science, 232,* 494–497.

McRobert, S. P., & Tompkins, L. (1983). Courtship of young males is ubiquitous in *Drosophila melanogaster*. *Behavior Genetics, 13,* 517–523.

Reddy, P., Jacquier, A. C., Abovich, N., Petersen, G., & Rosbash, M. (1986). The *period* clock locus of *Drosophila melanogaster* codes for a proteoglycan. *Cell, 46,* 53–61.

Reddy, P., Zehring, W. A., Wheeler, D. A., Pirrotta, V., Hadfield, C., Hall, J. C., & Rosbash, M. (1984). Molecular analysis of the *period* locus in *Drosophila melanogaster* and identification of a transcript involved in biological rhythms. *Cell, 38,* 701–710.

Schilcher, F. V. (1976a). The behavior of cacophony, a courtship song mutant in *Drosophila melanogaster*. *Behavioral Biology, 17,* 187–196.

Schilcher, F. V. (1976b). The function of pulse song and sine song in the courtship of *Drosophila melanogaster*. *Animal Behaviour, 24,* 622–625.

Schilcher, F. V. (1977). A mutation which changes courtship song in *Drosophila melanogaster*. *Behavior Genetics, 7,* 251–259.

Siegel, R. W., Hall, J. C., Gailey, D. A., & Kyriacou, C. P. (1984). Genetic elements of courtship in *Drosophila:* Mosaics and learning mutants. *Behavior Genetics, 14,* 383–410.

Smith, R. F., & Konopka, R. J. (1981). Circadian clock phenotypes of chromosome aberrations with a breakpoint at the *per* locus. *Molecular and General Genetics, 183,* 243–251.

Technau, G. M. (1984). Fiber number in the mushroom bodies of adult *Drosophila melanogaster* depends on age, sex, and experience. *Journal of Neurogenetics, 1,* 113–126.

Tompkins, L., Hall, J. C., & Hall, L. M. (1980). Courtship stimulating volatile compounds from normal and mutant *Drosophila*. *Journal of Insect Physiology, 26,* 689–697.

Tompkins, L., Siegel, R. W., Gailey, D. A., & Hall, J. C. (1983). Conditioned courtship in *Drosophila* and its mediation by association of chemical cues. *Behavior Genetics, 13,* 565–578.

Tully, T. (1984). *Drosophila* learning: Behavior and biochemistry. *Behavior Genetics, 14,* 527–557.

Tully, T. (1987). *Drosophila* learning and memory revisited. *Trends in Neuroscience, 10,* 330–335.

Young, M. W., Jackson, F. R., Shin, H. S., & Bargiello, T. A. (1985). A biological clock in *Drosophila*. *Cold Spring Harbor Symposium in Quantitative Biology, 50,* 865–875.

Yu, Q., Colot, H. V., Kyriacou, C. P., Hall, J. C., & Rosbash, M. (1987a). Behaviour modification by *in vitro* mutagenesis of variable region within the *period* gene of *Drosophila*. *Nature, 326,* 765–769.

Yu, Q., Jacquier, A. C., Citri, Y., Hamblen, M., Hall, J. C., & Rosbash, M. (1987b). Molecular mapping of point mutations in the *period* gene that stop or speed up biological clocks in *Drosophila melanogaster*. *Proceedings of the National Academy of Sciences USA, 84,* 784–788.

Zawistowski, S., & Richmond, R. C. (1985). Experience-mediated courtship reduction and competition for mates by male *Drosophila melanogaster*. *Behavioral Genetics, 15,* 561–569.

Zehring, W. A., Wheeler, D. A., Reddy, P., Konopka, R. J., Kyriacou, C. P., Rosbash, M., & Hall, J. C. (1984). P-element transformation with *period* locus DNA restores rhythmicity to mutant, arrhythmic *Drosophila melanogaster*. *Cell, 39,* 369–376.

# 7
# Development of the Nervous System: Genetics, Epigenetics, and Phylogenetics

ROBERT H. BENNO

No discussion of developmental behavior genetics in an evolutionary context can be considered complete without a thorough examination of the nervous system. It is all too obvious, however, that correlations between brain structure and function and the simplest behaviors are difficult to obtain. Therefore, any attempt to relate neuronal functioning to such a broad interdisciplinary framework is bound to encounter many obstacles. Fortunately, many scholarly works have addressed the problem of brain/behavior integration quite successfully, thus making my attempt to bridge these extremely diverse areas much easier. For instance, Sarnet and Netsky (1974) presented what is now considered the classic work on the evolution of the nervous system from an anatomical perspective. With ever-increasing technology, these early anatomical perspectives have been supplemented by an understanding of the functional evolution of the brain. Jacobson (1978), in his textbook survey of the literature on brain development, discussed the relationship between neuronal ontogeny and phylogeny both at the macroscopic and the microscopic levels. Evidence for merging of the disciplines of behavior and neurobiology across phylogenetic levels is evident in the rapid proliferation of scholarly papers and texts in the field on neuroethology (Camhi, 1984). Looking at neural development from a different perspective, one notes that the literature is filled with studies on the role of genetics in both normal and abnormal brain development (see Chapter 5 in this volume). Stent (1981) discussed the strengths and weaknesses of the genetic approach to the development of the nervous system. More recently, Purves and Lichtman (1985) have attempted to summarize the neurodevelopmental literature, bearing in mind the possible roles of genes and phylogeny. Drawing on these studies, as well as on many others, I have attempted to develop a framework for looking at developmental behavior genetics in the evolutionary context from the perspective of developmental neurobiology. In the last part of this chapter I present some data from our laboratory that illustrate differences in the rate of behavioral development in mice bred for differences in brain size (Fuller, 1979) and attempt to explain these data using the theoretical framework developed in this chapter.

## GENETICS VERSUS EPIGENETICS

It is now clear that the total genetic information available to an animal—perhaps $10^5$ genes in mammalian cells—is simply not sufficient to specify the total number of neuronal interconnections—perhaps $10^{15}$ in the mammalian nervous system—that are made. The development of the nervous system therefore requires epigenetic processes in which specific portions of the genome contained within developing cells are sequentially activated and modulated. It may be instructive at this point to define what is meant by *epigenesis*. As a process, epigenesis refers to that interaction of an organism with its environment that results in modification of both the organism and the environment. The resulting organism and its genome therefore interact somewhat differently with the surrounding environment than did the original organism. Thus it is difficult to determine the relative contributions of genetics/environment to a particular developmental event, since genes influence the environment and the developmental environment influences the genes.

A somewhat different use of the concept of epigenesis has been offered by Purves and Lichtman (1985). They defined epigenetic influences on cell development as those influences arising from factors other than genetic instructions. I find Purves and Lichtman's use of epigenesis somewhat easier to apply to my discussion on development of the nervous system and will therefore use their definition throughout the remainder of this chapter. According to Purves and Lichtman (1985), epigenetic influences arise either from within the embryo or from the external environment. The internal environment includes surface interactions between cells and the diffusion of chemical substances over long distances or between neighboring cells. The external environment includes nutritive factors, appropriate sensory and social experiences, and learning. A variety of internal and external factors impinge on a developing cell. The actions of several of these factors are thought to be critical for enabling a neuron to differentiate appropriately. Each signal is presumably not only chemically, but also temporally and possibly topographically, specific. To be effective, the signal often has to act on a cell at a specific stage of development.

## HISTORICAL VIEWPOINTS OF GENETIC VERSUS EPIGENETIC CONTROL OF NEURONAL DEVELOPMENT

### Loose Constructionist View

Psychologists of the 1920s and 1930s generally believed in a loose constructionist model. In this view there is only a limited developmental program for CNS development. The loose constructionists maintained that developmental forces leave the central nervous system an unorganized network capable of only random reactions. Out of this randomness, behavioral feedback from trial-and-error learning produces a coherent neural organization. According to this view, interconnections between cells are directed by the pattern of stimulation that the organism receives from its external environment.

### Strict Constructionist View

Some biologists maintain that the complexity of the brain demands a prescribed developmental program. They believe that there is an extensive genetic and developmental program in synapse formation and a more limited role for learning.

### System-Dependent Constructionist View

We might consider an alternative view—that epigenetic processes play different roles in shaping development depending on the neuronal circuit we are investigating. For instance, in muscles and autonomic ganglia, genetic information is probably sufficient to shape the appropriate pattern of neuronal innervation. Conversely, epigenesis probably is very important in determining the neuronal connections formed in association cortical areas. In the visual system, moderate amounts of both a prescribed genetic blueprint and epigenesis are probably required for development of a functional neuronal architecture.

## CRITICAL PERIODS IN DEVELOPMENT; VISUAL SYSTEM AS A MODEL

Neuronal development requires that various aspects of both structural and functional development be present and validated at a prescribed time. Absence of validation of the neuroethological substrate results in a mature system with compromised function. The time interval during which this "learning" takes place is called the *critical period*. It is readily apparent that the concept of critical periods takes on different meanings depending on the complexity of the function under consideration and of the organism being analyzed. In addition, critical periods for both cellular development and organismal learning exist but represent rather diverse perspectives of the concept. From a cellular neurobiological perspective, the critical period might refer to any or all of the processes of determination and/or differentiation, as will be discussed later.

Wiesel and Hubel addressed the problem of how much of the actual wiring in the visual system is preprogrammed and specific except for some fine tuning. They hypothesized that critical aspects of the basic wiring are present at birth and are not dependent on subsequent learning. They addressed this problem in the cat (Hubel & Wiesel, 1963) and found that in the retina, the lateral geniculate, and primary visual cortex, the types of response properties normally found in adults are largely present in newborns. Thus their data support the view of the strict constructionist that the major connections of the nervous system are established primarily under genetic control and that the initial establishment of the connection occurs in the absence of learning. If the visual system is manipulated experimentally during early gestation, the normal genetic developmental program may be disrupted and the animal may be functionally blind. Learning, however, is important for subsequent fine tuning and maintenance, as is seen in other experiments of Hubel and Wiesel (Wiesel & Hubel, 1963) in which cats deprived of normal visual experience due to suturing of one eyelid for a short specified period after birth do not develop binocular vision. Thus if the animal is isolated from the appropriate experience in development, the fine tuning of neuronal systems (e.g., binocular vision or pattern recognition) in the visual system might be absent, but the animal will not be functionally blind.

## PHYLOGENY: SIMPLE VERSUS COMPLEX ANIMALS

Simple (lower) animals are thought to have little impetus for behavioral flexibility because of the stereotyped nature of the developmental environment, their brevity of life, and the narrowness of biological purpose. This is reflected in the largely preprogrammed development of their nervous systems. Thus most of the cells in the developing nervous system of the invertebrate are only minimally affected by epigenetic influences, and the fate of every nerve cell in organisms of this sort is largely preordained. This type of developmental process is generally referred to as *mosaic* or *regulative* development (Roux, 1888). Thus when the neuronal developmental program of a simple organism (e.g., an invertebrate) is altered experimentally, the animal has only minimal ability to compensate for the pertubation. Tests of this developmental rigidity have been performed in the invertebrate *Caenorhabditis elegans* (Sulston & White, 1980). These authors, using a laser beam to destroy individual cells in the developing neural tube, showed that there was little change in the developmental program of surrounding cells. This finding supports the hypothesis that the early developmental program in invertebrates is largely preprogrammed. However, it has been challenged by numerous studies and is currently an area of great controversy. For example, Kimble (1981) has defined the concept of equivalence groups of developing neuroblasts in *C. elegans*. This study shows that equivalence cells may respond to manipulation of cells having similar cell lineages, although the ability of the equivalence cell to take over the function of the originally damaged cell has been questioned (Sulston & Horvitz, 1981). In general, however, most experimental studies on simple animals support the concept of preprogramming in neuronal development. Invertebrate neurons seem to differentiate almost immediately—not surprising since mosaicism is already present in the invertebrate egg (Conklin, 1932). This early developmental program, however, may be the very heart of the reason that the developmental program in invertebrates seems so rigid and stereotyped in experimental studies. One might consider that the developmental process has been accelerated so quickly in these organisms that our ability to define the appropriate experimental paradigm necessary to study their neuronal development is simply not possible. Thus the question of whether neuronal development in vertebrates and invertebrates is fundamentally different may remain a question of experimental procedure (Purves & Lichtman, 1985).

In vertebrates, on the other hand, it is quite apparent that neuronal development must be malleable over a much greater period of time than in invertebrates. A variety of experiments on vertebrates indicate that cells have a broader range of potential fates during the early developmental period. For example, vertebrate eggs (Gurdon, 1968) do not show the mosaicism that is inherent in invertebrate eggs (Conklin, 1932). Similarly, induction of presumptive neuroectoderm by mesoderm in vertebrates occurs relatively late and is therefore a process that shows a large degree of plasticity following experimental manipulation (Spemann, 1938). In longitudinal studies, experience is found to be an important factor in development of the vertebrate nervous system. The nervous system needs to remain modifiable for as long as possible so that it is best able to respond to the individual environment. Modifiability of behavior is most likely represented at the structural level by synaptic plasticity. As a general rule, the extent of synaptic plasticity is considered to correlate with the level of intelligence of the animal. For example, the establishment of patterned connections in the visual system of humans takes years, whereas the same process requires only weeks in the cat (Purves & Lichtman, 1985). The necessity for a prolonged neuronal modifiability is obviously most important in humans.

In summary, it is readily apparent that epigenesis plays a more important role in the shaping of neuronal development of complex animals than of simple organisms. An extended period of neuronal modifiability is necessary because of differences in the spheres of influence acting on the nervous system throughout development. This period of development might be considered (at least in humans) to last a lifetime. The epigenetic environment influencing neural cells continually enlarges during the lifetime of an organism (especially humans). Initially this environment is the cell's cytoplasm. As development proceeds, neurons send their axons to targets that they influence, and in turn they are influenced by these targets. Finally, at least in humans, the nervous system is shaped by experience in the postnatal world (Figure 7-1). The level of this neuronal modifiability is probably at the level of the modifiable synapse.

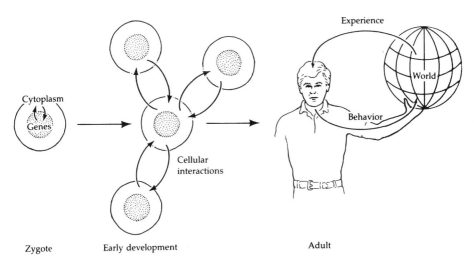

**Figure 7-1** The environment that shapes the structure of the nervous system continuously enlarges as the organism matures. Thus at each continual stage of development the genome interacts with a changing environment. The interaction of the genome with the environment results in epigenesis, whereby the structure and function of both the genome and the environment may be altered. (From Purves & Lichtman, 1985.)

## INDIVIDUAL NEURONAL CELLS—GENETICS AND EPIGENETICS

Due to the great diversity of neuronal cell types, it might be constructive to consider the relative roles that genetics and epigenetics have on the different types. Marcus Jacobson (1978) considered this question quite extensively. He made two broad categorizations of neurons and looked at them from several aspects, including the roles that genetics and epigenetics play in shaping the final adult structure (Table 7–1). His Type I neurons form the primary sensory and motor pathways, whereas his Type II are local circuit neurons that form central integrating circuits. The numbers of Type I neurons are closely correlated with body weight, whereas the numbers of Type II neurons are highly correlated with behavioral complexity. Thus from an evolutionary perspective, the ratio of Type II/Type I neurons increases as one ascends the phylogenetic scale. Similarly, increases in the ratio of brain to body weight, which one observes during phylogeny (see Chapter 8) probably result from this relative increase in the ratio of Type II to Type I cells. For example, the ratio of granule cells in the cerebellar cortex (Type II) to cerebellar Purkinje cells (Type I) is 1500:1 in humans, 600:1 in the cat, and 140:1 in the mouse (Blinkov & Glezer, 1968).

As mentioned earlier, there are not sufficient numbers of genes to code for all the possible neuronal types and connectivities in the vertebrate nervous system. Thus in his model Jacobson proposed that the Type I neurons (5% of the total number in humans) are programmed by genes whereas the Type II neurons (95% of the total number in humans) and their connections are defined on the basis of epigenetic processes. The Type II neurons are considered to have rather lax specification of function and a rather lengthy period of modifiability. Type I neurons, on the other hand, have rigid function specification and a very brief period of modifiability. We will make extensive use of this model when we look at specific aspects of neuronal determination and differentiation in the remainder of this chapter. As we shall see later, some of the variations in neuronal determination and differentiation between neurons are dependent on whether the cell is a Type I or Type II cell. However, just as

**Table 7–1** Type I and Type II neurons: Relative importance of genetics/epigenetics

| Type I | Type II |
|---|---|
| Severely constrained genetically | Loosely constrained genetically |
| Rigid specification | Lax specification |
| Generated early | Generated late into ontogeny |
| Mainly macroneurons (afferent and efferent neurons with long axons) | Mainly microneurons (interneurons with short axons), "local circuit neurons" |
| Invariant connectivity | Variable connectivity |
| Genetic specification sufficient | Genetic specification *not* sufficient—functional validation needed |
| Unmodifiable after early developmental specification | Modifiable until specification (late in development) |

*Note:* Ratio of Type II/Type I increases later in phylogeny.
*Source:* Adapted from M. Jacobson, 1978.

clear-cut differences between the processes of neuronal development in invertebrate and vertebrate systems are difficult to establish, distinctions between developmental processes in Type I and Type II neurons are also not well defined. For example, pyramidal neurons of layer five of the rostral two thirds of the cerebral cortex are well known to be the origin of the primary corticospinal projection in the adult rat. During early development, however, neurons in the occipital cortex have been shown by long-term retrograde dye labeling experiments to project to these same spinal motor neurons (a region they do not innervate in the adult rodent brain). These occipital neurons retain persistent connections with the superior colliculus and pontine nuclei throughout adulthood (Cowan et al., 1985). Thus, there seems to be extensive collateral elimination in this neuronal cell, which is a classic representative of the Type I neuron. Apparently it is necessary to provide even the most primary (Type I) neuron with rather general instructions (e.g., "project to brainstem and spinal cord") and to allow for the epigenetic selection of those connections that are appropriate for a given neuronal phenotype projecting to a specific region of the brain. In a similar vein, Lund (1978) argued against the validity of Jacobson's TypeI/ Type II classification scheme, stating that another classic Type I macroneuron, the retinal ganglion cell, has a very plastic developmental process. Thus, as stated earlier, there is insufficient genetic information to code for all the connections in the mammalian brain, and it is apparent that there must be some distinction between neurons as to how much of their development is genetic and how much is dependent on epigenesis. Perhaps some neurons are Type I, some are Type II, and still others are somewhere in between or mixtures of these two types.

## NEURONAL CONNECTIONS FROM A PHYLOGENETIC/ONTOGENETIC PERSPECTIVE

As a general rule, parts of the nervous system that appeared first in phylogeny also tend to appear early in ontogeny, and structures that arose late in evolution tend to arise late in ontogeny (M. Jacobson, 1978). For example, neurons in the outer layer of the cerebral cortex are known to have the latest birthdays of all cells in the developing cortex and they are also the most recent to develop in evolution. Thus, with knowledge of the phylogenetic background of an organism it should be possible to predict the structure of the organism's nervous system and the developmental program that produced it. Interestingly, Marin-Padilla (1972) showed that there are connections between cells in cortical layers in the cat that are present only during development, whereas similar cortical connectives are found throughout the lifetime of the reptile. The importance of understanding the relationship of ontogenetic principles to phylogeny is of the utmost importance if we are to understand developmental behavior genetics from an evolutionary perspective. Thus, not only must we understand how developmental behavioral repertoires of entire organisms are related to phylogeny (see Chapter 3), we must also identify how individual neuronal connections develop on the basis of their evolutionary background. After all, the behavior of the organism is ultimately dependent on the neuronal structure underlying this behavior.

## THE VERTEBRATE NEURON AS PART OF AN ECOSYSTEM

A particularly useful way to view a developing neuron in the vertebrate brain might be to consider it a part of an ecosystem. Recently, Edelman (1987) has used this ethological construct in an attempt to describe how the brain functions. Edelman's treatise, entitled *Neural Darwinism,* is based on a theory of neuronal group selection. From this perspective, the fate of a developing vertebrate neuron (such as a Type II neuron in M. Jacobson's model) cannot be foretold with any exactness, since the ultimate fate of a member of an ecosystem depends on competitive interactions with other members of the system vying for survival in the same niche. Analysis of the processes of determination and differentiation in the developing vertebrate nervous system can be greatly aided by considering this ecological concept, as we shall soon discover. Such a viewpoint may also prove useful in our attempt to understand such complex processes as learning and memory, which we might consider to be the result of synapses competing with one another for some specific connections and/or trophic substances. In the final analysis, however, we must remember that we are ultimately concerned with the role of the total organism in its ecological niche. It is the organism that is important for the evolution of the species and the survival of the genetic information into the next generation (see Chapter 3). We must, of course, be aware of the limitations of our attempt to understand the behavior of an organism in its ecological niche by considering the behavior of its component neurons.

## PROCESSES OF NEURONAL DEVELOPMENT

In the remainder of this chapter I will discuss the mechanisms involved in the process of neuronal development. In an attempt to relate these mechanisms to the overall theme of this symposium, I will relate them to the two subthemes previously discussed: genetics versus epigenetics and simple versus complex animals.

There are two major mechanisms involved in the development of the nervous system: determination and differentiation. *Determination* in general refers to the derivation of neurons and glia from neuroectoderm in a process known to involve induction by mesoderm. *Differentiation* is the resultant fate of the newly formed neuron or glial cell. Included in this differentiative life cycle are the stages of cell proliferation, generation of specific classes of neurons, migration, aggregation into cellular masses, expression of neurotransmitter, acquisition of cell shape, axonal growth, synaptogenesis, dendritic maturation, histogenic cell death, and synaptic rearrangement.

### Determination

Determination of a cell refers to that stage in its life cycle when it ceases to be totipotential and assumes a more limited fate. The precise time at which this process takes place cannot be told with exactitude. In fact, there is really no criterion for determining when this process has taken place. What might be considered an irrevocable process may prove not to be so under a new set of conditions (Harrison,

# GENETICS, EPIGENETICS, AND PHYLOGENETICS

1937). In general, though, the concept of determination has proven to be quite useful and experimentally testable in a number of experimental paradigms.

The amphibian has been the model of choice of numerous investigators interested in determination in the nervous system. Neural determination is known to take place during gastrulation in these simple vertebrates. This process has been elegantly studied by Spemann and Mangold (1924) in a series of classic experiments. By transplanting presumptive mesoderm into the blastocele of an early gastrula host embryo, these authors showed that ectoderm of the amphibian's trunk region can be influenced to develop into nervous tissue (Figure 7–2). Extending this model, Holtfreter (1944) demonstrated that hypertonic saline solutions, which prevent invagination of mesoderm toward the presumptive neuroectoderm, result in failure of development of the nervous system. Recent experiments involving separation with filters of explanted neuroectoderm and embryonic mesoderm have established that the inductive factor may be a peptide with a molecular weight greater than 1000 daltons.

Toivonen and Saxen (1968) examined the possibility that quantitative differences in the interactions between ectoderm and mesoderm could lead to regional specificity in the neuroectoderm. Their experiments showed that increasing the ratio of cultured

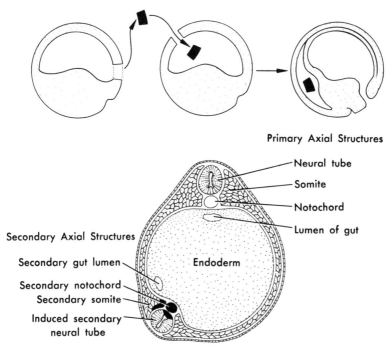

**Figure 7–2** The role of mesoderm in the process of neural induction was demonstrated by Spemann and Mangold (1924). These authors transplanted precursor mesoderm from one developing gastrula into the embryonic cavity of a second gastrula. The transplanted mesoderm came to underlie ectoderm that under control conditions would become epidermis covering the trunk. The transplanted mesoderm induced the presumptive trunk epidermis to become a secondary set of axial structures that, as can be seen, includes a secondary neural tube. (From Saunders, 1970.)

mesodermal cells in relation to ectodermal cells produced progressively more caudal neural structures. This regional specification along the anterior-posterior axis of the embryo was shown to be irreversible. Thus, if forebrain neuroectoderm is transplanted into the posterior part of a host embryo during the late gastrula stage (before the neural tube has formed and when only a few neural cells have differentiated), the nervous system will develop as already programmed. Thus the transplanted forebrain ectoderm will develop into forebrain even though it is now located in a posterior portion of the host embryo. The process of neural induction thus determines the overall regional organization of the nervous system. However, local details of this regional specification can be modulated by later signals that are critical for differentiation of specific neuronal populations and their interconnections.

## Differentiation

### Proliferation

Cell proliferation begins in the nervous system only after the neural tube has closed. The single layer of cells that formed the neural plate begins to undergo rapid cell proliferation and soon appears as a thick layer of cells. This proliferation is known to take place along the entire extent of the nervous system in regions adjacent to the ventricular surface. This proliferative region is known as the *germinal zone*. The actively dividing neuroblasts show characteristic movements of their nuclei and retraction of their cellular processes as they migrate toward and away from the ventricular surface (Fujita, 1962). Only during the mitotic (M) stage of the neuron's life cycle is the cell located at the ventricular surface with its characteristic rounded shape. One of the daughter cells formed in the M phase leaves the germinal zone and migrates to its appropriate position, permanently arrested in the G2 phase (gap period 2, which lasts from the end of DNA synthesis until the next cell division) as a neuron (or with potential to divide if it differentiates into a glial cell). The other daughter cell remains in the proliferative cycle and continues in this cycle until the developing neuroblast loses its ability to divide (Figure 7-3). In the mammalian brain, all neurons are believed to be generated before birth with the exception of the granule cells of the olfactory bulb (Hinds, 1968), the granule cells of the fascia dentata in the hippocampus (Altman 1966; Angevine, 1965), the granule cells of the rhombic lip in the brainstem (Taber Pierce, 1966), and the granule cells of the cerebellar cortex (Miale & Sidman, 1966).

There are a few exceptions to the general rule that neurons are unable to divide again after migration away from the ventricular zone. In the forebrain, for example, cells that give rise to small neurons in the basal ganglia and cerebral cortex are thought to proliferate in a subventricular zone found between the ventricular zone and the intermediate zone. In the cerebellum, cells of the external granular layer proliferate at the pial surface before beginning an uncharacteristic outside-in migration, during which they establish connections with Purkinje cell dendrites.

### Specific Classes of Neurons

Although the mechanisms responsible for the generation of specific classes of neurons are not known, some general rules have been shown to hold true for a variety

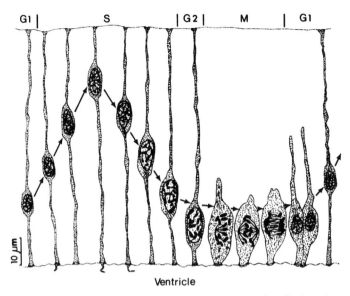

**Figure 7-3** Neurons are derived from germinal cells that line the ventricles of the developing brain. Individual germinal cells show a characteristic interkinetic movement toward and away from the ventricle during the various stages of the cell cycle [Gap period 1 (G1), Synthesis of DNA (S), Gap period 2 (G2), Mitosis (M)]. Following mitosis one daughter cell migrates away from the proliferative zone while the other daughter cell reenters the mitotic sequence. (From M. Jacobson, 1978.)

of different regions. Thus, as development proceeds temporally in any given region, large cells develop first, motor cells develop before sensory cells, interneurons develop last, and granule cells develop after neurons (M. Jacobson, 1970). There is a great deal of overlap among these events, and more recent evidence suggests that it is more accurate to discuss temporal gradients in the peak periods of proliferation of these classes of neurons (Holley, 1982).

It is now possible to follow the differentiation of descendents of individual cells rather than entire groups of neurons. These studies demonstrate important differences in the differentiation of vertebrate and invertebrate animals, as was discussed earlier in the section on phylogeny. Many invertebrates lend themselves easily to cellular lineage studies because of the small number of neuronal cells and the accessibility of these cells to visualization with the microscope. For example, the nervous system of the roundworm *C. elegans* has been almost completely reconstructed (White, Southgate, Thompson, & Brenner, 1976), and the lineage of virtually every somatic cell has been traced back to the zygote (Sulston, Schierenberg, White, & Thompson, 1983). These lineage studies have shown that cells with a common function share a common lineage rather than a particular precursor (Sulston & Horvitz, 1977). Thus cells that have a similar function in different parts of the worm are descendants of different cells. They may, however, have the same lineage. For instance, cells that are functionally homologous may be the anterior daughters of the posterior daughters of the posterior daughters of the posterior daughters of the primary neuroblasts. The question therefore arises as to how different portions of *C.*

*elegans* become specialized for discrete functions. Apparently, cells that are "programmed to die out" in some regions of the developing nervous system will survive in other regions, giving rise to regional specializations. It is possible to test the rigidity of this programming experimentally in invertebrates by systematically destroying individual cells and analyzing the effects in neighboring cells. When this is done using a laser beam, the neighboring cells are generally unaffected (Sulston & White, 1980). Thus individual neuronal cells in this simple organism are highly specialized for a particular function and even at early ages are unable to take over the function of surrounding cells. In general, this inflexibility holds for simple organisms, but evidence shows that some plasticity exists even in the neuronal development of *C. elegans*. For example, Sulston and Horvitz (1981) showed that ablation of individual cells in *C. elegans* can sometimes lead to proliferation of surrounding cells, although the surrounding cells will not necessarily take over the function of the destroyed cells.

While cellular lineage has long been known to be important in the development of the invertebrate nervous system, only recently has it been shown that cellular lineage occurs also in vertebrate models. For example, elegant studies with chimeric mice have proven the existence of lineage in vertebrate neuronal development (Herrup, Wetts, & Diglio, 1984), although its relative importance is not clear (see Chapter 5). In general, the more complex organism requires greater flexibility in its neuronal development and thus programmed lineage must be capable of being altered in the case of insult or changing environment. It seems likely that neuronal lineage is a vestigial function in complex organisms and that programmed cellular differentiation has been replaced by functional validation in the changing environment that the complex organism encounters.

## *Migration*

Developing neurons obviously have sophisticated guidance mechanisms that allow them the migrate to appropriate positions in the adult organism following cessation of proliferation. The signals that tell them where to go and when to stop are, however, poorly understood. As previously stated, the amount of genetic information is not sufficient to determine the migratory paths of each neuron. Therefore, for the great majority of cells, epigenesis probably determines their migratory routes (Figure 7–4). For example, granule cells migrating from the external surface of the cerebellum to their final resting place in the internal granular layer contact Purkinje cells along their route (Miale & Sidman, 1961) (Figure 7–5). The Purkinje cells, which represent a population of highly genetically specified Type I cells, might impart either migratory information or other epigenetic instructions to the Type II granule cells during contact. Some neural cells are thought to make contact with and migrate along the surface of previously established cells. For instance, neurons of the cerebral cortex in the monkey are known to migrate along the processes of radial glial cells on their way to final cortical destinations (Rakic, 1972). Migration of vertebrate neural crest cells illustrates similar principles, although operating through somewhat different mechanisms. It is believed that neural crest cells follow specific biochemical features of their surrounding extracellular matrix during their routes to the periphery (LeDouarin & Teillet, 1974; LeDouarin, 1980). Although these examples are not generalizable to all regions of the developing nervous system, they do serve to illus-

# GENETICS, EPIGENETICS, AND PHYLOGENETICS

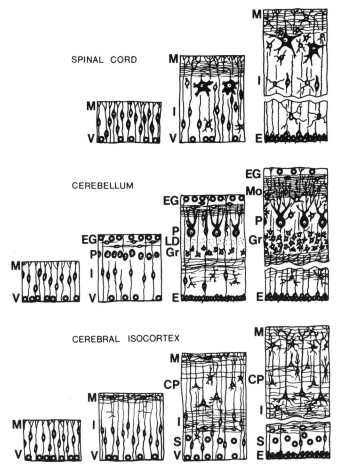

**Figure 7–4** Schematic diagram demonstrating several aspects of development in the spinal cord, cerebellum, and cerebral isocortex of mammals. Progressively later stages of development are demonstrated from left to right in each panel. Migration of cells and possible interactions among these cells are described in the text. Abbreviations: V, ventricular germinal zone; M, marginal zone; I, intermediate zone; P, Purkinje cell layer; EG, external granule layer; Gr, granule layer; LD, lamina dessicans; Mo, molecular layer; S, subependymal zone; CP, cortical plate. Cells undergoing mitosis are denoted by stars. (From M. Jacobson, 1978.)

trate a mechanism by which minimal genetic information can be used to specify a large amount of developmental processing.

## *Aggregation into Cellular Masses*

Following their migration toward the pial surface, neurons aggregate together into cellular masses known as nuclei. It is immediately obvious to any student of neuroscience that individual neurons cluster together as defined groups subserving apparently similar functions (albeit recent studies with tracer techniques and immunocy-

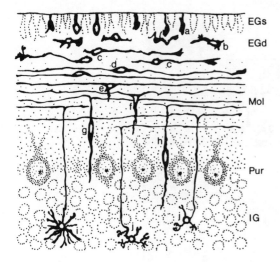

**Figure 7-5** External granular cells of the mammalian cerebellum undergo differentiation as they migrate from the superficial layer of the external granular layer (EGs) to the deep zone of the external granular layer (EGd), through the molecular layer (Mol) and Purkinje cell layer (Pur) to the internal granule layer (IG). (From M. Jacobson, 1978.)

tochemical double labeling for neurotransmitters (Hokfelt, Johansson, & Goldstein, 1984), call into question what were once believed to be homogeneous population of cells). The ability of individual cells to form recognizable nuclear masses that are practically nonvariant within the members of a species is indeed remarkable. The mechanisms for this cellular aggregation are uncertain, but probably involve some type of cell-cell contact, presumably mediated through cell surface glycoproteins (appropriately classified as cell-adhesion molecules, or CAMs), (Edelman, 1983). Clues for the mechanism of this association may be found in studies of dissociated cell cultures of regions of the developing nervous system. For example, neurons of the hippocampus are known to regain their characteristic hippocampal shape after being removed from the developing brain and placed into cell culture (DeLong, 1970).

### Expression of Neurotransmitter

One of the more unexpected findings in vertebrate neuronal development is the ability of certain classes of neurons to change their expression of neurotransmitters. This has been demonstrated most clearly using autonomic ganglion cells with both in vivo and in vitro paradigms. In both cases, sympathetic/adrenergic-secreting cells could be transformed into parasympathetic/cholinergic-secreting cells merely by changing the external environment that they encountered. LeDouarin (1980) transplanted neural crest cells destined to migrate and become sympathetic ganglia in the chick so that they would instead follow a migratory route of parasympathetic neurons. Although LeDouarin's evidence for change in transmitter expression in individual autonomic cells is compelling, these experiments do not clarify whether the environment had instructed the undifferentiated precursor cell in its future transmitter expression or if the environment had only functioned permissively, selecting for the survival of one of two populations of cells (sympathetic and parasympathetic) in the ganglion. Not until LeDouarin's paradigm was extended to tissue culture did it become clear that changes in autonomic transmitter expression are instructed by environmental factors. Several investigators (Furshpan, MacLeish, O'Lague, & Potter, 1976; Potter, Landis & Furshpan, 1981, Reichardt & Patterson, 1977) were able to follow individual ganglion cells in culture and examine the effect on transmitter expression of changing the culture medium. These authors were able to convert individual adrenergic neurons to cholinergic neurons by changing the percentage of heart-cell-conditioned medium to which the ganglion cells were exposed. These experiments clearly show the importance of epigenetic factors in the expression of neurotransmitter in vertebrate autonomic development. These experiments also shed light on one of the central themes of modern developmental biology: Are epigenetic factors instructive to the developing neuron, or merely permissive? Further elucidation of this issue will be a cornerstone in the field of developmental behavior genetics.

## Changes in Cell Shape

All neuroblasts demonstrate a similarly rounded shape when they are undergoing mitosis at the ventricular lumen. During their differentiation they undergo changes in shape that are under both genetic and epigenetic control. Experiments with dissociated neuronal cell cultures have proven valuable in addressing the relative importance of genes in this instance. Neurons dissociated in culture lose their processes and develop new ones. These dissociated neurons, which are thus removed from their normal environment, allow the relative roles of genetics and epigenetics in determining cell shape to be dissected. Several types of neurons—motor (Fischbach, 1970), sensory (Scott, Englebert, & Fisher, 1969), and hippocampal pyramidal cells (Banker & Cowan, 1979)—have developed appropriate phenotypes following dissociation in culture. It is clear, however, that although these cells showed broad similarity in shape they were by no means isomorphic, even though specific classes of neurons (e.g., $\alpha$ motor neurons) share identical genotypes. Thus, individual cells from similar populations are able to express different phenotypes, a fact of obvious importance if individuals in a neuronal population are to vary in synaptic connectivity and function. Thus, as the complexity of innervation changes in a neuron, so too does the complexity of its neuronal shape, especially its primary receptive elements—the dendrites. Alteration of the number and geometry of the postsynaptic dendrites allows the neuron to receive different populations of incoming sensory information. In general, more extensive dendritic arbors allow multiple innervation because competitive interactions between axonal inputs are minimal. On the other hand, neurons lacking dendrites are usually innervated by a single axonal input. This input proved to be the most viable in a highly competitive interaction for available receptive sites (Purves & Lichtman, 1985).

## Axonal Outgrowth

Axonal outgrowth can be considered to consist of two distinct phases under different control mechanisms. Initial outgrowth, Phase I, is apparently under rigidly defined genetic constraints, while the continued growth of the axon toward its target, Phase II, is greatly influenced by the substratum the axon encounters along its path. These phenomena are illustrated with a series of elegant experiments on Mauthner's giant neuron in the amphibian neurula (Hibbard, 1965; C. O. Jacobson, 1968; Stefanelli, 1950, 1951). In these experiments, rostrocaudal inversion of the presumptive Mauthner's neuron resulted in the development of an inverted Mauthner's neuron, with the initial axonal outgrowth proceeding from the normal pole of the neuron and hence in the wrong direction for making appropriate connections (Figure 7–6). In the majority of cases the axon grew out for a short distance, then reversed itself and proceeded toward its appropriate destination. Apparently the axon from this inverted neuron found itself in an inappropriate substratum and altered its course. Thus in this case, the neuron's polarity is intrinsically or genetically specified, whereas the subsequent pathway followed by the axon to its target is primarily influenced by epigenetic interaction of the axon with its environment.

## Synaptogenesis

The formation of synapses in the developing nervous system involves mechanisms that have long fascinated neurobiologists. Most scholars would agree that synaptogenesis is dependent on a two-way interaction between presynaptic and postsynaptic elements. Thus, some sort of trophic interaction is necessary for the appropriate matching of presynaptic and postsynaptic elements. The competition between axons for target sites may be modulated by several different influences in the course of an organism's development: (1) Different classes of neurons may respond to specific types and amounts of trophic agents, (2) Competing axons may exhibit variations in temporal patterns of neural activity, (3) Competing axons may encounter varying degrees of geometrical complexity in their target neurons. These three influences probably regulate convergence, synaptic strength, and neural unit configuration when put together in varying proportions. Synaptogenesis is known to begin as early in development as a few hours after neural tube closure (Kullberg, Lentz, & Cohen, 1977). Synapse formation is known to continue well into postnatal life in most vertebrate models studied (Huttenlocher, de Couten, Garey, & Van der Loss, 1982; Smolen, 1981). In fact, studies on synaptogenesis measure only the number of synapses present; they are not able to assess the turnover of existing synapses. It is perhaps instructive to consider synaptic modifiability in the adult as merely an extension of synaptogenesis during development. The rules for pattern formation in the neo-

---

**Figure 7–6** These drawings show the fate of Mauthner cell axons in the medulla of an amphibian *(Pleurodeles waltlii)* in which an extra segment of medulla has been grafted into a host. Thus a supernumerary pair of Mauthner cells has been grafted rostrally to the normal medulla. Panel A shows the resultant fate of the Mauthner cells when the graft is transplanted in the correct rostrocaudal orientation. In panel B, results of experiments in which the graft was placed in the reversed rostrocaudal orientation are illustrated. Most of the results in panel B show that the axons of the supernumerary pair of Mauthner cells grew rostrally with respect to the host, that is, in the wrong direction. These misguided axons eventually corrected themselves and grew down the spinal cord, decussating in the process. Thus we can conclude that the misdirected Mauthner axons sense their error in orientation when they emerge from the graft rostrally into the host tissue (an environment they would not normally encounter) and then redirect their growth into the appropriate environment. (After Hibbard, 1965.)

### (A) NORMAL ORIENTATION

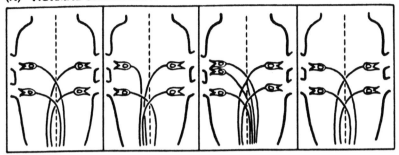

### (B) REVERSED ORIENTATION

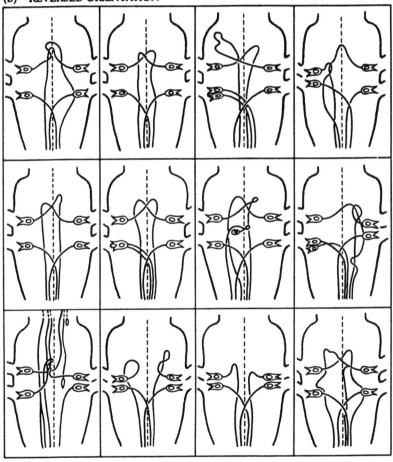

nate and the adult would be similar except that the target cell in the adult has probably been subjected to extensive environmental cues that alter what axonal connections and activities it will accept. Such changes in unit configuration are obvious in the visual system, where effects of experience on target cell function are easily seen. Lund (1978) has provided critical review of these developmental visual experiential studies. Since the experiential world of higher organisms is quite variable, synapses are thought to be modifiable throughout life, so that the organism will be able to change its behavior when faced with new environmental situations. Thus we now turn our attention to the mechanisms by which synapses are maintained and how they may be modified.

## Synaptic Modifiability

A prevailing view of efficient functioning in an organism is illustrated by the phrase, "Use it or lose it." Synaptic connections once formed are now believed to be dependent on active maintenance for continued viability. The key to this synaptic maintenance seems to be in the exchange of trophic factors between the presynaptic and postsynaptic elements. In the presence of a normal exchange of trophic materials, the current synaptic arrangement is maintained. With an abundance of trophic support, sprouting of presynaptic elements will take place; with a paucity of trophic support, retraction of synapses will result (Figure 7–7). Evidence for the importance of trophic exchange in the maintenance and modifiability of synapses comes from experiments in which trophic exchange is prevented. For example, colchicine, which blocks axoplasmic transport while leaving connections intact, results in synapse loss (Watson, 1974). Similarly, increases of nerve growth factor (NGF) at the synapse in mammalian sympathetic ganglia lead to increases in synapse sparing following axotomy (Nja & Purves, 1978). These mammalian ganglia cells, which are dependent on NGF during development, show substantial loss of synapses following treatment with anti-NGF (Nja & Purves, 1978).

## Dendritic Maturation

Dendritic maturation and outgrowth involve both genetic and epigenetic factors analogous to those observed earlier for axonal development. Factors that are intrinsic to the neuron (e.g., genetic endowment) are important for the orientation of the dendrites and the distribution of dendritic spines. This intrinsically determined dendritic outgrowth was noted by Van der Loos (1965) when he observed inverted pyramidal cells in the developing mammalian cerebral cortex. Van der Loos found that approximately 20 percent of all mammalian pyramidal cells in the normal cortex are misoriented. The dendritic arbors in these cells conform with the axis of the cell body and not with the axis radial to the cortical surface. Valverde and Ruiz Marcos (1969) observed the effects of inborn inversion of pyramidal cells in the visual cortex of the mouse on the distribution and number of dendritic spines. Although the absolute number of dendritic spines was diminished, normal distribution of spines on the apical dendrite was maintained. The decrease of dendritic spines these authors observed for the inverted pyramidal axons clearly demonstrates the importance of interaction of the neuron with its environment for normal dendritic maturation. Several lines of evidence demonstrate the failure of dendrites to develop fully in the absence of nor-

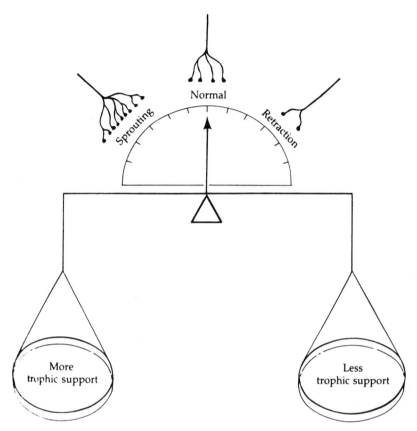

**Figure 7-7** This diagram shows the conditions under which neurons in mature animals may form new connections or retract existing ones. It is believed that the sprouting of axon terminals and synapse formation are stimulated by an increase in trophic support, while synapse retraction may be caused by a decrease in trophic support. An intermediate amount of trophic support might result in homeostasis of sprouting and retraction. (From Purves & Lichtman, 1985.)

mal axonal innervation. For example, removal of the frog eye during embryonic development (Larsell, 1931) results in failure of development of dendrites of the optic tectum neurons (targets of the optic nerves). Similarly, maturation of cerebellar Purkinje cell dendrites is dependent on afferent input from axons of cerebellar granule cells. If the granule cells fail to develop either due to irradiation (Altman, Anderson, & Wright, 1969) or because of a genetic abnormality (Sidman, 1968), Purkinje cell dendrites show an immature pattern in the adult consisting of reduction in size and number of dendritic spines.

Finally, if we consider that synaptic maintenance and modifiability result in ongoing rearrangements in dendritic structure, dendritic maturation may be considered to last the entire lifetime of an individual. The loss of dendritic spines and complexity with aging (Peters & Vaughn, 1981; Scheibel, 1982) reflect the importance of synaptic activity in the maintenance and modifiability of dendritic structure.

## Histogenic Cell Death

Cell death has long been recognized as a feature of normal development (Glucksmann, 1951). Histogenic cell death in the nervous system refers to a diffuse loss of neurons during the normal development of the system. Normally occurring death of neuronal precursors proceeds by one of two mechanisms: intrinsic instruction to die or dependence on interaction with other cells (Figure 7–8). Intrinsic instructions for programmed cell death are typically found in invertebrates. Experiments with the invertebrate nematode *C. elegans* have shown that fixed rules govern the survival of neuronal cells during development (Horvitz, Ellis, & Sternberg, 1982). The loss of specific neuroblasts (e.g., posterior daughter of anterior-posterior division, instead of the anterior daughter) during development in *C. elegans* illustrates that histogenic cell death in simple animals is largely under genetic control. On the other hand, histogenic cell death of neuroblasts in complex animals is apparently the result of dependence on interactions with other cells. Early experiments on amphibians showed that removal of a limb bud during development resulted in a decrease in the number of limb motor neurons (Shorey, 1909), whereas the addition of a supernumerary limb resulted in the increased survival of neurons (Detwiler, 1920). These early experiments established the dogma that neuron survival was dependent on contact with target organs, which could be manipulated experimentally, but these exper-

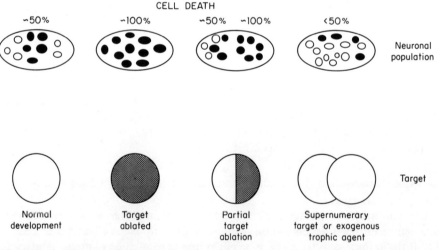

**Figure 7–8** Diagram demonstrating the phenomenon of naturally occurring neuronal death and experimental procedures that can alter this process. In most regions of the brain approximately 50 percent of the neurons that are initially generated die (left) (filled circles). This process happens about the time the neuronal population is forming connections with its target field. If either all or part of the target is removed over the same developmental period, proportionally more of the neuronal population will die (middle two ovals). Expansion of the target field or addition of an exogenous trophic factor (right) will rescue some of the neurons that might have been expected to die. (Adapted from Cowan et al., 1985.)

iments did not suggest neuronal cell death in the normal organism. Experiments by Hamburger and Levi-Montalcini (1949) were crucial in showing that there was histogenic cell death during normal development of the vertebrate nervous system. It is now generally believed that during normal vertebrate development, developing neurons compete for target organs and that their survival is dependent on the establishment of conections, although cell proliferation is probably not dependent on target influences (Currie & Cowan, 1974). Neurons appear to be competing for some trophic factor produced by the target organ. One such trophic factor, NGF, is apparently necessary for survival of mammalian and avian sensory and sympathetic ganglion cells (Gorin & Johnson, 1979; Levi-Montalcini, 1972). These and other experiments suggest that in the absence of peripherally derived NGF the innervating neurons die, and that in the presence of exogenous NGF neurons that were programmed for cell death are spared. In addition to affecting cell survival, NGF is capable of maintaining and modulating terminal arborizations in sympathetic ganglion cells grown in culture (Campenot, 1982).

## Synaptic Rearrangement

Although the exact mechanisms involved in the maintenance and modifiability of synaptic connections are unclear, synaptic rearrangement is necessary for learning in both simple and complex animals. The elegant experiments of Kandel and co-workers on habituation and sensitization of the gill withdrawal relex in *Aplysia* (Kandel & Schwartz, 1982) illustrate that synaptic functioning must be modifiable even in the most primitive behavioral systems. It is unclear in Kandel's work whether this change in synaptic functioning actually involves change in synaptic cytoarchitecture or simply a change in efficacy of existing synaptic structures. Recent experiments from the laboratory of Purves and Lichtman have shown that synaptogenesis can be studied in vivo. In these experiments the transition from synaptic multiple innervation to single innervation of muscle was observed microscopically in real time during development (Purves & Lichtman, 1985). Using this type of in vivo study of synaptogenesis it should be possible to determine to what degree synaptic modifiability is genetic and to what extent it is experiential.

The importance of changes in synaptic functioning in higher organisms is readily apparent in our own experiences. We are all too aware that our memory capacity decreases as we age, but we also realize that "even old dogs can learn new tricks."

## A MODEL SYSTEM FOR ANALYSIS OF THE PROCESS OF NEURONAL DEVELOPMENT AND ITS RELATIONSHIP TO DEVELOPMENTAL BEHAVIOR GENETICS

Much can be learned about the process of neuronal development with use of genetic models in which differences in behavioral development are expressed. Several types of genetic models have been used to look at the relationship of neuronal development to the onset of behavior. One such model, animals from a single inbred mouse strain, enables one to keep the genetics of an animal constant while manipulating experimental variables. The importance of inbred strains in animal research is obvious when one notes the difficulty involved in assessing differences in behavior as a function of neuronal development in humans (see Chapter 2). With inbred mice it is possible to assess the relative importance of genetics and/or epigenetics in neuronal development (Wahlsten, 1974). In addition, strain comparisons between inbred strains of mice have proven extremely fruitful in understanding how genetic differences may correlate with differences in developmental profiles (Oliverio, 1983). Studies from our laboratory using a diallel cross of inbred strains of mice have demonstrated correlations of the rate of behavioral development with the rate of somatic development and cerebellar foliation pattern (Cooper, Hahn, Hewitt, & Benno, 1987). Furthermore, studies with recombinant inbred strains of mice have the potential to demonstrate how a finite number of genes may be involved in behavioral differences observed among these animals. For example, recent studies by Roubertoux and his colleagues (1985) using recombinant C57BL/6By and BALB/cBy mice have illustrated that some of the differences in behavioral development in inbred mice may be due to individual genes that can be ascribed to particular chromosomes.

Another type of genetic analysis involves the use of mutant animals to determine the specific role of neuronal circuits in the behavioral repertoire of animals. Several strains of mice with cerebellar mutations have been used to study the complexities of the cerebellum and how this cellular cytoarchitecture relates to behavior (Sidman, 1968). These mutant models have proven useful not only in relating altered structure to function but also in aiding understanding of normal developmental mechanisms (see Chapter 5).

Still another genetic approach involves the use of selective breeding for specific behaviors or morphological characteristics. For instance, animals that are bred for differences in aggression have been analyzed for hormonal and brain differences (Lagerspetz, Tirri, & Lagerspetz, 1968; see also Chapter 3 in this volume). Similarly, several investigators have selected for differences in brain sizes in rodents and looked for differences in behaviors (Fuller, 1979, Roderick, Wimer, & Wimer, 1976). In keeping with this idea, the ratio of brain to body size has been argued to be of important consequence in the evolution of behavior (Jerison, 1973) and the role of genes in the evolution of brain and body size has been investigated (see Chapter 8, this volume). Brain-weight-selected (BWS) mice have been used to study the relationship between brain size and rate of motoric development (Benno, Desroches, Hahn, & Salinas, 1985; Fuller, 1979; Hahn, 1979) and learning (Jensen 1979; Wimer, 1979). We are currently investigating differences in the rate of reflexive motoric development in Fuller BWS mice as a function of neuronal development. Our preliminary studies and how these studies on BWS mice relate to the mechanisms of neuronal development and the field of developmental behavior genetics form the basis for the remainder of this chapter.

**Working Hypothesis**

Differences in the developmental rate of reflexive behaviors in Fuller low and high BWS mice can be shown to be correlated with differences in the rate of anatomical and chemical development of their nervous systems.

## Background and Test of Hypothesis

Fuller L line mice are known to develop faster behaviorally than their H line counterparts (Benno el al., 1985; Fuller & Geils, 1973; Hahn, 1979). The behavioral tasks analyzed in these BWS developmental studies include (among other behaviors) determination of the day on which a mouse is able to right itself when placed on its back (Figure 7-9) and the amount of time the mouse is able to stay on a rotorod. Thus one would presume that functional neuronal connections responsible for these behaviors are established earlier in the L line mice than in the H line mice. Although it is conceivable that functional development of a single neuronal network may be responsible for the onset of a particular behavior, direct proof of this is technically difficult to obtain. It is probably more realistic to hope to be able to analyze development of the neurological substrate in a single brain region known to be associated with but not totally responsible for a particular behavior.

Specifically, one might study the anatomical and biochemical maturation of the cerebellum in Fuller BWS mice in an attempt to understand the difference in the rate

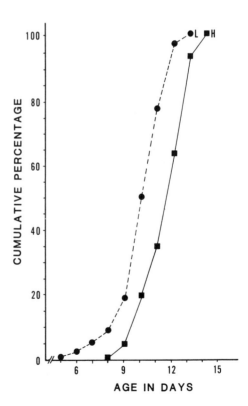

**Figure 7-9** Fuller high (H) and low (L) line mice ($N = 61$) were tested for their ability to right themselves when placed on their backs. Criterion for this test is the ability to become upright within 0.5 second for two consecutive days (age in days). The cumulative percentage represents the percentage of the total number of either H or L line mice that reached criterion by a particular day.

at which H and L line BWS mice obtain criterion on righting and rotorod running tasks (behaviors at least partially related to cerebellar maturation). Several anatomical and biochemical criteria might be used to assess the developmental stage of the BWS mice. In theory, to determine the relative neurodevelopmental stages between H and L line mice a chronological analysis of any of the processes of neuronal differentiation discussed in this paper could be used as a criterion. For example, anatomical maturation of the cerebellum has previously been shown to correlate with development and disappearance of cells in the external granular layer (Mares & Lodin, 1970; Wahlsten, 1974). Similarly, morphological maturation of cerebellar Purkinje cells has been used as an indicator of cerebellar maturation. Biochemical maturation of the cerebellum can also be assessed by several criteria. For example, the time course of development of the cerebellar specific peptides, cerebellins A and B, is believed to be associated with cerebellar maturation (Slemmon, Waleed, Hempstead, & Morgan, 1985). In addition, thymidine kinase levels are an index of biochemical maturation. Preliminary investigations to determine possible differences in these anatomical and biochemical markers for cerebellar development in Fuller BWS mice are currently under way in our laboratory.

The ultimate hope of a developmental behavior genetic study of the type described for the Fuller BWS mice is the possibility that the mechanism responsible for the neurobiological and behavioral differences might be uncovered. One possible mechanism for differences in the rate of neuronal and behavioral development as well as for differences in the size of brains in the H and L line mice might be variations in amount of or response to thyroid hormone. Thus, variations in thyroid hormone response may represent the primary locus on which selection operates in the Fuller BWS mice, while the phenotypic variations in brain weight between the lines are simply correlated characteristics. Previous studies have shown that thyroid hormone administration will increase the rate of both morphological (Clos, Crepel, Legrand, Legrand, Rabie & Vigouroux, 1974; Legrand, Selme-Matrat, Rabie, Clos, & Legrand, 1976) and behavioral development (Benno et al., 1985; Chen & Fuller, 1975) in rodents. Studies on thyroid (thyroxine)-injected Fuller BWS mice show that the behavioral development of both H and L line mice is sped up on a wide variety of developmental tasks (Figure 7-10). These studies also show that thyroid administration significantly lowers brain weights and body weights in both the H and L Fuller lines (Figures 7-11, 7-12). Although the magnitude of the effects of thyroid administration on somatic and behavioral development is similar in these experiments, we did find that daily thyroid injections resulted in about a 20% mortality rate in the L line whereas there were no such mortalities in the H line mice. This difference in mortality rate suggests that a minimal brain size, body size, or brain/body size ratio may be necessary for survival of an organism. Since thyroid hormone administration increases the metabolic rate, the increased mortality rate in the treated L line mice might also reflect a maximal metabolic rate for an organism that, if exceeded, would be detrimental for survival. Leamy (Chapter 8) suggests that the metabolic rate of the mother may be associated with the brain size of the offspring. Although there have been no direct metabolic studies on Fuller BWS mice, preliminary studies from our laboratory do not show significant differences in thyroid hormone levels between the lines on days 7, 9, 12, or 42 days of age (Table 7-2). We are

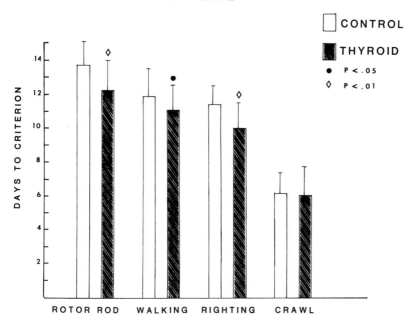

**Figure 7-10** Thyroxine (1 μg) was administered daily to 58 Fuller H line mice and the average days to criterion for four developmental behavioral tests were analyzed relative to saline-injected controls. Criteria for the behavioral tasks are as follows: crawl—all four limbs move in sequence; walk—the animal locomotes on his paws; rotor rod—the animal is able to remain on a rotorod moving at a speed of 1 revolution per minute; and righting—as described in Figure 7-9. □ control; ▨ thyroxine; ◊ $p \leq .01$; ● $p \leq .05$.

presently extending our investigation to analyze thyroid levels at earlier times of development in these mice. We are also considering how other factors such as glucocorticoid levels or maternal environmental factors may correlate with the developmental differences expressed in the Fuller BWS mice.

In summary, the Fuller BWS lines represent an excellent model for investigating how development of the nervous system correlates with development of behavior within the framework of the discipline of developmental behavior genetics. The primary objective of a neuroscientist is, of course, to understand how the brain works. How is it that behaviors are produced through activation of neuronal circuits? And what is the role of genes in this process? This task is formidable, and approaches to answering these questions range from reductionistic to holistic. The field of developmental behavior genetics provides an interdisciplinary framework within which we can begin to tackle the crucial issues of structure/function relationships in the neurosciences.

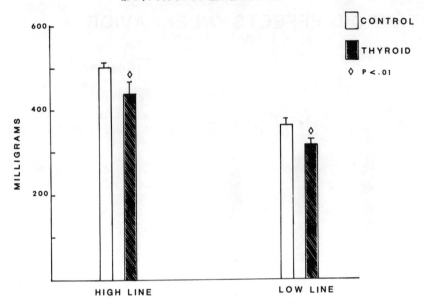

**Figure 7–11** Thyroxine (1 μg) was administered daily to 84 Fuller H and L line mice until day 21, when the animals were sacrificed using $CO_2$ inhalation and their brains were removed and weighed. □ control (saline); ▨ thyroxine; ◊ $p \leq .01$.

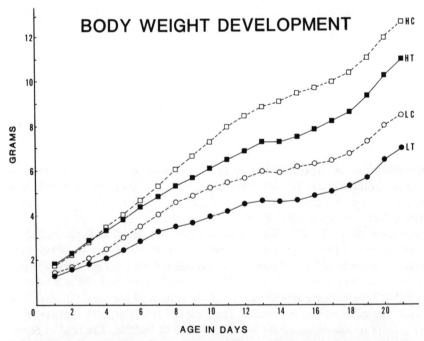

**Figure 7–12** Same experiment as described in legend to Figure 7–11. This graph shows the average weight of each of the four groups of animals recorded on each day of the experiment. HC, control high line; HT, thyroxine high line; LC, control low line; LT, thyroxine low line.

**Table 7-2** Thyroid hormone ($T_3$ and $T_4$) levels in serum of BWS mice

| Day | $T_3$ (ng/dl) | | $T_4$ (µg/dl) | |
|---|---|---|---|---|
| | High | Low | High | Low |
| 7 | 48.62 ± 17.29 | 31.15 ± 8.38 | 2.36 ± 0.84 | 2.81 ± 0.16 |
| 9 | 78.27 ± 17.81 | 65.71 ± 12.80 | 4.04 ± 0.97 | 4.43 ± 0.72 |
| 12 | 110.24 ± 9.26 | 95.19 ± 15.70 | 6.06 ± 1.01 | 5.65 ± 0.41 |
| Adult (M) | 15.85 ± 25.08 | 23.24 ± 19.81 | 0.94 ± 0.54 | 1.99 ± 1.17 |
| Adult (F) | 37.00 ± 21.52 | 14.37 ± 7.23 | 1.46 ± 0.95 | 0.79 ± 0.54 |

*Note:* Values for 7-, 9-, and 12-day-old animals are from pooled sera from four or five litters at each day for both strains. Adult M (male) and F (female) represent serum values from four or five animals from different litters for each condition. Mean ± S.D.

# REFERENCES

Altman, J. (1966). Autoradiographic and histological studies of postnatal neurogenesis. II. A longitudinal investigation of the kinetics, migration and transformation of cells incorporating tritiated thymidine in infant rats, with special reference to postnatal neurogenesis in some brain regions. *Journal of Comparative Neurology, 128,* 431–474.

Altman, J., Anderson, W. J., & Wright, K. A. (1969). Early effects of X-irradiation of the cerebellum in infant rats: Decimation and reconstitution of the external granular layer. *Experimental Neurology, 24,* 196–216.

Angevine, J. B. (1965). Time of neuron origin in the hippocampal region: An autoradiographic study in the mouse. *Experimental Neurology Supplement, 2,* 1–70.

Banker, G. A. & Cowan, W. M. (1979). Further observations on hippocampal neurons in dispersed cell culture. *Journal of Comparative Neurology, 187,* 469–494.

Benno, R. H., Desroches, D., Hahn, M., & Salinas, J. (1985). Brain growth and behavior in developing Fuller brain weight selected mice: Effects of thyroid hormone. *Neuroscience Abstracts, 11,* p. 529.

Blinkov, S. M., & Glezer, I. (1968). *The Human Brain in Facts and Figures.* New York: Plenum Press.

Camhi, J. M. (1984). *Neuroethology: Nerve Cells and the Natural Behavior of Animals.* Sunderland, MA: Sinauer.

Campenot, R. B. (1982). Development of sympathetic neurons in compartmentalized cultures. II. Local control of neurite survival by nerve growth factor. *Developmental Biology, 93,* 13–21.

Chen, C., & Fuller, J. (1975). Neonatal thyroxine administration, behavioral maturation, and brain growth in mice of different brain weight. *Developmental Psychobiology, 8(4),* 355–361.

Clos, J., Crepel, F., Legrand, C., Legrand, J., Rabie, A. & Vigouroux, E. (1974). Thyroid physiology during the postnatal period in the rat. A study of the development of thyroid function and of the morphogenetic effects of thyroxine with special reference to cerebellar maturation. *General Comparative Endocrinology, 23,* 178–192.

Conklin, E. G. (1932). The embryology of *Amphioxus. Journal of Morphology, 54,* 69–133.

Cooper, P., Hahn, M., Hewitt, J., & Benno, R. H. (1987). Biometrical genetic analysis of a cerebellar foliation pattern. *Neuroscience Abstracts, 13,* p. 254.

Cowan, W. M., Fawcett, J. W., O'Leary, D. D. M., & Stanfield, B. B. (1985). Regressive events in neurogenesis. In P. H. Abelson, E. Butz, & S. H. Snyder (Eds.), *Neuroscience* (pp. 13–29). Washington, DC: American Association for the Advancement of Science.

Currie, J., & Cowan, W. M. (1974). Some observations on the early development of the optic in the frog (Rena pipiens) with special reference to the effects of early eye removal on mitotic activity in the larval tectum. *Journal of Comparative Neurology, 156,* 123–142.

DeLong, G. R. (1970). Histogenesis of fetal mouse isocortex and hippocampus in reaggregating cell culture. *Developmental Biology, 22,* 563–583.

Detwiler, S. R. (1920). On the hyperplasia of nerve centers resulting from excessive peripheral loading. *Proceedings of the National Academy of Science, 6,* 96–101.

Edelman, G. M. (1983). Cell adhesion molecules. *Science, 219,* 450–457.

Edelman, G. M. (1987). *Neural Darwinism. The Theory of Neuronal Group Selection.* New York: Basic Books.

Fischbach, G. D. (1970). Synaptic potentials recorded in cell cultures of nerve and muscle. *Science, 169,* 1331–1333.

Fujita, S. (1962). Kinetics of cell proliferation. *Experimental Cell Research, 28,* 52–60.

Fuller, J., & Geils, H. (1973). Behavioral development in mice selected for differences in brain weight. *Developmental Psychobiology, 6(5),* 469–474.

Fuller, J. L. (1979). Fuller BWS lines: History and results. In M. E. Hahn, C. Jensen, & B. C. Dudek (eds.), *Development and Evolution of Brain Size, Behavioral Implications* (pp. 187–204). New York: Academic Press.

Furshpan, E. J., MacLeish, P. R., O'Lague, P. H., & Potter, D. D. (1976). Chemical transmission between rat sympathetic neurons and cardiac myocytes developing in microcultures: Evidence for cholinergic, adrenergic, and dual-function neurons. *Proceedings of the National Academy of Science USA, 73,* 4225–4229.

Glucksman, A. (1951). Cell deaths in normal vertebrate ontogeny. *Biological Review, 26,* 59–86.

Gorin, P. D., & Johnson, E. M. (1979). Experimental autoimmune model of nerve growth factor deprivation: Effects on developing peripheral sympathetic and sensory neurons. *Proceedings of the National Academy of Science USA, 76,* 5382–5386.

Gurdon, J. B. (1968). Transplanted nuclei and cell differentiation. *Scientific American, 219(6),* 24–35.

Hahn, M. E. (1979). Fuller BWS lines: Parental influences on brain size and behavioral development. In M. E. Hahn C. Jensen, & B. E. Dudek (Eds.), *Development and Evolution of Brain Size; Behavioral Implications.* New York: Academic Press, pp. 239–261.

Hamburger, V., & Levi-Montalcini, R. (1949). Proliferation, differentiation, and degeneration in the spinal ganglia of the chick embryo under normal and experimental conditions. *Journal of Experimental Zoology, 111,* 457–501.

Harrison, R. G. (1937). Embryology and its relations. *Science, 85,* 369–374.

Herrup, K., Wetts, R., & Diglio, T. J. (1984). Cell lineage relationships in the development of the mammalian CNS. II. Bilateral independence of CNS clones. *Journal of Neurogenetics, 1,* 275–285.

Hibbard, E. (1965). Orientation and directed growth of Mauthner's cell axons from duplicated vestibular nerve roots. *Experimental Neurology, 13,* 289–301.

Hinds, J. W. (1968). Autoradiographic study of histogenesis in the mouse olfactory bulb. I. Time of origin of neurons and neuroglia. *Journal of Comparative Neurology, 134,* 287–304.

Hokfelt, T., Johansson, O. & Goldstein, M. (1984). Chemical Anatomy of the Brain. In P. H. Abelson, E. Butz & S. H. Snyder (Eds.), *Neuroscience* (pp. 199–215). Washington, DC: American Association for the Advancement of Science.

Holley, J. A., Wimer, C. C. & Vaughn, J. E. (1982). Quantitative analyses of neuronal development in the lateral motor column of mouse spinal cord. III. Generation and settling patterns of large and small neurons. *Journal of Comparative Neurology, 207,* 333–343.

Holtfreter, J. (1944). Neural differentiation of ectoderm through exposure to saline solution. *Journal of Experimental Zoology, 95,* 307–343.

Horvitz, H. R., Ellis, H. M., & Sternberg, P. W. (1982). Programmed cell death in nematode development. *Neuroscience Commentary, 1,* 56–65.

Hubel, D. H., & Wiesel, T. N. (1963). Receptive fields of cells in striate cortex of very young, visually inexperienced kittens. *Journal of Neurophysiology, 26,* 994–1002.

Huttenlocher, P. R., de Couten, C., Garey, L. J., & Van der Loos, H. (1982). Synaptogenesis in the human visual cortex—evidence for synapse elimination during normal development. *Neuroscience Letters, 33,* 247–252.

Jacobson, C. O. (1968). Selective affinity as a working force in neurulation movements. *Journal of Experimental Zoology, 168,* 125–136.

Jacobson, M. (1970). *Developmental Neurobiology.* New York: Holt, Rinehart and Winston.

Jacobson, M. (1978). *Developmental Neurobiology.* New York: Plenum Press.

Jensen, C. (1979). Learning performance in mice genetically selected for brain weight: Problems of generality. In M. E. Hahn, C. Jensen, & B. C. Dudek (Eds.), *Development and Evolution of Brain Size: Behavioral Implications* (pp. 205–220). New York: Academic Press.

Jerison, H. J. (1973). *Evolution of the Brain and Intelligence.* New York: Academic Press.

Kandel, E. R., & Schwartz, J. H. (1982). Molecular biology of learning: Modulation of transmitter release. *Science, 218,* 433–443.

Kimble, J. E. (1981). Strategies for control of pattern formation in *Caenorhabditis elegans. Philosophical Transactions of the Royal Society of London (Biology), 295,* 539–551.

Kullberg, R. W., Lentz, T. L., & Cohen, M. W. (1977). Development of the myotomal neuromuscular junction in *Xenopus laevis:* An electrophysiological and fine structural study. *Developmental Biology, 60,* 101–129.

Lagerspetz, K. Y. H., Tirri, R., & Lagerspetz, K. M. J. (1968). Neurochemical and endocrinological studies of mice selectively bred for aggressiveness. *Scandinavian Journal of Psychology, 9,* 157–160.

Larsell, O. (1931). The effect of experimental excision of one eye on the development of the optic lobe and opticus layer in larvae of the tree frog *(Hyla regilla).* II. The effect on cell size and differentiation of cell processes. *Journal of Experimental Zoology, 58,* 1–20.

LeDouarin, N. M., & Teillet, M. -A. (1974). Experimental analysis of the migration and differentiation of neuroblasts of the autonomic nervous system and of the neuroectodermal mesenchymal derivatives, using a biological cell marking technique. *Developmental Biology, 41,* 162–184.

LeDouarin, N. M. (1980). The ontogeny of the neural crest in avian embryo chimeras. *Nature, 286,* 663–669.

Legrand, J., Selme-Matrat, M., Rabie, A., Clos, J., & Legrand, C. (1976). Thyroid hormone and cell formation in the developing rat cerebellum. *Biology Neonate, 29,* 368–380.

Levi-Montalcini, R. (1972). The morphological effects of immunosympathectomy. In G. Steiner & E. Schonbaum (Eds.), *Immunosympathectomy.* Amsterdam: Elsevier.

Lund, R. D. (1978). *Development and Plasticity of the Brain.* New York: Oxford University Press.

Marin-Padilla, M. (1972). Prenatal ontogenetic history of the principal neurons of the neocortex of the cat (Felis domestica). A Golgi study. II. Developmental differences and their significances. *Zeitschrift feur Anatomie und Entwicklungsgeschichte, 136,* 125–142.

Mares, V., & Lodin, Z. (1970). The cellular kinetics of the developing mouse cerebellum. II. The function of the external granular layer in the process of gyrification. *Brain Research, 23,* 343–352.

Miale, I. L., & Sidman, R. L. (1961). An autoradiographic analysis of histogenesis in the mouse cerebellum. *Experimental Neurology, 4*, 277–296.

Nja, A., & Purves, D. (1978). The effects of nerve growth factor and its antiserum on synapses in the superior cervical ganglion of the guinea pig. *Journal of Physiology (London), 277*, 53–75.

Oliverio, A. (1983). Genes and behavior: An evolutionary perspective. In J. S. R. Rosenblatt, A. Hinde, C. Beer, & M. C. Bushnell (Eds.), *Advances in the Study of Behavior.* New York: Academic Press.

Peters, A., & Vaughan, D. W. (1981). Central Nervous System. In J. E. Johnson (Ed.), *Aging and Cell Structure,* Vol. 1. New York: Plenum Press, pp. 1–34.

Potter, D. D., Landis, S. C., & Furshpan, E. J. (1981). Adrenergic-cholinergic dual function in cultured sympathetic neurons of the rat. In K. Elliot & G. Lawrenson (Eds.), *Development of the Autonomic Nervous System,* Ciba Foundation Symposium 83 (pp. 123–138). London: Pitman Books.

Purves, D., & Lichtman J. W. (1985). *Principles of Neural Development. Sunderland, MA: Sinauer.*

Rakic, P. (1972). Mode of cell migration to the superficial layers of fetal monkey neocortex. *Journal of Comparative Neurology, 145,* 61–84.

Reichardt, L. F., & Patterson, P. H. (1977). Neurotransmitter synthesis and uptake by isolated sympathetic neurons in microcultures. *Nature, 270,* 147–151.

Roderick, T. H., Wimer, R. E., & Wimer, C. C. (1976). Genetic manipulation of neuroanatomical traits. In L. J. Petrinovich & L. McGaugh (Eds.), *Knowing, Thinking, and Believing.* New York: Pergamon.

Roubertoux, P., Semal, C., & Ragueneau, S. (1985). Early development in mice. II. Sensory motor behavior and genetic analysis. *Physiology and Behavior, 35,* 659–666.

Roux, W. (1988). Beitrage zur Entwicklungsmechanik des embryo. *Virchows Archives feur Pathologische Anatomie und Physiologie und feur Klinishe Medizin. 114:* 113–153. H. Lauferedit, Trans. In B. H. J. Willier & M. Oppenheimer (Eds.), *Foundations of Experimental Embryology* (2nd ed.). New York: Hafner Press.

Sarnat, H.B., & Netsky, M. G. (1974). *Evolution of the Nervous System.* New York: Oxford.

Saunders, J. W., Jr. (1970). *Principles and Patterns of Animal Development.* New York: MacMillan Publishing.

Scheibel, A. B. (1982). Age related changes in the human forebrain. *Neurosciences Research Program Bulletin. 20,* 577–583.

Scott, B. E., Engelbert, V. E., & Fisher, K. C. (1969). Morphological and electrophysiological characteristics of dissociated chick embryonic spinal ganglion cells in culture. *Experimental Neurology, 23,* 230–248.

Shorey, M. L. (1909). The effect of the destruction of peripheral areas on the differentiation of the neuroblasts. *Journal of Experimental Zoology, 7,* 25–63.

Sidman, R. L. (1968). Development of interneuronal connections in brains of mutant mice. In F. D. Carlson (Ed.), *Physiological and Biochemical Aspects of Nervous Integration.* Englewood Cliffs, NJ: Prentice-Hall.

Slemmon, J., Waleed, D., Hempstead, J., & Morgan, J. (1985). Cerebellin: A quantifiable marker for Purkinje cell maturation. *Proceedings of the National Academy of Science, 82,* 7145–7148.

Smolen, A. (1981). Postnatal development of ganglionic neurons in the absence of preganglionic input: Morphological synapse formation. *Developmental Brain Research, 1,* 49–58.

Spemann, H. (1938). *Embryonic Development and Induction,* New Haven, CT: Yale University Press.

Spemann, H. & Mangold, H. (1924). Induction von embryonolangen durch implantation art-

fremder Organisatoren. *Archives Mikroskopy Anatomisches Entwicklungsmechung, 100,* 599–638.
Stefanelli, A. (1950). Studies on the development of Mauthner's cell. In P. Weiss (Ed.), *Genetic Neurology.* Chicago; University of Chicago Press, pp. 161–165.
Stefanelli, A. (1951). The Mautherian apparatus in the ichthyopside, its nature and function and correlated problems of neurohistogenesis. *Quarterly Review of Biology, 26,* 17–34.
Sulston, J. E. & Horvitz, H. R. (1977). Post-embryonic cell lineages of the nematode *Caenorhabditis elegans. Developmental Biology, 56,* 110–156.
Sulston, J. E. & Horvitz, H. R. (1981). Abnormal cell lineages in mutants of the nematode *Caenorhabditis elegans. Developmental Biology, 82,* 41–55.
Sulston, J. E., Schierenberg, E., White, J. G., & Thompson, J. N. (1983). The embryonic cell lineage of the nematode *Caenorhabditis elegans. Developmental Biology, 100,* 64–119.
Sulston, J. E. & White, J. G. (1980). Regulation and cell autonomy during postembryonic development of *Caenorhabditis elegans. Developmental Biology, 78,* 577–597.
Stent, G. S. (1981). Strength and weakness of the genetic approach to the development of the nervous system. *Annual Review of Neuroscience, 4,* 163–194.
Taber Pierce, E. (1966). Histogenesis of the nucleus griseum pontis, corporis pontobulbaris, and reticularis tegmenti pontis (Bechterew) in the mouse. An autoradiographic study. *Journal of Comparative Neurology, 126,* 219–239.
Toivonen, S. & Saxen, L. (1968). Morphogenetic interaction of presumptive neural and mesodermal cells mixed in different ratios. *Science, 159,* 539–540.
Valverde, F. & Ruiz-Marcos. (1969). Dendritic spines in the visual cortex of the mouse: Introduction to a mathematical model. *Experimental Brain Research, 8,* 269–383.
Van der Loos, H. (1965). The "improperly" oriented pyramidal cell in the cerebral cortex and its possible bearing on problems of neuronal growth and cell orientation. *Bulletin of Johns Hopkins Hospital, 117,* 228–250.
Wahlsten, D. (1974). A developmental time scale for postnatal changes in brain and behavior of B6D2F mice. *Brain Research, 72,* 251–264.
Watson, W. E. (1974). Cellular responses to axotomy and to related procedures. *British Medical Bulletin, 30,* 112–115.
White, J. G., Southgate, E., Thompson, J. N., & Brenner, S. (1976). The structure of the ventral nerve cord of *Caenorhabditis elegans. Philosophical Transactions of the Royal Society of London (Biology), 275,* 327–348.
Wiesel, T. N. & Hubel, D. H. (1963). Single cell responses in striate cortex of kittens deprived of vision in one eye. *Journal of Neurophysiology, 26,* 1003–1017.
Wimer, C. (1979). Correlates of mouse brain weight: A search for component morphological traits. In M. E. Hahn, C. Jensen, & B. C Dudek (Eds.), *Development and Evolution of Brain Size, Behavioral Implications* (pp. 147–162). New York: Academic Press.

# 8
# The Evolution of Brain and Body Size: Genetic and Maternal Influences

LARRY LEAMY

Interest in brain size and its relationship to body size in mammals, especially humans, has continued virtually unabated for a long time (see the historical review of Rosenzweig, 1979). Brain size was first expressed as a fraction of body size over 200 years ago, for example, although even then some of the problems associated with the use of such a ratio were recognized (Rosenzweig, 1979). Today we, of course, tend to scale brain size as a power ($k$) of body weight by making use of the standard allometric formula (Gould, 1966), that is

$$\text{Brain size} = a \, (\text{body size})^k$$

where $a$ is a constant and $k$ is the allometric coefficient. From evidence gathered in a large number of studies (Bauchot, 1978; Eisenberg, 1981; Gould, 1975, 1977; Jerison, 1979; Lande, 1979; Martin, 1981; Radinsky, 1977, 1978; Riska & Atchley, 1985), we now know that the scaling of brain size on body size within or between species or closely related taxa is low (typically $k = 0.2$ to $0.4$), whereas that between higher taxa such as orders of mammals is considerably higher (up to 0.77).

What might be the reason for this difference in scaling of brain size with body size among various taxa? What does it tell us, if anything, about the evolution of brain size? Up to about 1978, which coincidentally was the year of a major symposium (Hahn, Jensen, & Dudek, 1979) devoted specifically to the topic of brain size and its behavioral consequences, the answers to these and similar questions were speculative at best. It was thought that, among diverse taxa, brain size scaled to about the two-thirds power with body size, suggesting a connection between brain size and body surface area (Jerison, 1979). But intraspecific scaling of brain size with body size was totally unexplained, often being so low as to suggest that within species, "body size has been overemphasized as a factor in brain size" (Jerison, 1979). Further, in spite of the voluminous literature on brain size scaling, there simply was no model to account for the disparity in magnitude between intraspecific and interspecific allometry.

Given the state of knowledge at that time, it is understandable that arguments also were advanced that if we wish to fully understand the evolution of the brain, we must look beyond its size to its organization (Holloway, 1979). And so we should, for overall size is but one aspect of the developing brain, and perhaps only a manifestation of more important developmental processes (Hahn, Jensen, & Dudek, 1979). The size of the brain may be constrained across animal taxa by some developmental or physiological mechanism, but organization within this kind of constraint may be very important for our ultimate understanding of the brain, including the brain's behavioral consequences (Holloway, 1979).

In the few years since the 1978 symposium, however, there has been renewed interest in brain size and its scaling with body size, primarily because of several exciting developments. Three of these in particular seem especially significant and are discussed here.

First, reanalysis of the allometry of brain and body size across a large number of species of placental mammals has shown that the scaling coefficient is typically about three fourths rather than two thirds as originally had been found (Armstrong, 1983; Eisenberg, 1981; Hofman, 1982; Martin, 1981; Martin & Harvey, 1985). This difference is not trivial, for it immediately suggests that brain size is related to metabolism (rather than to body surface area), because basal metabolic rate also scales to body weight at this same magnitude (see Martin, 1981). Body size certainly is related to metabolic rate, which in turn is related to brain size, so metabolism ultimately may prove to be the causal mechanism linking the size of the brain and body.

Second, Lande (1979) provided the first genetic model for brain and body size evolution. This model basically suggests that brain size within and between closely related species has evolved as a correlated response to selection for body size. Using principles of population genetics, Lande also showed that in theory and with certain assumptions, evolutionary allometry of brain size can be predicted from genetic, but not phenotypic, within-species allometry (Lande, 1979). Further, he calculated the genetic scaling of brain size on body size and obtained a value (.36) that fell within the expected range of .2 to .4 (Lande, 1979). This model therefore is one way to account for the scaling of brain size at least among closely related taxa.

Third, and perhaps most significantly, the genetic model recently has been extended such that it may explain the higher brain–body allometric slopes among more distantly related taxa as well (Riska & Atchley, 1985). This new model is based in part on the fact that, in mice, the genetic allometry of brain on body size tends to be higher in prenatal and early postnatal life when brain and body are still growing rapidly. Growth occurs by increases in cell numbers more than by increases in cell sizes (Atchley, Riska, Kohn, Plummer, & Rutledge, 1984). Since the evidence also indicates that distantly related mammals differ more in cell numbers than in cell sizes, selection affecting changes at early ages could produce the steeper brain size/body size allometric slopes commonly seen among such groups (Riska & Atchley, 1985). Thus these genetic models suggest that selection for body size is capable of producing either lower or higher levels of brain scaling, depending on whether changes occur more by increases in cell size or by increases in cell numbers (Riska & Atchley, 1985).

## PERSPECTIVE AND PURPOSE

The genetic models just described would seem to offer a great deal to our understanding of the origin of brain and body size relationships. However, we certainly need more data (such as estimates of genetic correlations and regressions of brain and body size) to test these concepts adequately. Further, it is not clear at present how these models relate to metabolism, if indeed they do. The metabolic rate of the mother could be quite important in brain size development, for example, but if so, this should be manifested in genetic terms as a heritable, mostly prenatal, maternal effect. Thus beyond the direct genetic control of brain size and its scaling with body size, we need to discover if there is a significant indirect genetic control as well, operating through the maternal physiology.

This chapter is devoted to an assessment of genetic and maternal influences on brain size and the scaling of the brain with body size. The basic approach conventionally used in quantitative genetics is outlined first to provide the background necessary for proper interpretation of factors affecting brain size. Studies that have given rise to the genetic models are next reviewed. New empirical data on brain and body size are presented from random-bred and inbred/hybrid populations of mice. Finally, suggestions are made for future studies that could contribute to our understanding of the relationship of brain and body size.

## THE QUANTITATIVE GENETIC APPROACH

Both brain size and body size are typical quantitative or metric traits whose total phenotypic variance, $V_P$, can be partitioned into genetic and environmental components by the conventional approaches used in quantitative genetics (Falconer, 1981). The chief component of interest is the additive genetic variance ($V_A$), which, if expressed as a proporiton of the total phenotypic variance, is known as the heritability ($h^2$). Heritabilities are useful because they determine, in part, the expected response to selection for a given character. Specifically, the response to selection ($R$) is given by:

$$R = \frac{V_A}{V_P} S = h^2 S \qquad (8-1)$$

where $S$ is the selection differential, or difference between the means of the selected and unselected individuals (Falconer, 1981). Environmental sources of variation are many and varied; an important one in most mammals is maternal, nongenetic effects.

One method used for heritability estimation is parent-offspring regression. Regression of offspring on either parent estimates $\frac{1}{2}h^2$, and regression of offspring on midparent is a direct estimate of $h^2$. If maternal effects are present, however, they may inflate the female parent estimates, and thus estimates from the male parent are considered more reliable (Falconer, 1981). Differences in the regression estimates using male versus female parents are ascribable to contributions from a direct-maternal additive genetic covariance ($COV_{AoAm}$) plus one half of the additive maternal (or indirect) genetic variance ($V_{Am}$). The indirect genetic variance is produced by genes in the mother that indirectly affect her offspring, whereas $COV_{AoAm}$ is a genetic covariance between these indirect genetic effects that act through the female parent (Am) and direct genetic effects in the offspring themselves (Ao). This covariance is actually produced by pleiotropy of genes influencing growth both in the individuals and in their female parents (see Riska, Rutledge, & Atchley, 1985a).

Heritabilities also may be estimated from analyses of variance of half sibs and full sibs (see Falconer, 1981, for details) or of inbred strains. In analyses using inbred strains, the proportional contribution of the between-strain variance is equal to twice the heritability, provided maternal effects and epistatic interactions are absent or negligible.

The phenotypic *covariance* of two traits $x$ and $y$ ($\text{COV}_{Pxy}$) may be partitioned by any of the same methods used to partition the variance of individual traits. Twice the average of the two parent-offspring cross-covariances, for example, provides an estimate of the additive genetic covariance, $\text{COV}_{Axy}$ (Falconer, 1981). Genetic covariances also may be obtained by analyses of covariance of (full or half) sibs or inbreds. Both characters may be added together to create a new synthetic character ($x + y$) for which the analysis of variance can be run and covariances calculated by making use of the following:

$$\text{COV}_{x+y} = \text{VAR}_x + \text{VAR}_y + 2\text{COV}_{xy} \tag{8-2}$$

Genetic covariances may be used to predict the correlated response of a character $y$ ($\text{CR}_y$) to selection in $x$:

$$\text{CR}_y = (\text{COV}_{Axy}/V_{Px})S_x \tag{8-3}$$

where the terms are as previously described (Falconer, 1981).

Once the genetic covariance of two traits is estimated, the additive genetic correlation ($r_{Axy}$) of these traits is calculated as the ratio of this covariance over the square root of the product of the genetic variances of the separate traits, where the genetic variances are simply the numerators of their heritabilities. Genetic correlations reflect associations among traits resulting from pleiotropy and linkage disequilibrium (Falconer, 1981). Similarly, additive genetic regressions ($b_{Ayx}$) may be obtained as the ratio of the genetic covariance for two traits ($\text{COV}_{Axy}$) to the genetic variance of the independent trait ($V_{Ax}$). And if both traits are logarithmically transformed, these regressions predict allometric coefficients that would result from selection on $x$ (Lande, 1979). For brain and body size in logarithmic form, the standard allometric formula becomes

$$\log(\text{brain size}) = \log(a) + k(\log \text{body size})$$

Environmental correlations ($r_{Exy}$) may be calculated either directly from environmental sources of variance or indirectly by making use of the following:

$$r_{Pxy} = h_x h_y r_{Axy} + e_x e_y r_{Exy} \tag{8-4}$$

where $r_{Pxy}$ is the phenotypic correlation between $x$ and $y$, $h$ is the square root of the heritability, and $e^2 = 1 - h^2$ (Falconer, 1981). Environmental correlations reflect associations among traits arising from all sources of environmental influences, including maternal effects, although in some cases it is possible to further partition the environmental covariance in order to estimate maternal correlations themselves.

## REVIEW OF GENETIC MODELS

We are now ready to take a more detailed look at the studies that have given rise to the genetic models relating brain and body size within species. Nearly all of these studies have been done with laboratory mice and rats, for these animals have long proven amenable to the quantitative genetic approach (Green, 1966). Mice also exhibit a pattern of brain growth that apparently parallels that in humans (Hahn, Walters, Lavooy & DeLuca, 1983) and thus serves as a useful human model. Basically the mouse brain exhibits rapid, linear growth up to about 11 to 14 days, with much slower (although still linear) growth after this time (Hahn, Walters, Lavooy, & DeLuca, 1983; House, Berman, & Carter, 1985; Kobayashi, 1963). Postnatal growth in body size, on the other hand, typically is exponential up to about three weeks (point of inflection on the growth curve), becoming linear afterward and starting to level off at about six weeks (Riska, Atchley, & Rutledge, 1984). Thus both brain and body grow rapidly during prenatal and early postnatal life, but growth in body size continues long after growth in the brain essentially has ceased (Riska & Atchley, 1985).

Heritability estimates of brain size in mice and rats have generally been high, typically about .6 at most ages sampled (Atchley, 1984; Atchley, Riska, Kohn, Plummer, & Rutledge, 1985; Hahn & Haber, 1978; Leamy, 1985; Roderick, Wimer, & Wimer, 1976). This suggests that selection for brain size should be rapid and effective, and such has proven to be the case (Fuller & Geils, 1972; Roderick, et al., 1976). The relatively high heritability of brain size suggests that this is not an important fitness character in the sense that it probably has not been subjected to strong directional selection in the recent past (Falconer, 1981). This is not true for body size, however, for its heritability in mice typically is about .3 (Riska, et al., 1984; many other studies) and there is much evidence for its importance in evolution (Gould, 1966). It therefore seems reasonable to assume that selection has acted on body size in mice and most mammals; to the extent that it is genetically associated with body size, brain size exhibits a correlated response to this selection (Lande, 1979).

If we assume that body size has been selected for, what is the expected evolutionary scaling of brain size on body size? Lande (1979) clearly showed that, provided both variables have been suitably transformed (as by logging) to promote linearity, this scaling is given by the ratio of the correlated response of brain size ($CR_y$) to the direct response of body size ($R_x$). Using equations (8–1) and (8–3), this ratio is

$$\frac{CR_y}{R_x} = \frac{(Cov_{Axy}/V_{Px})S_x}{(V_{Ax}/V_{Px})S_x}$$

or

$$\frac{CR_y}{R_x} = \frac{COV_{Axy}}{V_{Ax}} \qquad (8-5)$$

Here the selection differentials and phenotypic variances cancel out, and evolutionary allometry is predicted by the ratio of the genetic covariance between brain and body size to the genetic variance of body size, or equivalently, by the genetic regression of brain size on body size (Lande, 1979).

Lande (1979) was able to estimate indirectly the genetic regression of brain size on body size by using mouse brain selection data of Roderick, Wimer, and Wimer (1976). To do this, he used estimates of several parameters, including the heritability of brain size (0.64) and of body size (.37) and the genetic correlation between the two variables (.68). As previously mentioned, the genetic regression of brain on body size calculated by Lande (1979) was .36, which falls within the .2 to .4 range commonly observed among closely related taxa. This in turn suggests that the selective forces that have resulted in the differentiation of such taxa have operated primarily on body size, with brain size responding in a predictable, correlated fashion (Lande, 1979).

Atchley and colleagues (1984) provided excellent corroboration of Lande's (1979) hypothesis in a study conducted with a large population of random-bred mice. In this study, a full-sib design with cross-fostering was used to estimate genetic and environmental (including maternal) components of variance and covariance for brain and body size in mice of several different ages. Their estimate of the genetic regression of brain size on body size for 70-day-old mice was .37 (Atchley et al., 1984), nearly the same as the .36 calculated by Lande (1979) and identical to the average within-genus allometry calculated by Martin and Harvey (1985) using over 200 species of mammals. For 38-day-old mice, however, this estimate increased to .53, with the phenotypic, genetic, maternal, and residual environmental correlations of brain and body size showing similar increases (Atchley, et al, 1984).

Riska and Atchley (1985) have postulated that the association of brain and body size should be greater at earlier ages when both brain and body are still growing rapidly, and should decline when brain growth slows and body continues to grow. They nicely illustrated this by showing that genetic correlations of 38-day brain size in random-bred mice are positive for body weight *gain* in the periods 0 to 14 days (+ .73) and 14 to 21 days (+.46), but negative after this (21 to 38 days, −.74). This suggests that brain and body share many of the same gene effects in their early growth stages, probably because both are influenced by general growth factors, or mitogens (Riska & Atchley, 1985). Later, growth in body size is controlled by gene effects that have, if anything, a negative effect on brain size.

It also therefore follows that selection on body size at early ages should result in steeper brain size/body size allometric slopes than selection at later ages (Riska & Atchley, 1985). If so, selection resulting in subspecies and species divergence has generally affected later growth, but to cause the divergence of more distant forms, has probably affected early growth stages. Evidence for this is indirect but reasonably compelling. It is known, for example, that early growth generally is hyperplastic, occurring largely by increases in the numbers of cells, whereas later growth is hypertrophic, occurring largely by increases in cell size (Cheek, 1975; Goss, 1966). Both numbers and sizes of cells are changed in the typical body size selection experiments in mice (Falconer, Gauld, & Roberts, 1978), indicating that selection has in fact affected both components of growth. But distantly related mammals differ far more in cell numbers than in cell size (Raff & Kaufman, 1983), suggesting that their differentiation has occurred by selection that has primarily affected hyperplastic growth (Riska & Atchley, 1985).

Although where we place the dividing line between "early" versus "later" growth is arbitrary, it is convenient to use the time of birth as the division. With this as our frame of reference, the genetic models also imply that the prenatal growth period should be far more important for the evolution of body size among distantly related species than among closely related species, for which the postnatal growth period should be more significant. Comparisons of the relative importance of prenatal and postnatal body size variation in closely related taxa (greater postnatal growth variation) versus distantly related taxa (greater prenatal growth variation) support this hypothesis (Riska, personal communiction, 1986). Further, scaling of neonatal to adult body size is higher among distantly related taxa compared with more closely related taxa (Leutenegger, 1976; Martin & MacLarnon, 1985), presumably again because of the relative importance of prenatal versus postnatal growth. Thus, although prenatal growth affects both neonatal and adult body size, postnatal growth affects adult body size only.

With regard to the evolution of brain size, it seems apparent that the prenatal growth period is crucial. Genetics comes much into play at this time, but part of this genetic influence on brain size could very well be mediated through the maternal physiology. If Martin (1981) and others are correct in suggesting that the metabolic rate of the mother is a key determiner of brain size in the offspring, this should be reflected in a significant, prenatal maternal effect. Atchley and colleagues (1984) have estimated that postnatal maternal effects in random-bred mice account for about 10% of the total variance in brain size, but there appear to be no comparable estimates for prenatal maternal effects.

Although estimates of heritability of brain size are generally high, as previously detailed, they give no indication of the relative importance of heritable maternal effects. It is possible, however, that most of these estimates include a maternal (indirect genetic) component as well. Thus the brain size heritability of .6 estimated by Atchley et al. (1984) in cross-fostered mice was obtained by doubling the genetic dam component, and this component also encompasses any prenatal maternal effects. If these effects amounted to 10% of the total variance, for example, then the true heritability estimate for brain size estimated by Atchley and colleagues (1984) would have been closer to .4 than to .6. It is suggestive in this regard that in these same mice, regressions of offspring on male parents yielded heritabilities that were lower than, although not significantly different from, .6 (Riska, personal communication, 1986).

Heritability estimates from between-strain components calculated in analyses of variance of inbred strains also contain additive direct and indirect (maternal) genetic variances (Henderson, 1979). In addition, such components theoretically estimate some epistatic genetic variance (Falconer, 1981), although this is conventionally assumed to be negligible. Roderick, Wimer, Wimer, and Schwartzkroin (1973) estimated the heritability of brain size in mice to be .67 in females and .62 in males, although the large number of inbred strains they sampled (25) should have maximized the range of genetic variability. In an analysis using three inbred strains, Leamy's (1985) estimates of the between-strain percentage for brain size at three different ages averaged .65, resulting in approximate heritabilities averaging .48. Full diallel designs that make use of both inbreds and hybrids provide better heritability estimates that are free of dominance variance and maternal effects (Henderson, 1979).

# BRAIN AND BODY SIZE: GENETIC AND MATERNAL INFLUENCES

Heritabilities of brain size estimated from such designs have varied from .46 (Henderson, 1979) to .68 (Hahn & Haber, 1978).

It is apparent that we need additional estimates of heritabilities of brain size that are free from the bias of maternal effects. Only then can we make a relative assessment of the importance of direct versus indirect genetic influences on the developing brain. We also need more estimates of genetic correlations/regressions of brain and body weight, especially at different ages, to adequately test the genetic models that explain scaling in these characters within and between species. We therefore next examine new empirical data derived from both random-bred and inbred/hybrid populations of mice, to gain additional insight into brain size and body size and their association.

## EXPERIMENTAL DATA ON BRAIN AND BODY SIZE

### Random-bred Population

The data on brain and body size presented here were derived from house mice of random-bred strain CV1; they are treated in greater detail in Leamy (1987). This population of mice was generated from 200 single-pair matings, which produced over 1000 total offspring (also see Leamy, 1974). Litter sizes in each of the families were reduced to six individuals, although when there were less than six, all were retained. In each litter, three sublitters of two mice each were sacrificed at 35 days of age (denoted 1 month), 90 days (3 months) and 150 days (5 months); all parents were sacrificed at 5 months. After sacrifice, each mouse was weighed and then skeletonized, brain sizes being measured by weighing the amount of alfalfa seeds necessary to fill the brain cavity. Brain weighings were made twice on each skull, and the mean of these two measurements was used in each case. Additionally, values for brain size and body size were logarithmically transformed (base 10) for subsequent analysis.

Heritabilities of brain size and body size were calculated for each of the three ages (Table 8–1) from regressions on male and female parents. As may be seen from the

**Table 8–1** Estimates of heritabilities ($h^2$) and their standard errors for brain and body size calculated from regressions of offspring in each age group on male and female parents, and maternal effects for each age calculated from full-sib correlations (see text)

|  | Age at sacrifice | | |
|---|---|---|---|
|  | 1 Month | 3 Months | 5 Months |
|  | Brain size | | |
| $h^2$ (male parent) | .26 ± .134 | .24 ± .109 | .16 ± .142 |
| $h^2$ (female parent) | .36 ± .085 | .47 ± .093 | .57 ± .125 |
| Maternal effects | .35 | .19 | .28 |
|  | Body size | | |
| $h^2$ (male parent) | −.34 ± .164 | .48 ± .122 | .20 ± .163 |
| $h^2$ (female parent) | .59 ± .127 | .35 ± .091 | .27 ± .138 |
| Maternal effects | .47 | .20 | .19 |

table, heritabilities for brain size derived from *male* parents (which should be the best estimators of direct genetic effects) are surprisingly low, averaging only about .2 over all ages. Heritabilities of body size (again from regressions on male parents) are more erratic over the ages, jumping from a negative value at 1 month to +.5 and +.2 at 3 and 5 months. We must be somewhat cautious in interpreting the values for the 1-month (and to some extent, 3-month) offspring, however, since they were obtained from regressions of offspring of one age on parents of a different (5-month) age. It could well be, for example, that the genes involved in determining brain and body size are simply not the same at different ages. If we therefore ignore the obviously poor estimate of heritability of body size at 1 month, the estimates at 3 and 5 months (average = 0.34) seem quite reasonable. However, the estimate of heritability for brain size is no higher at 5 months (or 3 months) than at 1 month, so overall we must conclude the heritability of brain size is relatively low in this population of mice.

The origin of the CV1 population of mice used here may help to explain the somewhat low level of heritability for brain size. This population is a random-bred derivative of inbred strain 101 and has generated its level of genetic variation through mutation alone over about 52 generations (Leamy, 1974). Previous estimates of heritability for various morphometric characters in this population have generally averaged about .4 (Leamy, 1974), but it is possible that the genetic variance accumulated for brain size itself has not been enough to generate a heritability level higher than that seen here. This is more likely if brain size is controlled by fewer genetic loci than the other morphometric characters (including body size), and the early selection plateaus reached in brain size selection experiments (Fuller, 1979) suggest that this may be the case.

Beyond the overall low level of heritability of brain size, the more interesting finding in Table 8–1 is the higher $h^2$ estimates obtained from regressions on female parents (roughly .4, .5, and .6 for the 1-, 3-, and 5-month offspring) compared with the male parents. This disparity between the male versus female parent regressions increases with age (.05, .12, .20), averaging about .12 overall. For body size (assuming zero for the negative regression at 1 month), the comparable average is .11. As previously detailed, this sort of difference is ascribable to the direct maternal additive genetic covariance ($COV_{AoAm}$) plus one half of the indirect genetic variance ($V_{Am}$). Whatever the relative importance of these two components, this suggests that the brain may be influenced by indirect as well as direct genetic effects.

The total maternal impact (including both prenatal and postnatal maternal effects) for brain and body size also was measured by subtracting full-sib correlations from $½h^2$ estimates obtained from regressions of offspring on male parents, and these values are included in Table 8–1. For both brain and body size, maternal effects contribute a rather high amount at 1 month (35% for brain size, 47% for body size), although in the same sense as the heritabilities, the estimates at this age are the least reliable. Maternal contributions decrease at the later ages, however, averaging about 23 percent for brain size and 20 percent for body size. The value of 23 percent for brain size is considerably greater than the average of 10 percent estimated by Atchley et al. (1984) for postnatal maternal effects in this same trait, suggesting that prenatal effects are as important as postnatal effects in contributing to the overall maternal impact on brain size. The 20-percent value for body size, however, is reasonably close to the postnatal maternal effects estimate of 17 percent made by Atchley and colleagues (1984). This implies that prenatal maternal effects are not nearly as important for body size as for brain size, a conclusion previously reached in a number of studies (Hanrahan & Eisen, 1974; Riska, Atchley, & Rutledge, 1984).

Phenotypic, genetic, and environmental correlations of brain and body size, as well as regressions of brain size on body weight, are given in Table 8-2. All estimates were made from covariances of offspring at each age on male and female parents, except that genetic (and therefore environmental) correlations and regressions involving the 1-month male parents could not be calculated because of the negative estimate of genetic variance (heritability) from the male parents at this age (see Table 8-1). Standard errors also are provided for all estimates except for the genetic and environmental regressions, because these have not yet been formulated.

As may be seen (Table 8-2), the phenotypic correlation of brain size and body size is higher at 1 month (.4) than at 3 or 5 months (.2). This sort of decline with age has previously been noted in mice (Atchley et al., 1984) and is indicative of the negative association of brain size with growth in body size after growth in brain size has largely ceased (Riska & Atchley, 1985). This association apparently also is responsible for the negative genetic correlations derived from male parents, and suggests that the genetic control of both brain size and body size is quite different at later, adult ages. The genetic correlations derived from the female parents, on the other hand, presumably are positive because they also estimate maternal as well as genetic sources of covariance. Their relative constancy in magnitude over the three ages also suggests a maternal involvement, for experience with most metric traits has shown that maternal correlations tend to be "locked in" at an early age and persist over later ages (Cheverud & Leamy, 1985). The environmental correlations show a pattern opposite to that of the genetic correlations, the male parent estimates being positive and the female parent estimates negative.

**Table 8-2** Phenotypic (P), genetic (G), and environmental (E) correlations (r) of brain and body size and regressions (b) of brain size on body size derived from covariances of offspring of each age with their 5-month male and female parents

|  |  | Age at sacrifice | | |
|---|---|---|---|---|
|  |  | 1 Month | 3 Months | 5 Months |
|  | $r_P$ | .40 ± .041 | .22 ± .042 | .22 ± .048 |
| Male parent | $r_G$ |  | −.23 ± .228 | −.43 ± .488 |
|  | $r_E$ |  | .34 ± .115 | .31 ± .119 |
|  | $r_P$ | .40 ± .040 | .23 ± .047 | .23 ± .042 |
| Female parent | $r_G$ | .48 ± .143 | .50 ± .121 | .56 ± .163 |
|  | $r_E$ | −.03 ± .167 | −.12 ± .109 | −.16 ± .161 |
|  | $b_P$ | .17 ± .018 | .10 ± .120 | .10 ± .022 |
| Male parent | $b_G$ |  | −.08 | −.19 |
|  | $b_E$ |  | .14 | .14 |
|  | $b_P$ | .17 ± .017 | .10 ± .019 | .10 ± .021 |
| Female parent | $b_G$ | .17 | .27 | .38 |
|  | $b_E$ | .01 | −.08 | −.16 |

Phenotypic, genetic, and environmental regressions of brain size on body size (Table 8–2) show the same general pattern as already described for the correlations. The magnitude of these regressions tends to be about one half that for the correlations, however, because body size is more variable than brain size. The genetic regressions estimate the expected evolutionary allometry, as previously explained, and it is interesting to note once again that only those values with a maternal component (female parent estimates) are positive. This suggests that selection for body size in adult (150-day-old) mice would produce a positive scaling of brain size only because of its indirect genetic link to body size through the maternal environment.

### Inbred/Hybrid Population

Let us now take a look at some brain and body size data derived from a different source—inbred and hybrid populations of mice. The data come from a recent study (Leamy, 1985) in which all possible crosses were made among three inbred strains (C57BL/6, C3HeB/Fe, and AKR). Three classes of inbreds and six classes of hybrids (including their reciprocals) of each sex were formed from these crosses, several litters being represented in each class. Altogether a total of 803 mice was sacrificed once again at 1, 3, and 5 months of age. Brain size and body size were measured as previously described, and all data were logarithmically transformed prior to analysis (Leamy, 1985).

The proportional contribution of variance among inbred lines for the 1-, 3-, and 5-month mice was .64, .63, and .67 for brain size and .28, .44, and .54 for body size (Leamy, 1985). If nonadditive genetic variation and maternal effects are considered negligible, these proportions represent heritabilities of approximately .47, .46, and .51 for brain size and .16, .28, and .37 for body size. These values compare favorably with the heritabilities calculated for brain size and body size from the random-bred population, provided the female parent estimates are used for brain size. This suggests that there may be a significant maternal component to the genetic variances estimated from inbreds for brain size, but not necessarily for body size.

Correlations (and regressions) of brain and body size in the inbreds and hybrids were calculated by making use of equation (8–2) and the procedures previously described. Variance and covariance components were obtained from the VARCOMP procedure of SAS (SAS Institute, Inc., 1985), and for the inbreds were estimated for differences between lines, litters within lines, within litters, and for the total. Whereas line differences are wholly genetic in origin, those between and within litters are strictly environmental, the between-litter variances also assessing maternal effects (Leamy, 1982). Genetic estimates obtained from variances among hybrid lines are complicated by heterosis (Henderson, 1979; Leamy, 1982), however, so variance and covariance components for hybrids were estimated only for differences among reciprocals within lines as well as between and within litters. Reciprocal differences are of particular interest because, assuming sex-linkage is negligible, they are ascribable to maternal effects (Henderson, 1979).

Correlations and regressions of brain and body size, calculated for each category in the 1-, 3-, and 5-month inbreds and hybrids, are given in Table 8–3. If we look first at inbreds, it is apparent that the between-line (genetic) correlations are essen-

**Table 8-3** Correlations of brain and body size and regressions (in parentheses) of brain size on body size for 1-month, 3-month, and 5-month inbred and hybrid mice

|  | Age at sacrifice | | |
|---|---|---|---|
|  | 1 Month | 3 Months | 5 Months |
|  |  | Inbreds |  |
| Between lines | 1.00 (0.67) | 1.02 (0.92) | 1.00 (0.53) |
| Between litters | 0.95 (0.22) | 0.58 (0.26) | 0.87 (0.36) |
| Within litters | 0.46 (0.18) | 0.57 (0.34) | 0.29 (0.10) |
| Total | 0.78 (0.37) | 0.79 (0.56) | 0.86 (0.41) |
|  |  | Hybrids |  |
| Between reciprocals | 1.17 (0.29) | 0.92 (1.05) | 1.10 (0.63) |
| Between litters | 0.92 (0.43) | 1.01 (0.35) | −0.20 (−0.01) |
| Within litters | 0.36 (0.16) | 0.13 (0.07) | 0.25 (0.09) |
| Total | 0.74 (0.33) | 0.58 (0.33) | 0.60 (0.26) |

tially 1.0 at all ages. Thus, in contrast to the comparable estimates of about .5 for the genetic correlations obtained in the random-bred population from covariances of (female) parents and offspring (see Table 8-2), those obtained here from between-strain variation indicate a nearly perfect association between brain size and body size. The genetic regression estimates also are higher than before, averaging about .7, or nearly the magnitude attained in the scaling of brain and body size among all mammals.

Why is the genetic association of brain size and body size so high among the inbred lines here? As pointed out by Hegmann and Possidente (1981), genetic correlations calculated from variation among inbred strains must be viewed with considerable caution. In the inbreeding process, genes at various loci get fixed purely by chance and thus may impose genetic correlations that would be strictly transient in randomly mating populations, if they occurred at all. Further, the chance of obtaining spurious correlations is increased if the traits involved are controlled by only a few loci, and the indirect evidence presented earlier suggests that this may be the case at least for brain size. In addition, we have used only three inbred strains here, and a greater number of strains might ensure a more general estimate (Hegmann & Possidente, 1981). It is noteworthy in this regard that the genetic correlation of brain and body size calculated by Lande (1979) from 25 inbred strains was high (+.77), but was not 1.0.

Beyond the between-line correlations and regressions for inbreds (Table 8-3), those generated from between- and within-litter variation are somewhat less in magnitude. The average of the correlations between litters (about .8) is greater than that within litters (about .4), presumably because of maternal covariation among litters that simply has no effect within litters (Falconer, 1981). Maternal influences are also evident in the very high correlations generated from reciprocal differences in the hybrids. Between-litter correlations again are higher than within-litter correlations (except at 5 months) in hybrids, although their overall average is somewhat less than that for the hybrids. Similar patterns are seen for the regressions, except that as expected, their overall average is less than that for the correlations.

## Synopsis

What do these data from the random-bred and inbred/hybrid populations tell us about brain and body size? First, they tend to support the genetic models that relate brain size and body size, but suggest more of a role for maternal influences than has previously been supposed. In both populations, for example, heritabilities are higher for brain size than for body size. But the results from the random-bred population suggest that a significant portion of this higher heritability for brain size stems from indirect genetic (maternal) variance. Further, especially by comparison with other studies, it would appear that this maternal component affecting brain size is largely prenatal in origin. This contrasts with body size, which is far more influenced by postnatal, rather than prenatal, maternal effects. Genetic associations between brain size and body size also are augmented by maternal influences, these influences persisting well into postnatal life.

## CONCLUDING REMARKS AND SUGGESTIONS

It is gratifying that the experimental data just presented are basically compatible with genetic models that try to explain the evolution of brain and body size. But it would be remiss not to point out that we need many more genetic data of this sort to see how the models hold up under further scrutiny. The random-bred population used here after all is but one population of many, and a rather unique one at that. It would be particularly useful to discover, for example, whether the obvious differences in the male versus female parent covariances found earlier occur in other outbred populations. It also would be instructive to know whether the use of parents of a younger age might change some of the patterns observed in these data. Similarly, we need additional estimates of genetic and environmental variances and covariances from inbred/hybrid populations, especially those that consist of a large, representative sample of lines.

Given the obvious involvement of maternal influences on brain size and the assocation of brain size with body size, we certainly need more estimates of maternal variances and covariances derived from various models. Besides assessing the total maternal impact for brain size, we particularly need specific estimates of the additive maternal (indirect) genetic variance as well as the direct additive maternal genetic covariance. These parameters have been estimated for such characters as body weight (Riska et al., 1985b) and length of the mandible (Riska et al., 1985a), so in practice they can be calculated for brain size as well. Their estimation obviously does require specific breeding designs (see Eisen, 1967; Riska et al., 1985a), designs that thus far have simply not been used for brain size. Further, if these estimates are to be reasonably precise, the experiments must use rather large numbers of individuals. Such experiments must be done if we are to assess the relative importance of maternally mediated genetic effects on both the variance of brain size and its covariance with other traits such as body size.

We also need an assessment of the relative importance of prenatal versus postnatal maternal influences that affect brain size. Postnatal maternal influences have been estimated (Atchley et al., 1984), but as we have seen, prenatal maternal effects appear to be as important for brain size. Embryo transfer techniques (for example, Aitken, Bowman, & Gauld, 1977) thus could be useful in assessing this. Another approach is that of cross-fostering between two inbred strains, as done by Tenczar and Bader (1966) for the widths of the second ($M_2$) and third ($M_3$) mandibular molars in house mice. Interestingly, they showed that prenatal maternal effects contributed more to the total variance of the earlier developing $M_2$ than to the the later developing $M_3$, which in turn was more affected by the postnatal environment. Brain size should show a pattern similar to that of early developing structures such as the $M_2$, but this needs testing.

Artificial selection for brain size may be another effective way of assessing prenatal and postnatal maternal influences. Riska and Atchley (1985) have suggested, for example, that the highest brain/body size association should be achieved when there is extensive change in early (mostly prenatal) growth but little or no change in later (postnatal) growth. This is equivalent to restricted index selection (Turner & Young, 1969), the outcome of which should be predicted by the partial regression of brain size on body size, holding later growth constant (Riska & Atchley, 1985). With this sort of selection regime, therefore, it should be possible to produce scaling coefficients for brain and body size equivalent to that found among all mammals (.75). Again, this would require considerable organization and labor, but it should be done.

Direct attempts also should be made to assess the genetic association of brain size with the metabolism of the mother. To do this, basal metabolic rates of all females would need to be taken at intervals during the gestation period so that they could later be related to the brain size of the offspring. We very much need to know whether metabolism is the proximate link between brain and body size, and a genetic correlation of 1.0 would provide powerful evidence of this. Brain size certainly scales to body size just as metabolism does (Martin, 1981), and Armstrong (1983) also has shown that among a number of mammalian species, brain size scales isometrically ($k = 1.03$) to body size when body size is adjusted for basal metabolic rate. Finally, metabolic rate is but one parameter of development, and development itself is the frontier to which no doubt we will increasingly turn for future progress in understanding brain size.

## ACKNOWLEDGMENTS

I wish to thank Bruce Riska, Martin Hahn, and Norman Henderson for helpful comments on earlier versions of this paper. This work was supported in part by a grant from the University Research Committee, California State University, Long Beach.

## REFERENCES

Aitken, R. J., Bowman, P., & Gauld, I. (1977). The effect of synchronous and asynchronous egg transfer on foetal weight in mice selected for large and small body size. *Journal of Embryology and Experimental Morphology, 37,* 50–64.

Armstrong, E. (1983). Relative brain size and metabolism in mammals. *Science, 220,* 1302–1304.

Atchley, W. R. (1984). The effect of selection on brain and body size associations in rats. *Genetical Research, 43,* 289–298.

Atchley, W. R., Riska, B., Kohn, L. A. P., Plummer, A. A., & Rutledge, J. J. (1984). A quantitative genetic analysis of brain and body size associations, their origin and ontogeny: Data from mice. *Evolution, 38,* 1165–1179.

Bauchot, R. (1978). Brain allometry in vertebrates. *Brain Behavior and Evolution, 15,* 1–18.

Cheek, D. B. (Ed.). (1975). *Fetal and Postnatal Cellular Growth.* New York: John Wiley.

Cheverud, J. M., & Leamy, L. J. (1985). Quantitative genetics and the evolution of ontogeny. III. Ontogenetic changes in correlation structure among live-body traits in randombred mice. *Genetical Research, 46,* 325–335.

Eisen, E. J. (1967). Mating designs for estimating direct and maternal genetic variances and direct-maternal genetic covariances. *Canadian Journal of Genetics and Cytology, 9,* 13–22.

Eisenberg, J. F. (1981). *The Mammalian Radiations.* Chicago: University of Chicago Press.

Falconer, D. S. (1981). *Introduction to Quantitative Genetics.* (2nd ed.). New York: Longman.

Falconer, D. S., Gauld, I. K., & Roberts, R. C. (1978). Cell numbers and cell sizes in organs of mice selected for large and small body size. *Genetical Research, 31,* 287–301.

Fuller, J. L. (1979). Fuller BWS lines: History and results. In M. E. Hahn, C. Jensen, and B. C. Dudek (Eds.), *Development and Evolution of Brain Size: Behavioral Implications* (pp. 190–204). New York: Academic Press.

Fuller, J. L. & Geils, H. D. (1972). Brain growth in mice selected for high and low brain weight. *Developmental Psychobiology, 5,* 307–318.

Goss, R. J. (1966). Hypertrophy versus hyperplasia. *Science, 153,* 1615–1620.

Gould, S. J. (1966). Allometry and size in ontogeny and phylogeny. *Biological Reviews, 41,* 587–640.

Gould, S. J. (1975). Allometry in primates, with emphasis on scaling and the evolution of the brain. *Contributions to Primatology, 5,* 224–292.

Gould, S. J. (1977). *Ontogeny and Phylogeny.* Cambridge, MA: Belknap Press.

Green, E. L. (Ed.). (1966). *Biology of the Laboratory Mouse.* New York: McGraw-Hill.

Hahn, M. E., & Haber, S. B. (1978). A diallel analysis of brain and body weight in male inbred laboratory mice *(Mus musculus). Behavioral Genetics, 8,* 251–260.

Hahn, M. E., Jensen, C., & Dudek, B. C. (Eds.). (1979). *Development and the Evolution of Brain Size: Behavioral Implications.* New York: Academic Press.

Hahn, M. E., Walters, J. K., Lavooy, J., & DeLuca, J. (1983). Brain growth in young mice: Evidence on the theory of phrenoblysis. *Developmental Psychobiology, 16,* 377–383.

Hanrahan, J., & Eisen, E. J. (1974). Genetic variation in litter size and 12-day weight in mice and their relationships with post-weaning growth. *Animal Production, 19,* 13–23.

Hegmann, J. P., & Possidente, B. (1981). Estimating genetic correlations from inbred strains. *Behavioral Genetics, 11,* 103–114.

Henderson, N. D. (1979). Genetic correlations between brain size and some behaviors of housemice. In M. E. Hahn, C. Jensen, & B. C. Dudek (Eds.), *Development and Evolution of Brain Size: Behavioral Implications* (pp. 348–371). New York: Academic Press.

Hofman, M. A. (1982). Encephalization in mammals in relation to the size of the cerebral cortex. *Brain Behavior and Evolution, 20,* 84–96.

Holloway, R. L. (1979). Brain size, allometry, and reorganization: Toward a synthesis. In M. E. Hahn, C. Jensen, and B. C. Dudek (Eds.), *Development and Evolution of Brain Size: Behavioral Implications* (pp. 61–88). New York: Academic Press.

House, D., Berman, E., & Carter, H. B. (1985). Description and implications for analysis of brain growth in suckling mice. *Growth, 49,* 426–438.

Jerison, H. J. (1979). The evolution of diversity in brain size. In M. E. Hahn, C. Jensen, and B. C. Dudek (Eds.), *Development and Evolution of Brain Size: Behavioral Implications* (pp. 30–57). New York: Academic Press.

Kobayashi, T. (1963). Brain-to-body ratios and time of maturation of the mouse brain. *American Journal of Physiology, 204,* 343–346.
Lande, R. (1979). Quantitative genetic analysis of multivariate evolution applied to brain: body size allometry. *Evolution, 33,* 402–416.
Leamy, L. (1974). Heritability of osteometric traits in a randombred population of mice. *Journal of Heredity, 65,* 109–120.
Leamy, L. (1982). Morphometric studies in inbred and hybrid house mice. II. Patterns in the variances. *Journal of Heredity, 73,* 267–272.
Leamy, L. (1985). Morphometric studies in inbred and hybrid house mice. VI. A genetical analysis of brain and body size. *Behavioral Genetics, 15,* 251–263.
Leamy, L. (1987). Genetic and maternal influences on brain and body size in randombred house mice. *Evolution, 42,* 42–53.
Leutenegger, W. (1976). Allometry of neonatal size in eutherian mammals. *Nature, 263,* 229–230.
Martin, R. D. (1981). Relative brain size and basal metabolic rate in terrestrial vertebrates. *Nature, 293,* 57–60.
Martin, R. D., & Harvey, P. H. (1985). Brain size allometry: Ontogeny and phylogeny. In W. L. Jungers (Ed.), *Size and Scaling in Primate Biology* (pp. 147–173). New York: Plenum Press.
Martin, R. D., & MacLarnon, A. M. (1985). Gestation period, neonatal size and maternal investment in placental mammals. *Nature, 313,* 220–223.
Radinsky, L. (1977). Brains of early carnivores. *Paleobiology 3,* 333–349.
Radinsky, L. (1978). Evolution of brain size in carnivores and ungulates. *American Naturalist, 112,* 815–831,
Raff, R. R., & Kaufman, T. C. (1983). *Embryos, Genes, and Evolution.* New York: MacMillan.
Riska, B., & Atchley, W. R. (1985). Genetics of growth predict patterns of brain-size evolution. *Science, 229,* 668–671.
Riska, B., Atchley, W. R., & Rutledge, J. J. (1984). A genetic analysis of targeted growth in mice. *Genetics, 107,* 79–101.
Riska, B., Rutledge, J. J., & Atchley, W. R. (1985a). Genetic analysis of crossfostering data with sire and dam records. *Journal of Heredity, 76,* 247–250.
Riska, B., Rutledge, J. J., & Atchley, W. R. (1985b). Covariance between direct and maternal genetic effects in mice, with a model of persistent environmental influences. *Genetical Research, 45,* 287–297.
Roderick, T. H., Wimer, R. E., & Wimer, C. C. (1976). Genetic manipulation of neuroanatomical traits. In L. Petrinovich and J. L. McGaugh (Eds.), *Knowing, Thinking and Believing* (pp. 143–178). New York: Plenum Press.
Roderick, T. H., Wimer, R. E., Wimer, C. C., & Schwartzkroin, P. A. (1973). Genetic and phenotypic variation in weight of brain and spinal cord between inbred strains of mice. *Brain Research, 64,* 345–353.
Rosenzweig, M. R. (1979). Responsiveness of brain size to individual experience: Behavioral and evolutionary implications. In M. E. Hahn, C. Jensen, and B. C. Dudek (Eds.), *Development and Evolution of Brain Size: Behavioral Implications* (pp. 264–295). New York: Academic Press.
SAS Institute, Inc. (1985). *SAS user's guide: Statistics, version 5 edition.* Cary, NC: SAS Institute, Inc.
Tenczar, P., & Bader, R. S. (1966). Maternal effect in dental traits of the house mouse. *Science, 152,* 1398–1400.
Turner, H. N., & Young, S. S. Y. (1969). *Quantitative Genetics in Sheep Breeding.* Ithaca, NY: Cornell University Press.

# III
# BIOMETRICAL APPROACHES TO EVOLUTION AND BEHAVIORAL DEVELOPMENT

JOHN K. HEWITT

Biometry is concerned with the measurement and characterization of biological variation. It deals with populations, their average characteristics, and the properties of distributions around the average. Biometricians are interested in the distributions of individual variables and in the joint distributions of two or more variables. Since variation is the raw material on which the forces of evolution operate, biometry has always been closely associated with the study of evolution. Sir Ronald Fisher, a biometrician of outstanding genius, brought together the fields of statistics, genetics, and evolutionary theory in his class treatise *The Genetical Theory of Natural Selection* (Fisher, 1930, 2nd ed. 1958). Prior to this Fisher had shown how the properties of continuously varying characteristics (like height and intelligence), as well as those of characteristics that fall into discrete classes (like eye color), could be explained by the theory of particulate inheritance put forward by Mendel. The key was the realization that if many genes, each of small effect, contributed to the determination of a characteristic, this would result in the continuous unimodal distributions for height, weight, strength, and so on that we observe within a population (for a given sex). More important, this *polygenic* hypothesis would predict particular patterns of variation and correlation between relatives based on their genetic relatedness.

Thus was born the scientific discipline of biometrical or quantitative genetics. By comparing appropriate data—means and variances of particular genetic crosses in animals and plants, or patterns of variation and covariation in human families—against hypotheses about genetic and environmental influences it was possible to infer the relative importance of these influences and their modes of action. Did the genes act additively, or were there nonadditive influences such as genetic dominance? Were there maternal influences or other influences common to offspring reared in the same family that affected the characteristic under study? Were *genes* reponsible for the observed correlations between two different characteristics (through pleiotropy, genetic linkage, or genetic association), or was the correlation caused by environmental influences that affected both characteristics? Questions of this kind could be answered only by a biometrical genetical approach (Falconer, 1981; Mather &

Jinks, 1983). Understanding of the sources of individual variation, in intelligence (IQ) for example, is a central concern of social science (see Jencks et al., 1972) as well as of psychology and human biology, and thus scientists in these areas have turned increasingly to biometrical genetics for methods of study and guidance in the interpretation of population and family data on human behavior (Jinks & Fulker, 1970).

This recognition of the necessity of a biometrical genetic framework for theorizing has been given additional impetus by the increasing familiarity with and acceptance of the methods of path analysis by social scientists. Path analysis, developed originally by the American biometrical geneticist Sewall Wright (Wright, 1921), provides an algorithm for specifying and estimating the influence of multiple causes on a series of outcome or manifest variables. The flexibility of the approach remains part of its appeal, and in the context of developmental phenomena it permits the exploration of changing patterns of genetic and environmental control. Perhaps most important, the accessibility of this technique to social scientists, for whom the application of the general principles of structural equation modeling (or path analysis) required no knowledge of genetics whatsoever, has confronted them head on with the question: What paths of influence do we need to consider when accounting for the correlations between parents and their children or between other pairs of relatives? In this context it is clear even to the most biologically illiterate scientist that genetic influences have to be considered, at least provisionally. Once considered, their effects have to be estimated, and typically this gives rise to a whole new set of interesting questions. How is it, for example, that there appear to be genetic influences on a person's income (Taubman, 1976)? What combination of heritable characteristics—intelligence, personality, physical appeal, prowess—contribute to a person's earning potential, and how do these interact with nongenetic inheritance and social and individual environments to produce the outcome? These kinds of behavior genetic questions are now being discussed, not by behavioral geneticists but by economists. Such questions are a far cry from the problems that have traditionally fascinated evolutionary biologists, although intraspecific competition among beasts should not be intrinsically more interesting to a biologist than competition among people.

Aside from the revolution in thinking about *human* variation that biometrical genetic approaches are bringing about, these approaches are contributing considerably to our understanding of animal behavior and its evolution. We can identify two important principles that are influencing this field. The first is that natural selection will not only alter the mean level of expression of a phenotype in the population but also affect the predominant pattern of genetic control. In the simplest case, deleterious alleles will remain in the population only if they are recessive; that is, we will typically find that there is genetic dominance toward the fitter phenotype. In polygenic systems, the same principle will apply wherever there has been continued directional selection toward a higher (or lower) optimum in the range of phenotypes expressed in the population, providing of course that the phenotype is heritable. Describing the patterns of additive and nonadditive genetic control influencing a phenotype has been likened to describing the genetic architecture of a trait. Behavioral geneticists have used these descriptions to draw inferences about the likely adaptive significance of behaviors, in the absence of the direct information about their influence on fitness that is so difficult to obtain (Broadhurst, 1979; Hahn, 1983; Henderson, 1978; Hewitt, Fulker & Broadhurst, 1981).

A second principle now emerging as significant in the study of behavioral evolution is that the *genetic correlation* between traits imposes constraints on the progress of evolution. That is, if two traits are genetically correlated, selection applied to one trait will necessarily have consequences for the correlated trait; if the correlated trait was already at its optimum value, this will slow the progress of selection on the primary trait. Lande (1979) has elaborated some of the consequences of this for evolutionary changes. Of course, an important early observation by Mather was that genetic correlations are themselves subject to change under selection. In particular, artificial selection for a characteristic, abdominal bristle number in *Drosophila melanogaster,* for example, will invariably reach a plateau when natural selection against all the correlated characters opposes artificial selection on the primary character to the extent that the selection lines faces extinction. Given several generations of relaxation of artificial selection and the opportunity for genetic recombinations to occur, even within tightly linked groups of "supergenes," renewed artificial selection may surpass the earlier plateau. The original genetic correlation between high bristle number and lower fitness has been reorganized through genetic recombinations. The small number of generations over which Mather observed this change (Mather, 1973; Mather & Harrison, 1949) reminds us that genetic correlations, like other aspects of the genetic architecture, are themselves responsive to selection in ways that can be understood only by reference to the genetic properties of the organism, rather than to its statistical properties.

Armed with these kinds of principles, biometrical geneticists have turned their attention to developmental questions. For this volume, we asked five scientists to approach evolution and behavioral development from their own perspectives within the broad context of biometrical or quantitative genetics.

Steve Arnold's chapter on the inheritance and evolution of behavioral ontogenies introduces recent work on evolutionary theory and in particular on the implications of genetic correlations for evolution. He presents examples from animal behavior—niche ontogeny, male mating tactics, and birdsong—to illustrate insights into the evolutionary process that quantitative genetics may be able to provide. He is dissatisfied with a univariate approach to genetic architecture that has typified much of biometrical genetic work and emphasizes the probable importance of the role of genetic correlations, in particular those from age to age, in understanding the constraints on the evolution of ontogenic patterns. He argues that, given relatively weak selection on ontogenetic patterns, the genetic parameters—correlations and heritabilities—may be assumed to remain constant long enough for one to make predictions without worrying about the kinds of changes in the genetic correlations that Mather observed over a relatively short number of generations.

Norman Henderson approaches the questions from a somewhat different perspective. He is concerned with what we can really say about genetic architecture and genetic correlations, given the kinds of experiments it is feasible for us to carry out. He provides a detailed analysis of the power of an animal breeding design, the diallel cross, that is widely used to inform us about the parameters of the genetic system—additive and nonadditive effects, maternal effects and sex-linkage, and genetic correlations. Like Arnold, he is more optimistic about predictions to populations descended from the animals being studied than he is about the inherences made to a hypothetical ancestral population. Henderson cautions us not to expect too much too quickly from the study of developmental behavior genetics, a caution that echoes that of Hahn earlier in this volume (Chapter 4).

John Hewitt takes his starting point in the literature on the genetics of animal learning. Here, as elsewhere, the genetic architecture is related to the adaptiveness of the behaviors studied, but most intriguingly, *changes* in the genetic architecture during learning and across the life span point to changes in the direction of selection at different ages or stages of development. This clearly has implications for the kinds of analysis advocated by Arnold. Taking up the developmental theme, Hewitt approaches the problem of modeling the developmental process, first in terms of the transitions postulated by learning theorists and then in terms of the kinds of developmental path models being formulated by Lindon Eaves and his colleagues. Although the approaches are different, they converge on some common hypotheses that may serve as guidelines in developmental behavior genetic research.

Corley and Fulker consider the use of adoption studies in research on human development. The study of children adopted away from their biological parents provides, along with the study of identical and fraternal twins, a direct way to disentangle the genetic and family environmental influences on human behavior that are confounded in the normal nuclear family. Combining the adoption design with a longitudinal study permits us to answer questions about the development of behavior, and in particular about the relationship between early and later manifestations of personality and cognitive variation.

Finally, Lindon Eaves and his colleagues take up where Hewitt left off in exploring approaches to the quantitative genetic modeling of development and age-related changes. They first show how quantitative models of development and change may be extended to the multivariate case. They then introduce three new considerations for the study of behavioral development, which are suggested by quantitative genetics: What is the role of the genetic control of homeostasis (and its potential breakdown)? How is liability to exhibit a particular behavior or to develop a disease related to its age of onset? How should statistical survival models (which are used to predict age of onset) be modified to allow for genetic and environmental variation in vulnerability? These considerations take us beyond the traditional concerns of biometrical genetics, but they may well be just the kinds of new questions that will set developmental behavior genetics apart from traditional behavior genetics.

## REFERENCES

Broadhurst, P. L. (1979). The experimental approach to behavioral evolution. In J. R. Royce & L. P. Moss (Eds.), *Theoretical Advances in Behavior Genetics*. Germantown, MD: Sijthoff and Noordhoff.
Falconer, D. S. (1981). *Introduction to Quantitative Genetics* (2nd ed.). London: Longman.
Fisher, R. A. (1930) (2nd ed., 1958). *The Genetical Theory of Natural Selection*. Oxford: Clarendon Press (New York: Dover).
Hahn, M. (1983). Genetic "artifacts" and aggressive behavior. In E. C. Simmel, M. E. Hahn, & J. K. Walters (Eds.), *Aggressive Behavior: Genetic and Neural Approaches*. Hillsdale, NJ: Lawrence Erlbaum.

Henderson, N. D. (1978). Genetic dominance for low activity in infant mice. *Journal of Comparative and Physiological Psychology, 92,* 118–125.

Hewitt, J. K., Fulker, D. W., & Broadhurst, P. L. (1981). Genetic architecture of escape-avoidance conditioning in laboratory and wild populations of rats: A biometrical approach. *Behavior Genetics, 11,* 533–544.

Jencks, C., Smith, M., Acland, H., Bane, M. J., Cohen, D., Gintis, H., Heyns, B., & Michelson, S. (1972). *Inequality: A Reassessment of the Effect of Family and Schooling in America.* New York: Basic Books.

Jinks, J. L., & and Fulker, D. W. (1970). Comparison of the biometrical genetical, MAVA and classical approaches to the analysis of human behavior. *Psychological Bulletin, 75,* 311–349.

Lande, R. (1979). Quantitative genetic analysis of multivariate evolution, applied to brain:body size allometry. *Evolution, 33,* 402–416.

Mather, K. (1973). *The Genetical Structure of Populations.* London: Chapman and Hall.

Mather, K., & Harrison, B. J. (1949). The manifold effect of selection. *Heredity, 3,* 1–52, 131–162.

Mather, K., & Jinks, J. L. (1983). *Biometrical Genetics* (3rd ed.). London: Chapman and Hall.

Taubman, P. (1976). The determinants of earnings: Genetics, family, and other environments. *American Economic Review, 66,* 858–870.

Wright, S. (1921). Correlation and causation. *Journal of Agricultural Research, 20,* 557–585.

# 9
# Inheritance and the Evolution of Behavioral Ontogenies

STEVAN J. ARNOLD

Developmental behavior genetics is the intersection of developmental and genetic studies of behavior. The principle focus in this new field is on how gene action changes during development and on how heritabilities and genetic correlations change with age (De Fries & Fulker, 1986; Plomin, 1983, 1986).

The evolutionary genetics of behavioral ontogenies is a still newer and relatively unexplored subject. Here the focus is on the inheritance and evolution of the program for behavioral development. My thesis is that this new field will grow faster if it fosters contacts with surrounding disciplines. In this chapter I review some relevant concepts of quantitative genetics and evolutionary theory and then examine three phenomenological examples of possible contacts: alternative male mating tactics, ecological niches, and birdsong. My aim, then, is to consider how our understanding of the evolution of these phenomena might be enhanced by focusing on the inheritance of behavioral ontogenies. These examples also have a transcendant interest to the student of developmental genetics for they suggest how an evolutionary perspective can be used to focus genetic work.

## BACKGROUND

To conceptualize the genetic issues, it will be helpful to use an idealized behavior with a simple ontogeny. Such a case is shown in Figure 9-1. In this hypothetical example the average value of the behavioral trait increases linearly with age, but any regular pattern of change could be used to illustrate the following points. Suppose we score the behavior in a large sample of individuals at each of three ages. We could then define three age-specific traits (e.g., behavior at age 1). Plotting our hypothetical data, we find that each of these age-specific traits shows a continuous distribution, as illustrated in Figure 9-1.

### Two Concepts from Quantitative Genetics

Some important conceptual tools from quantitative genetics will enable us to visualize the inheritance of behaviors that change during ontogeny. The first important

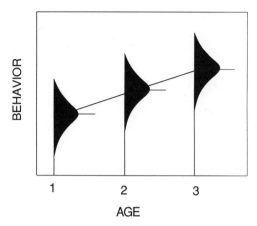

**Figure 9-1** A hypothetical behavior whose average value (horizontal lines) increases with age. The bell-shaped figures are frequency distributions portraying the behavioral variation among individuals at each age.

tool is *genetic variance,* which describes the degree of hereditary resemblance between parents and offspring in some phenotypic trait. In Figure 9-2 we see the average scores of offspring at age 1 plotted against the average scores of their parents at that same age. The data points in the example show a linear trend that might reflect Mendelian inheritance as well as the effects of environmental features that are shared by families. Let us imagine that such common environmental effects have been eliminated or disrupted by experimental design (e.g., by cross-fostering of offspring) so that the observed resemblance between offspring and parents can be wholly attributed to inheritance. In this case, we can show with a little algebra (Falconer,

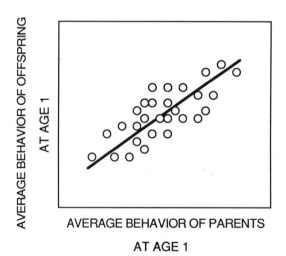

**Figure 9-2** Hypothetical data showing the average behavior of offspring at age 1 plotted against the average behavior of their two parents at that same age. The slope of the least squares best-fit line estimates the heritability of behavior at age 1.

1981, Chapter 9; Fisher, 1918) that the resemblance is due to a particular additive property of gene action. The property that measures offspring-parent resemblance in a particular trait is known as *additive genetic variance*. *Heritability* is simply a standardized genetic variance that varies between zero and one. The slope of the least-squares line in Figure 9-2 estimates the heritability of behavior at age 1.

Extending our approach to the behavior as it is expressed at other ages, we could likewise estimate the heritability of behavior at any age. We could then summarize our findings by graphing genetic variance or heritability as a function of age. Such profiles of genetic variance are the first critical type of genetic information needed to predict response to deliberate or natural selection. But before we consider the role of genetic variance in evolutionary prediction, let us examine the second critical type of genetic information.

The second important genetic issue is hereditary correspondence between behaviors expressed at different ages. Two extreme possibilities are shown in Figure 9-3, in which the average behavior of offspring at age 2 is plotted against the average behavior of their parents at age 1. As before, nongenetic causes of resemblance

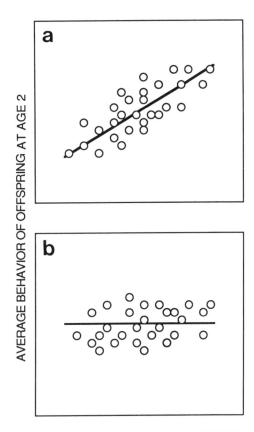

**Figure 9-3** Hypothetical data illustrating two extremes of genetic covariance of behavior as it is expressed at two different ages. (a) Strong positive genetic covariance between behavior at age 1 and the same behavior at age 2. (b) Lack of genetic covariance.

between offspring and parents have been disrupted by experimental design. One possibility is that the resemblance between relatives in behaviors expressed at different ages is as strong as the resemblance in behaviors expressed at the same age (compare Figure 9-3a with Figure 9-2). This circumstance could arise if virtually the same genes were responsible for genetic variation in behavior at the two ages. Alternatively, a wholly different set of genes might affect behavioral differences at age 2. In this case there might be no average resemblance between parent and offspring behavior at the two ages (Figure 9-3b). Eaves and colleagues (Chapter 13) refer to this distinction as a contrast between situations in which genes act all the time versus situations in which genes are occasion or age specific.

The genetic parameter that describes hereditary correspondence between behaviors expressed at different ages is called *genetic covariance*. These covariances are often standardized so that they vary between minus one and plus one; they are then known as *genetic correlations*. The situation in Figure 9-3a with strong hereditary resemblance reflects a large, positive genetic covariance between genetic values for behavior at ages 1 and 2. The lack of resemblance in Figure 9-3b reflects a zero genetic covariance between genetic values for behavior at the two ages. Genetic coupling, described by genetic covariance, can arise from pleiotropy or linkage disequilibrium (Falconer, 1981; Hazel, 1943).

Genetic variances and covariances are convenient descriptors of hereditary resemblance in behavior. Because these genetic parameters describe behavioral transmission from one generation to the next, they are also the critical hereditary variables needed to make evolutionary predictions. The elementary evolutionary issue, after all, is to make a prediction about how behavior will change from one generation to the next. To make such a prediction we need, aside from inheritance, a specification of selection. Selection describes how the parentage of the next generation is biased; inheritance (genetic variances and covariances) describes how that bias is transmitted.

## Evolutionary Response to Selection on Behavior

Selection acting at a particular age can cause evolution not only of behaviors expressed at that age but also of behaviors that are expressed at other ages (Falconer, 1981, Chapter 19). Let us consider the simple case of so-called truncation selection. Imagine that selection acts on behavior at age 1 so that only a subset of potential future parents actually survives to produce offspring. For example, suppose that only parents above some critical behavioral score actually produce offspring. If behavior at age 1 is heritable, those parents will leave offspring whose behavior at age 1 has higher scores than the behavior of average offspring of all potential parents (Figure 9-4a). That behavioral discrepancy between the offspring of selected parents and of all potential parents is known as the *direct response to selection* on behavior at age 1. Likewise, if the genetic covariance between behavior at age 1 and 2 is nonzero, selection on behavior at age 1 will produce a *correlated response* in behavior at age 2 (Figure 9-4b). By computing the direct and correlated responses to selection for each generation we can deduce the long-term evolutionary trajectory of the population and inquire about the long-term effects of genetic correlation.

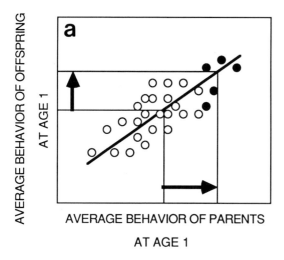

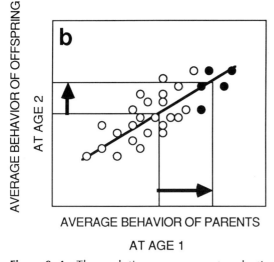

**Figure 9–4** The evolutionary response to selection on behavior. (a) Same hypothetical data as in Figure 9–2, with heritability equal to the slope of the heavy line. The average behavior of all potential parents is indicated by the vertical line at the center of the figure. The expected behavior of the average of all potential parents is indicated by the horizontal line at the center of the figure. If selection acts on the parental generation we can readily predict the evolutionary response. If only the five sets of parents with the highest average behavioral scores become the actual parents of the next generation, the expected behavior of their offspring is indicated by the uppermost horizontal line. The horizontal arrow indicates the strength of selection (the difference between potential and actual parents). The vertical arrow indicates the response to selection (the consequence of the imposed selection, assessed in the next generation). (b) Evolutionary response of behavior at age 2 to selection imposed on behavior at age 1. Same data as in Figure 9–3a, with the slope of the heavy line reflecting positive genetic covariance. The strength of selection on behavior at age 1 is indicated by the horizontal arrow. The correlated response in behavior at age 2 to selection imposed at age 1 is indicated by the vertical arrow.

## Evolution of Behavioral Ontogeny When Selection Acts at a Specific Age

The long-term consequence of genetic correlation is to cause the evolution of temporary maladaptation in the behavioral ontogeny. Imagine a situation in which the environment suddenly changes so that a new behavioral ontogeny is favored by selection. In Figure 9-5 we consider just the initial portion of the behavioral ontogeny, from age 1 to age 2. We suppose that in the old environment the population had evolved an optimal ontogeny (lowest line) but in the new environment there is a new optimal behavioral ontogeny (heavy line), one that slopes downward rather than upward. The only difference between these optimal ontogenies is in the behavioral score at age 1. In the new environment a higher behavioral score is favored at age 1, but the original behavioral score is optimal at age 2 in the old and new environments. Using this simple scenario we can visualize the effects of genetic correlation on the behavioral ontogeny as it evolves under the challenge of a new environment. To construct these visualizations I have used evolutionary theory based on the elementary concept of genetic correlation as well as on the concepts of direct and correlated response to selection (Lande, 1979, 1980).

The population will evolve directly to the new optimal ontogeny in the absence of genetic correlation. In other words, if gene effects are age specific, so that different sets of genes affect behavior at the two ages, the evolution of behavior at age 1 will not affect behavior at age 2. This uncoupled case is illustrated in Figure 9-5. Behavior at age 1 evolves rapidly at first and then slows as it reaches the new optimum, but behavior at age 2 remains at its optimum. Genetic coupling between behavior at the two ages will change this picture of independent evolution.

Genetic correlation will cause behavior at age 2 to evolve temporarily away from its optimum. This effect occurs because genetic correlation causes selection on behavior at age 1 to induce a correlated response in behavior at age 2. If the genetic correlation is positive, as in Figure 9-3a, the ontogeny will evolve in two phases, as shown in Figure 9-6. In the first phase (Figure 9-6a), behavior at age 1 rapidly evolves to higher scores, toward the new optimum. Behavior at age 2 also evolves toward higher scores (at a slower rate than behavior at age 1), away from its optimum. During the second evolutionary phase (Figure 9-6b), behavior at age 1 con-

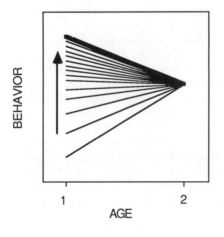

**Figure 9-5** Evolution of a simple behavioral ontogeny when there is no genetic correlation between behavior at ages 1 and 2. The ancestral behavioral ontogeny is indicated by the lowermost line. If a new selective regime prevails that favors a markedly increased value for behavior at age 1 but the same behavior at age 2 (heavy line), the behavioral ontogeny will gradually evolve toward that optimum without any change in behavior at age 2.

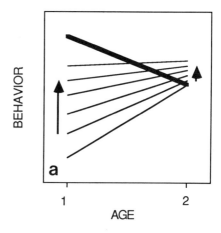

**Figure 9-6** Evolution of a simple behavioral ontogeny when there is a strong positive genetic correlation between behavior at ages 1 and 2. As in Figure 9-5, a new selective regime has been imposed so that the new optimum ontogeny is one in which high behavior scores are expressed at age 1 and low scores at age 2 (heavy line). (a) During the first evolutionary phase, behavior at both ages 1 and 2 increases in value. Behavior at age 2 evolves as a correlated response to selection on behavior at age 1. (b) During the second evolutionary phase, behavior at age 1 continues to evolve upward, toward the optimum, and behavior at age 2 evolves downward, toward the optimum.

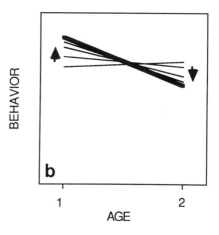

tinues to evolve upward, decelerating as it reaches its optimum. Behavior at age 2 reverses its evolutionary direction and gradually returns to its optimum.

Another visualization of the two evolutionary phases is shown in Figure 9-7. Here we plot the average ontogeny of the population each generation as a point in a two-dimensional space. The population evolves directly to the new optimum along a straight trajectory in the absence of genetic correlation. Genetic coupling between behaviors at the two ages causes the population to evolve along a curved trajectory. During the initial phase (to the left of the dotted line), the behavior at age 2 evolves away from its optimum, in a maladaptive direction. During the second phase (to the right of the dotted line), behavior at age 2 evolves back toward its optimum.

Selection acting at a specific age can produce evolutionary reverberations along the whole length of the ontogeny. The reverberations will be greatest at ages adjacent to the age(s) at which selection acts. For example, we consider first just the long-term correlated responses of behaviors that are expressed before and after the age that experiences strong directional selection. In Figure 9-8 we consider environmental

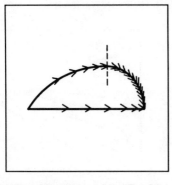

**Figure 9-7** Effect of genetic correlation on evolutionary trajectory. The straight line shows the evolutionary path to a new optimum consisting of a higher behavior score at age 1 and an unchanged score at age 2 when there is no genetic correlation between behaviors at ages 1 and 2. When there is a strong positive genetic correlation, the population evolves toward the new optimum along the curved path. The arrow heads are spaced at approximately equal intervals in number of elapsed generations.

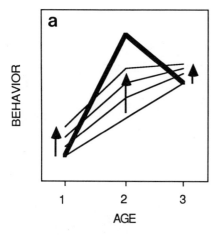

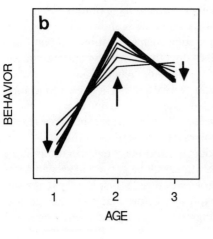

**Figure 9-8** Evolution of a behavioral ontogeny toward a new optimum when selection acts directly on behavior only at an intermediate age. The new optimum is indicated by a heavy line and is characterized by the highest behavioral score occurring at an intermediate age. (a) During the first evolutionary phase, behavior at all three ages evolves toward higher scores, with behavior at ages 1 and 2 evolving away from optimal behavior. The ancestral behavioral ontogeny is indicated by the lowermost line. (b) During the final evolutionary phase, behavior at all three ages evolves toward optimal behavior.

change that favors a new optimal ontogeny. That new optimum differs from the old one only in that high scores are now favored at the intermediate age. The population will evolve directly to the new ontogeny if there is no genetic coupling between behaviors at the three ages. However, the evolution of behavior at the intermediate age will induce evolution of behavior at the other two ages if the behaviors are genetically coupled. Figure 9-8 illustrates the case of positive genetic correlations between behaviors. As before, two evolutionary phases can be recognized: an initial phase of maladaptive evolution by behaviors at the two ends of the ontogeny (Figure 9-8a) and a final phase of approach to the optimum (Figure 9-8b). If we had considered behaviors much earlier and much later than ages 1 and 3, they would have shown less extreme departures from their optima. Maladaptive reverberations would have been less at ages further from the age experiencing selection because of our expectation that genetic correlations will be smaller between behaviors expressed at more disparate ages. The smaller the genetic correlation, the smaller the correlated response to selection (consult Figure 9-3).

In nature an environmental change is unlikely to induce selection at a specific age only. Nevertheless, by focusing on this extreme case we can visualize the consequences of selection pressures that act differently at different ages. As in the extreme case we considered, the evolutionary effect will be to cause temporary maladaptation of the ontogeny.

## Genetic Studies of Morphological and Behavioral Ontogeny

Because the morphological literature, while not large, is more extensive than the behavioral, it will be informative for us briefly to review the principal conclusions arising from the morphological studies of ontogeny. First, the heritability of a particular character often varies with the age at which the character is measured. For example, in house mice, the subjects of much of the developmental genetic work, heritability may increase or decrease with age, show a maximum at an intermediate age, or remain roughly constant during ontogeny, depending on the trait (Cheverud, Rutledge, & Atchley, 1983a; Rutledge, Robison, Eisen, & Legates, 1972). Likewise, genetic correlations can change with age and show age profiles that differ from one pair of traits to another (Leamy & Cheverud, 1984; Riska & Atchley, 1985; Riska, Rutledge, & Atchley, 1985).

Second, genetic correlations between morphological traits expressed at different ages are usually highest for traits expressed at adjoining ages. Genetic correlation usually decreases gradually from high positive values as the interval between the two age-specific traits increases (Cheverud, Leamy, Atchely, & Rutledge, 1983b; Leamy & Cheverud, 1984). Likewise, traits that are homologous or contiguous in space often show strong positive correlations among individuals (Bader & Hall, 1960; Berg, 1960; Kurtén, 1953; Lande, 1980b; Olson & Miller, 1958; Pearson, 1903). In all these cases, strong correlations probably reflect pleiotropy (shared gene effects) (Lande, 1980b).

The same trends—heritability changing with age and high genetic correlation between the same trait expressed at adjoining ages—can be seen in the behavioral literature, although the data base is not so large. For example, heritability is constant or increases with age in various behavioral scores for humans (Plomin, 1986; Eaves, Long, & Heath, 1986). Estimates of genetic correlation across ages are scarce because most behavior genetic studies have been cross-sectional rather than longitudinal (Plomin, 1986). Longitudinal studies in house mice, however, have revealed high genetic correlations between activity level at different ages, even between widely different ages (Henderson, 1986).

## Meaning and Consequences of Ontogenetic Trends in Genetic Parameters

Developmental change in genetic variances and covariances has been viewed from two perspectives: as a reflection of history of selection on the trait(s) or as given with focus on their evolutionary consequences.

Genetic parameters may reflect the history of selection, but the reflection may be distorted by the effects of other processes. Thus one may argue that characters that have experienced a history of strong directional selection will show low heritabilities. Directional selection will tend to cause fixation of genes with additive effects, and so will gradually erode genetic variance for the trait (Falconer, 1981; Fisher, 1958). Broad comparisons of the heritabilities of different traits tend to support this interpretation (Broadhurst, 1979; Falconer, 1981). The heritabilities of fitness components (e.g., fertility) tend to be lower than the heritabilities of traits less directly related to fitness (Falconer, 1981). Nevertheless, the argument should be used cautiously. Recent work indicates that genetic variation of polygenic traits may be appreciably restored each generation by mutation (Lande, 1975, 1980b, 1984; Turelli, 1984, 1985). Consequently, a trait that shows low genetic variance may have experienced strong selection or may be less prone than other traits to mutational input because, for example, it is affected by fewer loci. Furthermore, in virtually all heritability comparisons used to test the selection history hypothesis, differences in selection are simply inferred rather than measured (but see Gustafson, 1986). Fitness components undoubtedly experience stronger directional selection than most behavioral traits, but most trait comparisons are not so unambiguous. Henderson (1986) reviewed other qualifications of the method and illustrated how it can be cautiously applied to behavioral data. Empirical verification of the selection history hypothesis awaits simultaneous estimates of both selection (Endler, 1986; Lande & Arnold, 1983) and genetic architecture for a variety of traits in a spectrum of taxa.

Genetic parameters can be used to predict immediate responses to selection and perhaps even long-term evolutionary response, but prediction of very long-term consequences is problematic. Embracing this outlook, the second school focuses on the evolutionary consequences of genetic variances and covariances instead of using them to deduce selection that acted in the past. The departure point for this viewpoint is Lande's (1975, 1980b, 1984) result that genetic variances and covariances may evolve to an equilibrium under the opposing action of selection, mutation, and other forces. Once the inheritance parameters have achieved equilibrium, the average behavioral ontogeny or some other array of trait means could continue to evolve (Lande, 1976, 1979). A key but defensible assumption in this view is that the behavioral ontogeny experiences only weak selection, so that it evolves gradually and the genetic parameters are not jarred from their equilibrium values. Some comparative data support the supposition of evolutionary constancy in genetic parameters, at least on the time scale required for the differentiation of geographical races (Arnold, 1981a,b; Atchley, Rutledge, & Cowley, 1981; Ayres & Arnold, 1983; Lofsvold, 1986).

Over vast stretches of evolutionary time, the genetic parameters themselves will evolve and the supposition of constancy will no longer hold. A current challenge is to determine the time scale over which constancy is likely to hold.

Even without invoking long-term constancy of genetic parameters, we can use the results of developmental studies of morphological and behavioral genetics to predict important short-term consequences (e.g., Cheverud et al., 1983a). Thus the result that a particular behavior is likely to show a high genetic correlation when expressed at adjacent ages has immediate evolutionary consequences. To visualize those consequences, it will be helpful to consider some behavioral examples.

## PROSPECTS FOR ANALYZING THE INHERITANCE AND EVOLUTION OF BEHAVIORAL ONTOGENIES

Ontogenetic change in the niche, alternative male mating tactics, and bird song share several common denominators that make these behaviors particularly promising subjects for genetic analysis. First, the ecological setting for all three types of behavior is well understood, so that the actual agents of selection are accessible to the investigator. Often when behaviors are studied only in the laboratory, the selection pressures confronting the population in nature are unknown or only dimly perceived. Without an ecological reference, the investigation can easily drift away from substantive evolutionary issues. Second, in each case comparative studies have revealed the aspects of behavior that are most prone to evolutionary change. By comparing songs in different bird species, for example, we can determine which aspects of song are most variable among species and so identify the aspects most liable to evolutionary modification. Such a diagnosis can be used to focus genetic studies on recurrent evolutionary themes. Third, all three types of behavior have a strong developmental dimension. Patterns of resource use, mating tactics, and bird song can all show pronounced change during ontogeny. Perhaps because of this developmental complexity, these behaviors share the fourth characteristic: virtual absence of genetic analysis. The developmental dimension seems to have thwarted even the first stage of conceptualizing the relevant genetic issues. The literature on each of these three topics has often bogged down in the retrogressive pattern of treating ontogenetic modification and inheritance as alternatives. How can we move forward?

### Evolution of Ontogenetic Shifts in the Niche

Werner and Gilliam (1984) emphasized the point that the ecological niche has an ontogenetic dimension and stressed its importance for population dynamics and species interactions. Very commonly the pattern of resource utilization changes with age and size, so that juveniles exploit different foods and face different hazards and competitors than do adults. Many fish, for example, progress through several abrupt shifts in diet during ontogeny. To illustrate the genetic phenomena that may play a role in the evolution of such niche shifts, I will focus on a simpler example, the striped swamp snake.

The striped swamp snake *(Regina alleni)* is a small, aquatic natricine snake commonly residing in the floating water hyacinths found in Florida and southern Georgia. Juveniles feed predominantly on the aquatic larvae (naiads) of damselflies, but adults shift over to a diet of recently molted, soft crayfish (Godley, 1980). The dietary shift is accompanied by a change in prey-handling behavior. Crayfish are swallowed tail first but damselfly naiads are swallowed head first (Franz, 1977; Godley, 1980).

Comparative studies suggest that damselfly feeding is a more recent evolutionary innovation than crayfish feeding. The ancestral *Regina* was undoubtedly a crayfish feeder, since all four extant species in the genus have this habit (Rossman, 1963) and no other North American natricine snake regularly feeds on these prey. *R. alleni* is the only species in the genus that also preys on damselflies. A secondary addition of damselflies to the diet is also suggested by the fact that both juvenile and adult *R. alleni* will readily feed on crayfish in the laboratory. Godley (1980) suggested that the feeding propensities of juveniles are broad, so that the diet varies with seasonal availability of prey. Small crayfish are rare during the seasons when *R. alleni* juveniles are growing to maturity, but damselfly naiads are common.

The ontogenetic diet shift in *R. alleni* may reflect an ontogenetic change in chemoreceptive response to prey. In other species of *Regina,* naive newborn snakes give an active tongue-flicking response to crayfish odors but not to prey that do not occur in the diet (Burghardt, 1968). The flicking tongue transports odiferous particles to a chemosensor, the Jacobson's organ, in the roof of the mouth (Burghardt & Pruitt, 1975). The ontogeny of chemoreceptive responses to prey have not been studied in *R. alleni* or in any other *Regina,* but the ontogenetic dimension has been studied in a related natricine snake. The red-bellied water snake *(Nerodia erythrogaster)* shifts from a juvenile diet of fish to an adult diet of frogs (Mushinsky, Hebrand & Vodopich, 1982). This snake shows no ontogenetic change in chemoreceptive response to fish, but chemoreceptive response to frog odor increases suddenly as snakes reach maturity. Remarkably, the ontogenetic patterns for responses to fish and frog odor were unaffected by rearing snakes on fish versus frogs (Mushinsky & Lotz, 1980).

In other natricine snakes, populations show heritable variation in the chemoreceptive prey responses of naive juveniles. Unfortunately no genetic studies have been conducted with *Regina* or *Nerodia*. In related garter snakes *(Thamnophis),* analyses of full-sib families have revealed heritable variation in responses to the odors of a number of different prey (Arnold, 1981a,b). Those studies suggest that the chemoreceptive responses can respond to selection, and indeed, population differences in chemoreceptive response correspond to population differences in diet (Arnold, 1981a).

Returning to our damselfly-eating *Regina,* for which phylogenetic inferences are the strongest, we can try to view evolution of the ontogenetic niche from a genetic perspective. Selection for juvenile predation on damselflies in the ancestral crayfish-eating population of *R. alleni* may have produced a maladaptive evolutionary phase characterized by increased adult predation on damselflies. If, for the sake of concrete discussion, we take chemoreceptive response to damselflies as a possible target for selection, then we imagine selection favoring increased juvenile responsiveness to damselflies, but no corresponding selection on adults (juveniles but not adults are faced with a food shortage). If juvenile responsiveness were genetically coupled to adult responsiveness, as seems possible, adults would experience a correlated response to selection as juveniles. Over a period of many generations, adult response to damselflies would increase and then decrease, as ontogenetic differentiation in chemoreceptive response was gradually achieved (see Figure 9-6).

This example of genetic influence on the evolution of niche ontogeny is necessarily hypothetical because critical ontogenetic and genetic information is missing. Nevertheless, a quantitative genetic perspective raises issues not previously addressed in the ecological literature. Furthermore, by making geographic comparisons within *R. alleni* we might catch populations still engaged in the final slow stages of ontogenetic differentiation and so test the prediction of temporary maladaptation.

## Evolution of Alternative Male Mating Tactics

In a wide variety of animals, males use two or more tactics to gain mating success (Austad & Howard, 1984; Dunbar, 1985). Often these alternative tactics represent a contrast between offensive and defensive maneuvers in male-male competition for mates (Arnold, 1976). For example, the primary tactic may be to guard the female from rivals, while a secondary, offensive tactic may be to steal the female away from another guarding male. Alternatives such as these may involve polymorphism (in which males show persistent individual differences in behavior), facultative shifts in tactics (tactics vary with circumstances), and ontogenetic change in tactics. These are not mutually exclusive; in some of the most thorough studies, polymorphism, facultative shifts, and ontogeny change occur in the same population. Here I will focus on one such case, North American bass and sunfish. Caro and Bateson (1986) have recently emphasized the ontogenetic dimension of alternative tactics. I hope to extend their discussion by concentrating on the genetic issues.

Comparative studies by Gross and his colleagues highlight some behavioral transitions that must have occurred during the evolution of mating systems in North American sunfish and their relatives (Figure 9-9). Rock bass *(Amblopistes rupestris)* males use only a single mating tactic (Gross & Nowell, 1980). They build shallow cuplike nests in the shallow water along lake margins and defend them. Females visit these nests, spawn, and immediately depart. Males tend the eggs until they hatch and defend them against predators and egg-eating conspecifics.

Pumpkinseed sunfish *(Lepomis gibbosus)* males facultatively shift between two tactics, and employment also varies with age (Gross, 1982). Larger, older males build and defend nests, as do rock bass, but they will also sneak into the nests of a male in an adjoining territory while that male attempts to spawn with a female. The sneaking male releases sperm over the spawning couple, fertilizing some of the eggs. The sneaker immediately returns to his territory and the rival rears his offspring. Younger, smaller males use the sneaking tactic. Perhaps they also use the nest-building parental tactic on rare occasions when nest sites are available.

The mating system of male bluegill sunfish *(Lepomis macrochirus)* consists of two ontogenetic pathways as well as facultative shifts in tactics (Gross, 1982). Some males mature relatively late in life. These large males adopt the parental nest-building tactic at about 7 years or age and use it consistently for three or four years, until they die. These parental males will also use the sneaking tactic on occasion when a rival is spawning in an adjoining territory. Other males in the bluegill population mature at an earlier age. These smaller males reach sexual maturity at 2 years of age and adopt

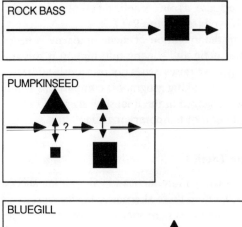

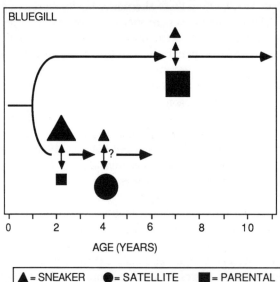

**Figure 9-9** Ontogeny of alternative male mating tactics in rock bass, pumpkinseed sunfish, and bluegill.

sneaking as their primary tactic. The tactic is facultative, for these small, young males will build and defend nests when population density is low and nest sites are available. Later, at the age of 4 years, the small males have grown to the size of mature females and adopt a new, so-called satellite mating tactic. Satellite males defend territories in mid-water above the territories of the larger, parental males. Satellite males do not build nests. Instead, they slowly move into the nests of spawning, parental males and steal fertilizations. Facultative shifts between sneaking and satellite tactics have been observed in males that grow along this second ontogenetic pathway have been observed.

In summary, four kinds of evolutionary change are needed to accout for the differences among just these three centrachid species. For the sake of concrete discussion, I will consider the mating system of the bluegill to be the most derived. One could equally well consider the following changes as occurring in the opposite direction (e.g., substituting loss of tactics for gain of tactics). Additional comparative work would be needed to settle the issue of evolutionary polarity; the present discussion relies only on the necessity of change that is implied by the species differences. The needed transitions are (1) evolution of an earlier onset of sexual maturity, (2) evolution of a facultative shift to new tactics (sneaking, satellite behavior), (3) evolution of ontogenetic changes in facultative shifts between tactics, and (4) bifurcation in the ontogeny of behavior and morphology. By focusing on these elements of evolution, we can see some of the genetic issues.

Consider, for example, the evolution of a new mating tactic, sneaking. In one of the ontogenetic pathways in bluegill, the tactic is expressed in young males, and to a much lesser extent in older males. It is conceivable that older males are simply not exposed to the environmental circumstances that elicit sneaking. It is also possible that young males are more prone to sneak than older males, given the same eliciting stimuli. In this case we must consider the evolution of age-specific differences in the expression of sneaking. From our preceding discussion of genetic correlation, we can anticipate that selection for sneaking in young males would have caused a correlated evolutionary response in older males as well as in the young targets of selection. The critical genetic parameter giving the effect is a genetic correlation between sneaking expressed at early and later ages. Following our earlier theoretical discussion we would expect the elaboration of age-specific differences in behavior to have been a time-consuming process perhaps lasting many thousands of generations.

The evolution of the bifurcation in ontogeny seen in bluegills probably involved some additional genetic phenomena. Let us explore, for example, the possibility that the bifurcation represents a genetic polymorphism. In the simplest case, the inheritance of alternative ontogenies could behave as simple Mendelian inheritance. Such inheritance is known for many polymorphisms, including the primary sex differences and polymorphisms involving coloration in many organisms. In such cases it is unlikely that the polymorphism evolves by saltation, with mutation to a gene that instantaneously produces all the differences associated with the present-day dimorphism. Instead, the polymorphism is probably built up gradually by selection for numerous genes that modify and augment the effects of a gene that acts as a developmental switch (Fisher, 1958, pp. 181–185). A good working hypothesis for the bluegill case is that the bifurcation was gradually elaborated from some ancestral ontogeny so that the present dimorphism behaves as a Mendelian factor. Under this scenario we would predict that populations of bluegill, as well as closely related species, might show variations in the completeness of developmental bifurcation.

## Inheritance and the Evolution of Birdsong

Comparative studies of birdsong have revealed tremendous variation in repertoire size among species (Krebs & Kroodsma, 1980; Marler, 1981). Individuals in some species may sing a single type of song, whereas individuals of other species may sing over 100 different songs. At one extreme, the average repertoire of an individual marsh warbler *(Acrocephalus palustris)* includes imitations of songs from many species of birds; the species as a whole is known to imitate the songs of 212 other species (Dowsett-Lemaire, 1979). Differences in age-specific tendency to incorporate new songs into the repertoire are an important component of species variation in repertoire size. Some species learn songs only during a critical period, whereas other species add songs to their repertoire throughout life.

Two lines of evidence indicate that repertoire size affects male mating success (Krebs & Kroodsma, 1980). In red-winged blackbirds *(Agelaius phoenicus),* males with larger repertoires have larger harems of females, but the correlation is apparently due to common correlation age (Yasukawa, Blank, & Patterson, 1980). In mockingbirds *(Mimus polyglottos),* males with larger repertoires breed earlier in the year, although the effect may be due to common correlation with territory quality (Howard, 1974). In the best study of fitness effects, McGregor, Krebs, and Perrins (1981) found that great tits *(Parus major)* with intermediate repertoire sizes left the most descendants. In this species repertoire size does not change with age. Kroodsma (1976), in an experimental study with canaries, found that repertoire size affected female nest building. While some of these results are equivocal, it seems likely that sexual selection exerted by mate choice has been an important force in producing diversification of repertoire size.

Comparative and experimental studies have identified onset and length of a sensitive learning period, tendency to incorporate songs into the repertoire during the sensitive period, and the size of the pool of tutors as key variables in the ontogeny and diversification of repertoires (Marler, 1981). Auditory input and feedback are critical for song development.

Males deafened early in life produce only the rudimentary elements of song, and birds deprived of tutors or model songs develop only simplified versions of species-characteristic songs. In some species songs are learned during a sensitive period lasting only a few weeks, but in others incorporation of new songs continues throughout life. The bird is not a *tabula rasa* during the sensitive period. Many song birds, for example, selectively learn the songs of conspecifics and reject the songs of even closely related species. Finally, the timing of the sensitive period in relation to dispersal can greatly affect the adult's repertoire size. The males of some species learn songs from their fathers. In other species the sensitive period does not begin until after the young male has left his natal territory. In such species the male may be exposed to a large pool of different conspecific songs.

To understand how repertoire size evolves, we need genetic studies of the developmental and learning processes. To date only one genetic study has focused on the key issues, and that study gave an unexpected result. McGregor et al. (1981) were unable to detect any father-son resemblance in the repertoire sizes of great tits. Is this an anomalous result (the sample size was small) or is there typically a low or even zero father-son resemblance in repertoire size? It may be that song characteristics are transmitted from one generation to the next by cultural channels that do not involve lines of descent. McGregor et al.'s (1981) results notwithstanding, let us consider some different models of transmission that could produce father-son resemblance in repertoire size, to see the genetic issues more clearly (Figure 9-10).

A nongenetic conception of song and repertoire transmission is implicit in much of the bird-song literature. Two varieties of this idea are shown in Figure 9-10. In the first model, sons do not learn songs from their fathers. Nevertheless sons resemble their fathers because they inherit or take over the father's territory and are conse-

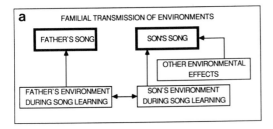
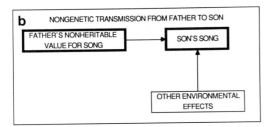
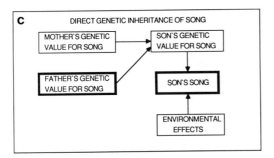
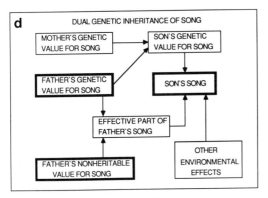

**Figure 9–10** Four models for father-son resemblance in songbirds. Causal paths are indicated by single-headed arrows. Correlations are indicated by double-headed arrows.

quently exposed to a common set of tutors (Figure 9–10a). Alternatively, sons might resemble their fathers because they learn their father's songs (Figure 9–10b). As before, the transmission of songs and repertoire is nongenetic. Cavalli-Sforza and Feldman (1981) and Boyd and Richerson (1985) have modeled and discussed these and other modes of purely cultural transmission.

Genetic transmission of songs is both conceivable and compatible with song learning. For example, the son might inherit a genetic propensity to sing a particular song or to develop a large repertoire. (Such genetic propensities could be transmitted both by fathers, who express the tendencies, and by mothers, who do not. Many sex-limited traits are affected by autosomal genes). Genetic transmission could promote father-son resemblance even though sons do not learn their fathers' songs and instead use unrelated tutors (Figure 9–10c).

Finally, father-son resemblance in repertoire could arise from dual inheritance in which the son inherits his father's propensities and also learns the father's songs. Models of dual inheritance have been used by quantitative geneticists for many years to analyze maternally affected traits in mammals (e.g., offspring body size) (Cheverud, 1984; Dickerson, 1947; Riska et al., 1985; Willham, 1963). A simplified version of the scheme is shown in Figure 9–10d. The key feature in this model is that the aspects of the father's song that are effective in teaching the son may themselves be heritable. Consequently father-son resemblance arises from two genetic routes: direct genetic effects on singing propensities and the genetic part of the father's singing.

Controlled breeding and manipulative experiments will be needed to distinguish between these and other models of song transmission. For example, cross-fostering could be used to determine whether tutoring by the father augments father-son resemblance. To cleanly estimate the many genetic parameters in the dual-inheritance model one needs a correspondingly large number of different kinds of relatives (e.g., paternal half sibs, full sibs, double first cousins, etc.) (Eisen, 1967; Willham, 1963) or a combination breeding and cross-fostering design (Atchley, Plummer, & Riska, 1985, Table 2). For example, it would be particularly useful to have many sets of paternal half-sib brothers and to know their song characteristics and those of their sons.

Correlation between direct and paternal genetic effects in the dual-inheritance model is of particular evolutionary interest. The important postulate of that model is that an individual's phenotypic value for song includes two genetic parts. One part acts directly in the individual, perhaps by affecting that individual's responsiveness to particular song types. The other part acts through the paternal song tutor. This paternal genetic part of song will influence the tutoring effect of the individual's song upon his sons. A particular gene may have effects on both of these genetic parts of song phenotype. The aggregate effect of many such pleiotropic genes can be to produce a genetic coupling or correlation between the two parts of song. The resulting genetic correlation might be positive or negative. If the correlation is positive, selection favoring high values for one or the other part of song will augment the evolution of the other part. But if the correlation is negative, evolution may be counter-intuitive (Cheverud, 1984; Dickerson, 1947). Selection for high values of one part of song could override selection for high values of another part and cause that part to evolve toward lower values. Thus, selection acting on the tutoring aspect of song could interfere with the evolution of the student part of song learning.

## CONCLUSIONS

Ontogeny is a dimension of the phenotype and genotype rather than an alternative to inheritance. To confront the key issues of how behavioral ontogenies are inherited and how they evolve, it is essential to see inheritance and ontogeny as complementary phenomena rather than as alternative explanations.

Quantitative genetics is a natural tool for analyzing the inheritance of behavioral ontogenies. The analytic machinery of this discipline is designed to deal with traits whose genetic variation is affected by many genes of individually small effect. Such genetic variation is common for morphological traits and probably for behaviors as well. In addition, a tradition of genetic analysis of morphological ontogenies provides a series of useful conceptual guideposts for the behaviorist. For example, the simple step of defining age-specific behaviors can make the ontogenetic dimension manageable.

Another important lesson from morphological studies is that longitudinal data are crucially important. Selection acting on ontogeny works on individual variation in ontogenetic trajectories, so it is crucial to describe that variation in order to understand selection. From a genetic standpoint one cannot compute genetic covariances between behaviors expressed at different ages unless individuals or families are followed through time and their behaviors scored at regular intervals.

To make incisive evolutionary studies of behavioral ontogeny, the feasibility of ecological work is as important as the feasibility of genetic analysis. An important criterion for genetic work is that individuals of known relationship can be identified in the field or bred in the laboratory. To test evolutionary hypotheses, we have the additional criterion that the behavior and its ontogenetic dimension can be placed in an ecological context. At a minimum we need observations that suggest how the behavior functions in a natural context. Ideally, we want careful demographic work that enables us to evaluate the fitness consequences of behavioral and ontogenetic variation. Thus an ideal behavior is one that can be observed in the field, is performed by an organism amenable to genetic work, and occurs in populations in which selection pressures can be directly measured.

Comparative studies can be of critical importance in identifying which aspects of behavioral ontogeny are most prone to evolutionary modification. Of the examples explored in this chapter, we have by far the best grasp of important evolutionary issues in the case of birdsong. Here the evolutionary issues are apparent because of in-depth studies of developmental processes and an abundance of comparative studies. From these two sources, length of the sensitive period for song learning emerges as a key variable, one that is particularly prone to evolution. In the cases of niche ontogeny and alternative mating tactics, the identification of key variables is more speculative because comparative studies of behavioral ontogeny are rare.

Genetic studies of facultative shifts in behavior are greatly needed. Behavioral shifts are particularly evident in studies of alternative mating tactics. Particular males commonly change their mating tactics according to circumstances: guarding their mate when another male intrudes, playing the role of intruder when they themselves are not guarding. By slowly changing the probabilities of using one or the other tactic under particular circumstances, we could easily model the evolution of complex systems of tactics. Unfortunately, we lack concrete data on the inheritance of facultative shifts. As a first step we need to know whether there is phenotypic variation in the parameters of behavioral plasticity in natural populations.

Genetic correlations are of crucial importance to the evolution of behavioral ontogenies, and yet they are seldom discussed outside the quantitative genetic literature. The importance of heritability for evolution is now routinely acknowledged outside the technical literature, some 50 years after the concept was first employed in animal breeding (Lush, 1937). The importance of genetic correlations was first stressed by Hazel in 1943, and such correlations have been employed ever since in schemes maximizing gains from deliberate selection on domesticated species (Falconer, 1981). The evolutionary importance of genetic correlations has recently been explored by Lande in a series of papers (Lande, 1979, 1980a, 1981, 1982; Via & Lande, 1985). Despite this long history of application and the recent surge in evolutionary inquiry, behaviorists and psychologists have tended to overlook the importance of genetic correlation. The first estimates of genetic correlations for behavioral traits were published only within the last two decades (DeFries & Hegmann, 1970; Hegmann & DeFries, 1970a,b; Henderson, 1972).

The present review has revealed two general situations in which genetic correlations are likely to play an important role in behavioral evolution. First, genetic correlation between the same behavior expressed at different ages is likely to impede the evolution of age-specific behaviors. Such correlations are likely to slow both the process of ontogenetic differentiation of the niche and the elaboration of ontogenetic differences in male mating tactics, for example. Second, in systems of dual inheritance, genetic correlation between direct effects on behavior and effects mediated through learning can cause counterintuitive evolutionary effects. A dual system of inheritance is conceivable for singing in birds and in many other behavior circumstances, and yet formal behavior genetic tests of the model have not been attempted. These two situations are examples of the general rule that genetic correlations cause curved evolutionary trajectories and promote temporary maladaptation (Lande, 1980a).

## ACKNOWLEDGMENTS

I am grateful to James Cheverud, Mart Gross, Russell Lande, and Patrick Phillips for stimulating discussions. The preparation of this manuscript was supported by National Science Foundation grant BSR 85-06766 and National Institutes of Health grant 1 RO1-GM-35492-01.

## REFERENCES

Arnold, S. J. (1976). Sexual behavior, sexual interference and sexual defense in the salamanders *Ambystoma maculatum, Ambystoma tigrinum* and *Plethodon jordani. Zeitschrift für Tierpsychologie, 42,* 247–300.

Arnold, S. J. (1981a). Behavioral variation in natural populations. I. Phenotypic, genetic and environmental correlations between chemoreceptive responses to prey in the garter snake, *Thamnophis elegans. Evolution, 35,* 489–509.

Arnold, S. J. (1981b). The microevolution of feeding behavior. In A. Kamil and T. Sargent (Eds.), *Foraging Behavior: Ecological, Ethological and Psychological Approaches* (pp. 409–453). New York: Garland STPM Press.

Atchley, W. R., Plummer, A. A., & Riska, B. (1985). Genetics of mandible form in the mouse. *Genetics, 111,* 555–577.

Atchley, W. R., Rutledge, J. J., & Cowley, D. E. (1981). Genetic components of size and shape. II. Multivariate covariance patterns in the rat and mouse skull. *Evolution, 35,* 1037–1055.

Austad, S. N., & Howard, R. D. (1984). Introduction to the symposium: Alternative reproductive tactics. *American Zoologist, 24,* 307–308.

Ayres, F. A., & Arnold, S. J. (1983). Behavioural variation in natural populations. IV. Mendelian models and heritability of a feeding response in the garter snake, *Thamnophis elegans. Hereditary, 51,* 405–413.

Bader, R. S., & Hall, J. S. (1960). Osteometric variation and function in bats. *Evolution, 14,* 8–17.

Berg, R. (1960). The ecological significance of correlation pleiades. *Evolution, 14,* 171–180.

Boyd, R., & Richerson, P. J. (1985). *Culture and the Evolutionary Process.* Chicago: University of Chicago Press.

Broadhurst, P. L. (1979). The experimental approach to behavioral evolution. In J. R. Royce & L. P. Mos *(Eds.) Theoretical Advances in Behavior Genetics* (pp. 43–95). Alphen van den Rijn, Netherlands: Sijthoff and Noordhoff.

Burghardt, G. M. (1968). Chemical preference: Studies on newborn snakes of three synpatric species of *Natrix. Copeia, 1968,* 732–737.

Burghardt, G. M., & Pruitt, C. H. (1975). Role of the tongue and senses in feeding of naive and experienced garter snakes. *Physiology and Behavior, 14,* 185–194.

Caro, T. M., & Bateson, P. (1986). Organization and ontogeny of alternative tactics. *Animal Behavior, 34,* 1483–1499.

Cavalli-Sforza, L. L., & Feldman, M. W. (1981). *Cultural Transmission and Evolution: A Quantitative Approach.* Princeton, NJ: Princeton University Press.

Cheverud, J. M. (1984). Evolution by kin selection: A quantitative genetic model illustrated by maternal performance in mice. *Evolution, 38,* 766–777.

Cheverud, J. M., Rutledge, J. J., & Atchley, W. R. (1983a). Quantitative genetics of development: Genetic correlations among age-specific traits and the evolution of ontogeny. *Evolution, 37,* 895.

Cheverud, J. M., Leamy, L., Atchley, W. R., & Rutledge, J. J. (1983b). Quantitative genetics and the evolution of ontogeny. I. Ontogenetic changes in quantitative genetic variance components in random bred mice. *Genetical Research, Cambridge, 42,* 65–75.

DeFries, J. C., & Hegmann, J. P. (1970). Genetic analysis of open-field behavior. In G. Lindzey & D. D. Thiessen (Eds.), *Contributions to Behavior-Genetic Analysis: The Mouse as a Prototype* (pp. 23–56). New York: Appleton-Century-Crofts.

DeFries, J. C., & Fulker, D. W. (1986). Multivariate behavioral genetics and development: An overview. *Behavioral Genetics, 16,* 1–10.

Dickerson, G. E. (1947). Composition of hog carcasses as influenced by heritable differences in rate and economy of gain. Ames, IA: Iowa Agricultural Experimental Station Research Bulletin 354, pp. 489–524.

Dowsett-Lemaire, F. (1979). The imitative range of the song of the marsh warbler *Acrocephalus palustris,* with special reference to imitations of African birds. *Ibis, 121,* 453–468.

Dunbar, R. I. M. (1985). Intraspecific variations in mating strategy. In P. P. G. Bateson & P. Klopfer (Eds.), *Perspectives in Ethology* (pp. 385–431). New York: Plenum Press.

Eaves, L. J., Long, J., & Heath, A. C. (1986). A theory of developmental change in quantitative phenotypes applied to cognitive development. *Behavioral Genetics, 16,* 143–162.

Eisen, E. J. (1967). Mating designs for estimating direct and maternal genetic variances and maternal covariances. *Canadian Journal of Genetics and Cytology, 9,* 13–22.

Endler, J. (1986). *Natural Selection in the Wild.* Princeton, NJ: Princeton University Press.

Falconer, D. S. (1981). *Introduction to Quantitative Genetics* (2nd ed.). London: Longman.

Fisher, R. A. (1918). The correlation between relatives on the supposition of Mendelian inheritance. *Transactions of the Royal Society of Edinburgh, 52,* 399–433.

Fisher, R. A. (1958). *The Genetical Theory of Natural Selection* (2nd ed.). New York: Dover.

Franz, R. (1977). Observations on the food, feeding behavior, and parasites of the striped swamp snake, *Regina alleni. Herpetologica, 33,* 91–94.

Godley, J. S. (1980). Foraying ecology of the striped swamp snake, *Regina alleni,* in southern Florida. *Ecological Monographs, 50,* 411–436.

Gross, M. R. (1982). Sneakers, satellites and parentals: Polymorphic mating strategies in North American sunfishes. *Zeitschrift für Tierpsychologie, 60,* 1–26.

Gross, M. R., & Nowell, W. A. (1980). The reproductive biology of rock bass, *Ambloplistes rupestris* (Centrarchidae), in Lake Opinicon, Ontario. *Copeia, 1980,* 482–494.

Gustafson, L. (1986). Lifetime reproductive success and heritability: Empirical support for Fisher's fundamental theorem. *American Naturalist, 128,* 761–764.

Hazel, L. N. (1943). The genetic basis of constructing selection indexes. *Genetics, 28,* 476–490.

Hegmann, J. P., & DeFries, J. C. (1970a). Are genetic correlations and environment correlations correlated? *Nature, 226,* 284–286.

Hegmann, J. P., & DeFries, J. D. (1970b). Maximum variance linear combinations from phenotypic, genetic and environmental covariance matrices. *Multivariate Behavioral Research, 5,* 9–18.

Henderson, N. D. (1972). Relative effects of early rearing environment and genotype on discrimination learning in house mice. *Journal of Comparative and Physiological Psychology, 79,* 243–253.

Henderson, N. D. (1986). Predicting relationships between psychological constructs and genetic characters: An analysis of changing genetic influences on activity in mice. *Behavioral Genetics, 16,* 201–220.

Howard, R. D. (1974). The influence of sexual selection and interspecific competition on mockingbird song *(Mimus polyglottos). Evolution, 28,* 428–438.

Krebs, J. R., & Kroodsma, D. E. (1980). Repertoires and geographical variation in birdsong. *Advances in the Study of Behavior, 2,* 143–177.

Kroodsma, D. E. (1976). Reproductive development in a female songbird: Differential stimulation by quality of male song. *Science, 192,* 574–575.

Kurtén, B. (1953). On the variation and population dynamics of fossil and recent mammal populations. *Acta Zoologica Fennica, 76,* 1–122.

Lande, R. (1975). The maintenance of genetic variability in a polygenic character with linked loci. *Genetical Research, Cambridge, 26,* 22–235.

Lande, R. (1976). Natural selection and random genetic drift in phenotypic evolution. *Evolution, 30,* 314–334.

Lande, R. (1979). Quantitative genetic analysis of multivariate evolution, applied to brain: body size allometry. *Evolution, 33,* 402–416.

Lande, R. (1980a). Sexual dimorphism, sexual selection and adaptation in polygenic characters. *Evolution, 34,* 292–305.

Lande, R. (1980b). The genetic covariance between characters maintained by pleiotropic mutations. *Genetics, 94,* 203–215.

Lande, R. (1982). A quantitative genetic theory of life history evolution. *Ecology, 63,* 607–615.

Lande, R. (1984). The genetic correlation between characters maintained by selection, linkage and inbreeding. *Genetical Research, Cambridge, 44,* 309–320.

Lande, R., & Arnold, S. J. (1983). The measurement of selection on correlated characters. *Evolution, 37,* 1210–1226.

Leamy, L., & Cheverud, J. M. (1984). Quantitative genetics and the evolution of ontogeny. II. Genetic and environmental correlations among age-specific characters in random bred house mice. *Growth, 48,* 339–353.

Lofsvold, D. (1986). Quantitative genetics of morphological differentiation in *Peromyscus.* I. Tests of the homogeneity of genetic covariance structure among species and subspecies. *Evolution, 40,* 559–573.

Lush, J. L. (1937), *Animal Breeding Plans.* Ames, IA: Iowa State University Press.

Marler, P. (1981). Birdsong: The acquisition of a learned motor skill. *Trends in Neuroscience, 3,* 88–94.

Maynard Smith, J. (1982). *Evolution and the Theory of Games.* Cambridge, U.K.: Cambridge University Press.

McGregor, P. K., Krebs, J. R., & Perrins, C. M. (1981). Song repertoires and reproductive success in the great tit. *American Naturalist, 188,* 149–159.

Mushinsky, H. R., Hebrand, J. J., & Vodopich, D. S. (1982). Ontogeny of water snake foraging ecology. *Ecology, 63,* 1624–1629.

Mushinsky, H. R., & Lotz, K. H. (1980). Chemoreceptive responses of two sympatric water snakes to extracts of commonly ingested species: Ontogenetic and ecological considerations. *Journal of Chemical Ecology, 6,* 523–535.

Olson, E. C., & Miller, R. L. (1958). *Morphological integration.* Chicago: University of Chicago Press.

Pearson, K. (1903). On the influence of natural selection on the variability and correlation of organs. *Philosophical Transactions of the Royal Society of London, A200,* 1–66.

Plomin, R. (1983). Developmental behavioral genetics. *Child Development, 54,* 253–259.

Plomin, R. (1986). Multivariate analysis and developmental behavioral genetics: Developmental change as well as continuity. *Behavioral Genetics, 16,* 25–43.

Riska, B., & Atchley, W. R. (1985). Genetics of growth predict patterns of brain-size evolution. *Science, 229,* 668–671.

Riska, B., Rutledge, J. J. & Atchley, W. R. (1985). Covariance between direct and maternal genetic effects in mice, with a model of persistent environmental influences. *Genetical Research, Cambridge, 45,* 287–297.

Rossman, D. A. (1963). Relationships and taxonomic status of the North American natricine snake genera *Liodytes, Regina,* and *Clonophis. Occasional Papers of the Museum of Zoology, Louisiana State University, 29,* 1–29.

Rutledge, J. J., Robison, O. W., Eisen, E. J., & Legates, J. E. (1972). Dynamics of genetic and maternal effects in mice. *Journal of Animal Science, 35,* 911–918.

Turelli, M. (1984). Heritable genetic variation via mutation-selection balance: Lerch's zeta meets the abdominal bristle. *Theoretical Population Biology, 25,* 138–193.

Turelli, M. (1985). Effects of pleiotropy on predictions concerning mutation-selection balance for polygenic traits. *Genetics, 111,* 165–195.

Via, S., & Lande, R. (1985). Genotype-environment interaction and the evolution of phenotypic plasticity. *Evolution, 39,* 505–522.

Werner, E. E., & Gilliam, J. F. (1984). The ontogenetic niche and species interactions in size-structured populations. *Annual Review of Ecology and Systematics, 15,* 393–425.

Willham, R. L. (1963). The covariance between relatives for characters composed of components contributed by related individuals. *Biometrics, 19,* 18–27.

Yasukawa, K., Blank, J. L., & Patterson, C. B. (1980). Song repertoires and sexual selection in the red-winged blackbird. *Behavioral Ecology and Sociobiology, 7,* 233–238.

# 10

# Genetic Analysis as a Route to Understanding the Evolution of Animal Behavior: Examples Using the Diallel Cross

NORMAN D. HENDERSON

For over 50 years a body of theoretical and empirical literature has accumulated supporting the view that the genetic architecture underlying a particular trait provides some clues regarding the evolutionary history of the trait. The clues are often limited and somewhat ambiguous and thus must be supplemented by other observations based on nongenetic methods. Nevertheless, the prospect of providing supporting evidence for the existence of a functional relationship between level of expression of a given trait and Darwinian fitness is attractive. The assumption that the mere presence of a behavior pattern or other trait in a population indicates that the trait is a product of natural selection has received much criticism concerning the ease of generating post hoc fitness explanations (e.g., Gould & Lewontin, 1979). A growing collection of such explanations in the literature has prompted some critics to suggest equally plausible arguments for the fitness of behaviors quite opposite to those described and to dismiss post hoc statements of the fitness value of the behavior as "just so stories." It would therefore be desirable to be able to bolster conclusions from field observations and other research with compatible results from a genetic analysis.

## GENETIC ARCHITECTURE AND DARWINIAN FITNESS

### Nonadditive Genetic Variance

Traits that have undergone a history of sustained selection pressure exhibit nonadditive genetic variance, such as genetic dominance or overdominance at one or more loci, or epistasis—the interaction of alleles at different loci (Falconer, 1981). Finding nonadditive gene action contributing to individual differences in a behavior thought to be related to Darwinian fitness constitutes genetic support for the fitness hypothesis. If the presence of nonadditive gene action is established, the magnitude and directionality of such action provides the next level of information concerning selection history. For some behaviors we might expect that, within some reasonable range, fitness is monotonically related to the degree of expression of the behavior—high fitness is associated with high levels of the behavior. For such traits the population mean is below its optimal value for fitness, and although natural selection favors higher scoring individuals, it is unable to increase the mean of the trait. The situation arises when effective selection limits have been reached either because deleterious recessive alleles are being maintained at low frequencies in the population or because alleles at more or less intermediate frequencies are overdominant with respect to the fitness-related behavior. In these cases some of the genotypic variance will be nonadditive, and dominance or epistasis will be directional, favoring higher scoring genotypes.

Behaviors monotonically related to fitness exhibit classic inbreeding depression. This suggests a simple and powerful test for determining if a trait is indeed related to fitness. For such traits, individuals with high coefficients of inbreeding should exhibit decrements in the behavior. This test may be done in natural populations with systematic inbreeding in the laboratory. Usually the reverse breeding experiment is carried out, wherein two or more inbred strains are crossed to produce $F_1$ and perhaps later generations of hybrid offspring. The inbred–hybrid comparisons on the behavior of interest are then used to make inferences about the intensity and direction of selection in an ancestral population from which the genetic material was derived. Inbreeding depression effects on a large variety of behaviors are now well documented, suggesting that these behaviors are more or less monotonically related to fitness (e.g., Ehrman & Parsons, 1981; Fuller & Thompson, 1978). Furthermore, the overall pattern of results makes intuitive sense, with mating and sexual behaviors tending to show the greatest degrees of dominance or overdominance favoring high activity; followed by learning measures, which typically exhibit dominance favoring rapid learning; followed by other behaviors such as locomotor activity and consummatory activity, which often show largely additive genetic variance or highly situation-specific nonadditive effects (e.g., Broadhurst, 1979; Henderson, 1979, 1986).

Sometimes nonadditive variance involving bidirectional genetic influences is found. Both trait-increasing and trait-decreasing dominant alleles exist for the behavior. In these cases some of the genotypic variance is nonadditive but there is little net directional effect so that inbred–hybrid differences are small. This genetic architecture is associated with characters for which intermediate phenotypes close to the population mean have the highest fitness value. The genetic results suggest that the trait is being subjected to stabilizing selection, with intermediates being favored because extreme expression of the phenotype in either direction leads to a fitness decline. The fitness reduction at the two extremes is usually for different reasons, such as in the balance between number of offspring and limitations of food supplies. Stabilizing selection favors genotypes with the least variability and thus tends to fix alleles that confer the greatest developmental stability (Waddington, 1957). Stabilizing selection may also reduce genetic variance by building up balanced combinations of positive and negative alleles, thus preserving genotypic variance while minimizing phenotypic variance (e.g., Lewontin, 1964; Mather, 1941, 1943). Genetic variance may be further reduced through changes in allele frequencies at loci that affect the character.

## Correlated Responses to Selection

Finding bidirectional dominance for a trait does not necessarily mean that what has been measured is functionally related to fitness itself. The behavior or trait may appear to have an intermediate optimum because it is correlated with other characteristics that affect fitness in opposite directions. Falconer (1981), for example, argued that body size is positively correlated with number of offspring produced and negatively related to ability to escape predators, its intermediate optimum thus being the result of directional selection on these two traits. There are also conditions in which the fitness profile for the observed trait may reflect gene pleiotropy. For example, the number of sternopleural bristles in *Drosophila* appear to be under stabilizing selection, yet the apparent selection that gives rise to this intermediate optimum takes place at the larval stages prior to the development of these bristles (Kearsey & Barnes, 1970).

## THE DIALLEL CROSS—SOME WORKED EXAMPLES

The relationships that exist between prior selection pressures and genetic architecture suggest that quantitative genetic approaches to the study of evolutionary history may be viable. Unfortunately, interpretations of genetic outcomes can be misleading and are often based on inconclusive data. To illustrate some of the advantages and limitations of a genetic analysis for understanding the evolutionary history of a behavior, we will examine several possible outcomes from one type of breeding experiment—the diallel cross design. I have chosen to illustrate this comprehensive genetic design in some detail for several reasons: (1) Diallel breeding designs have been used widely with many plant and animal species and many phenotypes, and also in a large number of behavioral studies. (2) Many different forms of analysis have been developed for the diallel; to a large degree these reflect the differing goals and constraints of quantitative genetic procedures. We can thus raise issues applicable to most breeding experiments used to assess the genetic architecture of behavioral traits. (3) Despite the use of diallel data in behavioral research for over 25 years, their potential and interpretive limitations are still not clearly understood, even by some users of the diallel design.

Most important, however, is that the diallel is one of the most useful breeding designs for behavioral research, especially as it pertains to evolution. Unlike selection experiments or designs such as the triple test cross, the diallel is not based on extreme scoring groups in a single behavioral dimension. There are few behavioral constructs for which a single operational definition or measure suffice. Behavioral constructs are manifested in multivariate measurements. To choose subjects on the basis of their performance on a single narrowly defined measure subsequently labeled learning, memory, activity, exploration, feeding, foraging or sexual, maternal, thermoregulatory, agonistic, predatory, or whatever behavior is to succumb to blind operationalism at the expense of biological reality. Unlike classic breeding designs that begin with a pair of inbred strains, the diallel design calls for the genotype to be sampled widely, which increases both the chance that the major alleles influencing the behavior are represented and the likelihood that the genetic sample is representative of

# THE DIALLEL CROSS IN GENETIC ANALYSIS OF ANIMAL BEHAVIOR

some natural breeding ancestral population. The diallel also permits assessment of maternal effects and, to some degree, their interaction with genotype, both useful biological considerations. Finally, the diallel provides an important safeguard in behavioral research that is highly vulnerable to measurement bias—the opportunity for multimethod analysis. A genetic conclusion obtained from diallel cross data is most properly one derived from several converging lines of evidence from several different analyses of the data. Methodological flaws rarely influence all such lines of evidence in a consistent manner.

## The Diallel Matrix

The full diallel design is conceptually simple—a series of $N$ parent strains are crossed in all $N^2$ possible combinations, producing a diallel matrix. The parents may represent heterogeneous stocks or may be highly inbred strains, although use of the latter has advantages in terms of power and the understanding of gene action. The half diallel, which omits reciprocal crosses, is an alternative design that permits wider genetic sampling for the same testing effort as the full diallel, but at the expense of data on maternal and sex-linked effects. Several contrived diallels will be illustrated to convey a sense of the kind of information, and its limitations, that can be obtained in a genetic analyses. For clarity in the examples, genetic factors are assumed to operate in the absence of maternal or other environmental influences.

We begin by constructing a set of hypothetical inbred parent strains that will serve as progenitors of a series of diallel cross matrices. For simplicity we assume that genes at only three loci are involved in the behavior and that only two alleles can exist at each locus—one that increases the expression of the behavior by one half unit $(+)$ and one that decreases expression of the behavior by one half unit $(-)$. Since inbred strains will be homozygous $(++ \text{ or } --)$ at each locus, only eight different strains can exist with respect to these three loci. These eight strains are represented along the diagonal of the diallel matrix in Tables 10–1, 10–2, and 10–4. Strain $S$ is homozygous for $(++)$ alleles at all three loci, strain $T$ possesses $(++)$ alleles at loci $a$ and $b$ and $(--)$ alleles at locus $c$, and so on for the eight possible combinations of alleles.

**Table 10–1** Generation of an 8 × 8 diallel matrix involving additive genetic variance with no dominance (Case I)

|  | Parent strain | | | | | | | | Array statistics | | |
|---|---|---|---|---|---|---|---|---|---|---|---|
|  | S | T | U | V | W | X | Y | Z | Mean | Var | Cov |
| S | 6 | 5 | 5 | 5 | 4 | 4 | 4 | 3 | 4.50 | 0.86 | 1.72 |
| T | 5 | 4 | 4 | 4 | 3 | 3 | 3 | 2 | 3.50 | 0.86 | 1.72 |
| U | 5 | 4 | 4 | 4 | 3 | 3 | 3 | 2 | 3.50 | 0.86 | 1.72 |
| V | 5 | 4 | 4 | 4 | 3 | 3 | 3 | 2 | 3.50 | 0.86 | 1.72 |
| W | 4 | 3 | 3 | 3 | 2 | 2 | 2 | 1 | 2.50 | 0.86 | 1.72 |
| X | 4 | 3 | 3 | 3 | 2 | 2 | 2 | 1 | 2.50 | 0.86 | 1.72 |
| Y | 4 | 3 | 3 | 3 | 2 | 2 | 2 | 1 | 2.50 | 0.86 | 1.72 |
| Z | 3 | 2 | 2 | 2 | 1 | 1 | 1 | 0 | 1.50 | 0.86 | 1.72 |
| Means: | P = 3.00 | $F_1$ = 3.00 | | | | | | | 3.00 | 0.86 | 1.72 |
| Var: | P = 3.43 | $F_1$ = 1.33 | | | | | | | 0.86 | 0 | 0 |

The behavioral phenotype of each parental strain is simply the sum of (+) and (−) effects at each locus, plus a constant of 3.0 to avoid negative numbers in our analysis. The eight inbred parent strains thus have scores on a hypothetical trait ranging from 0 to 6, with a mean of 3.00 and a variance of 3.43.

## Additive Genetic Variance

The eight parent strains are mated to produce all 56 $F_1$ offspring combinations of a diallel cross. Table 10-1 shows the cell means one would obtain if the genetic architecture of the trait studied consisted only of additive genetic variance. In this and subsequent examples there is a symmetry of the cell means above and below the diagonal. In the presence of maternal or paternal effects or other reciprocal $F_1$ effects, the means of the reciprocal pairs of cells, such as $S \times T$ and $T \times S$, would not have been identical as in the examples. Systematic maternal effects related to the strain of the female parent would have produced differences in comparable column and row means. Thus, if the maternal effect of being reared by strain $S$ had been to increase the phenotypic score of offspring, the overall mean of the cell means of row 1 would have been greater than the mean of column 1. In addition to systematic column and row differences created by strain-related maternal influences, differences between specific pairs of reciprocal $F_1$s due to idiosyncratic environmental effects or maternal-by-genotype interactions could have occurred. If any of these reciprocal effects varied differentially as a function of the sex of the offspring, sex linkage might be involved. Such complications are absent in the examples shown. Had they existed, analyses such as those described by Walters and Gale (1977) and by Crusio (1987) could be used.

Table 10-1 shows the idealized means that would have been obtained for each genotype if the genes determining the behavior were acting in an additive manner. In each case the $F_1$ cross between a pair of inbred parent strains falls at the average of the two strains, reproduced along the diagonal of the matrix. The symmetry of a completely additive genetic system extends beyond the cell means. The grand mean of the diallel matrix is identical with the inbred parent mean and the variance of the means of each row (i.e., the array of eight possible offspring of each parent strain) is identical. In each row of the matrix the $F_1$ offspring mean is perfectly correlated with its nonrecurring parent mean, with each unit increase in the nonrecurring parent leading to one-half unit increase in the $F_1$ offspring. The covariance between $F_1$ means and their nonrecurring parent means is thus half the variance of the parental strain means for each array. Finally, the variance of row means is identical to the variance of cell means within each row and equal to one quarter of the inbred strain variance. It is these systematic relationships between row variances and covariances, and parental and $F_1$ means and variances, that form the basis of genetic scaling tests and estimates of genetic components in some analyses of diallel data.

## Directional Dominance

Let us contrast the symmetrical relationships found in the completely additive diallel of Case I with those shown in Case II (Table 10-2). Case II was constructed under the assumption that (+) alleles are always dominant over (−) alleles, producing

# THE DIALLEL CROSS IN GENETIC ANALYSIS OF ANIMAL BEHAVIOR

**Table 10-2** Two diallel matrices showing genetic dominance: Directional dominance favoring high scores (Case II) and additive variance plus bidirectional dominance (Case III)

| | Case II | | | | | | | | | | Case III | | | | | | | | | |
|---|---|---|---|---|---|---|---|---|---|---|---|---|---|---|---|---|---|---|---|---|
| | Parent strain | | | | | | | Array statistics | | | Parent strain | | | | | | | Array statistics | | |
| | S | T | U | V | W | X | Y | Z | Mean | Var | Covar | S T U V W X Y Z | | | | | | | Mean | Var | Covar |
| S | 6 | 6 | 6 | 6 | 6 | 6 | 6 | 6 | 6.00 | 0 | 0 | 6 5 4 6 3 5 4 3 | | | | | | | 4.50 | 1.43 | 1.72 |
| T | 6 | 4 | 6 | 6 | 4 | 4 | 6 | 4 | 5.00 | 1.14 | 1.14 | 5 4 3 5 2 4 3 2 | | | | | | | 3.50 | 1.43 | 1.72 |
| U | 6 | 6 | 4 | 6 | 4 | 6 | 4 | 4 | 5.00 | 1.14 | 1.14 | 4 3 4 4 3 3 4 3 | | | | | | | 3.50 | 0.29 | 0.58 |
| V | 6 | 6 | 6 | 4 | 6 | 4 | 4 | 4 | 5.00 | 1.14 | 1.14 | 6 5 4 4 3 3 2 1 | | | | | | | 3.50 | 2.57 | 2.86 |
| W | 6 | 4 | 4 | 6 | 2 | 4 | 4 | 2 | 4.00 | 2.28 | 2.28 | 3 2 3 3 2 2 3 0 | | | | | | | 2.50 | 0.29 | 0.58 |
| X | 6 | 4 | 6 | 4 | 4 | 2 | 4 | 2 | 4.00 | 2.28 | 2.28 | 5 4 3 3 2 2 1 0 | | | | | | | 2.50 | 2.57 | 2.86 |
| Y | 6 | 6 | 4 | 4 | 4 | 4 | 2 | 2 | 4.00 | 2.28 | 2.28 | 4 3 4 2 3 1 2 1 | | | | | | | 2.50 | 1.43 | 1.72 |
| Z | 6 | 4 | 4 | 4 | 2 | 2 | 2 | 0 | 3.00 | 3.43 | 3.43 | 3 2 3 1 2 0 1 0 | | | | | | | 1.50 | 1.43 | 1.72 |
| Means: | P = 3.00  $F_1$ = 4.71 | | | | | | | | 4.50 | 1.72 | 1.72 | P = 3.00  $F_1$ = 3.00 | | | | | | | 3.00 | 1.43 | 1.72 |
| Var: | P = 3.43  $F_1$ = 1.84 | | | | | | | | 0.86 | 1.11 | 1.11 | P = 3.43  $F_1$ = 1.93 | | | | | | | 0.86 | 0.74 | 0.74 |

complete directional dominance favoring high-scoring phenotypes. The inbred parental scores remain identical to those in Case I but the $F_1$ phenotypes now reflect the dominance favoring (+) alleles. The result is a striking change in some of the statistics of each row, or parental array. The average array covariance remains 1.72, or half the parent strain variance, but the average array variance has now increased from one quarter to one half the parent strain variance and is equal to the average covariance. No longer are the variances of each array identical, nor are the parent-offspring covariances equal across all arrays. Offspring of parent strain S, which possesses dominant alleles at all three loci, are all identical to their S parent; hence there is no variance among the offspring of the S array. Since the phenotype of each $F_1$ of the S parent is identical to the S parent, there is also no correlation between scores of the $F_1$s in the S array and the scores of their nonrecurring T to Z parents. The covariance between $F_1$s of the S array and their nonrecurring parents is thus zero. In contrast to the S parent, strain Z, possesses only (−) alleles that are recessive at each locus. As a result, $F_1$ offspring in the Z array are identical to their nonrecurring S to Y parents, resulting in an array variance equal to that of the inbred parent strains and identical to the covariance of $F_1$s with their nonrecurring parents. Arrays of offspring from parent strains containing a mix of dominant (+) and recessive (−) alleles fall in a predictable manner between those of parent S and parent Z.

## Bidirectional Dominance

In Case III shown on the right in Table 10-2, the mode of gene action has been altered so that the (+) allele is dominant at locus $a$, the (−) allele is dominant at locus $b$, and the alleles act in an additive manner at locus $c$. This is a case of partial bidirectional dominance, with some dominant alleles that increase the phenotype, some dominant alleles that decrease the phenotype, and some additive gene action. Genetic architecture of this type is expected to evolve for traits in which the mean of the natural parent population is more or less at the optimum phenotype. In such populations, balanced combinations of increasing and decreasing dominant alleles may be favored by selection, especially in fluctuating environments. In this way a great deal of genetic diversity can be maintained in the population, which is accessible to natural selection in the case of changing environmental conditions, and yet phenotypic variation is minimized around a current optimum for the population. If selection favored balanced combinations of increasing and decreasing dominants in the natural breeding population, and this gametic phase disequilibrium was maintained during inbreeding, the inbred strains generated would predominantly look like strains $U$, $V$, $W$, and $X$ with balanced $(+)$ and $(-)$ combinations at the $a$ and $b$ loci. Phenotypic variation of the offspring of these four strains is small, ranging only from 2 to 4, despite the existence of maximal genetic diversity at each locus.

Something to note about the case of bidirectional dominance is how misleading and contradictory genetic conclusions would be if they were based on only pairs of strains and their $F_1$ hybrids. Results of the cross between strains $S$ and $T$ suggest intermediate inheritance; the cross between strains $S$ and $U$ suggests dominance favoring the lower scoring parent; the cross between $S$ and $V$ suggests dominance favoring the higher scoring parent; the cross between $S$ and $W$ suggests partial dominance favoring low scorers; the cross between $S$ and $X$ suggests partial dominance favoring high-scoring phenotypes; phenotypically similar strains $T$ and $U$ produce $F_1$s with lower scores than either, suggesting overdominance or epistasis favoring low scores; while the cross of strains $W$ and $Y$ produces the opposite, suggesting overdominance or heterosis for the high-scoring phenotype. Individual pairings among the parent strains in the diallel produce every possible genetic outcome, and subsets of three or four strains drawn from the eight also produce wide variations in the apparent genetic architecture of the character. Such results illustrate problems concerning genetic sampling and correlated gene distributions—issues to which we shall return.

With respect to the array statistics of Case III involving partial dominance, the average array variance falls between the completely additive Case I and the completely dominant Case II. The average array covariance, however, remains equal to half the parental variance. $F_1$ offspring of strains $U$ and $W$ are the least variable in the diallel and have the smallest covariance with their nonrecurring parents, indicating these other parents' small degree of influence on $U$ and $W$ offspring. Strains $U$ and $W$ are the two that possess dominant $(+)$ alleles at locus $a$ and dominant $(-)$ alleles at locus $b$, and thus possess the largest proportion of dominant alleles among the eight parent strains. Strains $S$, $T$, $Y$, and $Z$ each possess dominant alleles at only one locus, and strains $D$ and $X$ possess only recessive alleles at the $a$ and $b$ loci. As in Case II, both the variation of offspring of a parental array and the nonrecurring parent versus offspring covariance reflect the proportion of dominant alleles possessed by the parent strain. Unlike Case II, however, variances and covariances differ within each array.

## Relationships Among Array Statistics

The relationships of array variance ($V_r$) and covariances with nonrecurring parents ($W_r$) are shown in Figure 10-1 for Cases I, II, and III. Case I, which consisted of additive gene action only, produces only a single point, since all arrays had the same $W_r = 2V_r$ values. In Case II, with complete dominance favoring high scores, $W_r$ and

# THE DIALLEL CROSS IN GENETIC ANALYSIS OF ANIMAL BEHAVIOR

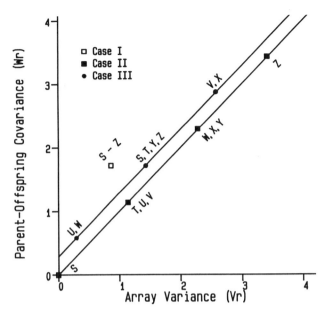

**Figure 10-1** Relationship between array variances and parent-offspring covariances for Cases I, II, and III. In Case I, involving only additive genetic variance, array variances and covariances cluster at a single point, $V_r = \frac{1}{4}$ parent variance, $W_r = \frac{1}{2}$ parent variance. In Cases II and III, involving genetic dominance, $V_r, W_r$ points are distributed on a line of unit slope. Parent strains are ordered along the regression line in terms of relative dominance, with those possessing the most dominant alleles at the lower left.

$V_r$ fall on a straight line passing through the origin with a slope of unity, since $W_r = V_r$ in each array. High-scoring strain $S$, with dominant alleles at all three loci, produced offspring most like itself, thus minimiziing the array $V_r$ and $W_r$. Parent strains possessing fewer dominant alleles systematically appear further out on the $V_r, W_r$ regression line, indicating progressively less genetic influence over their offspring. In Case III, the slope of the regression line is still unity but the intercept has shifted upward, reflecting the decreased variance of the average array due to the decreased degree of dominance. The position of arrays along the regression line again corresponds to the parent strain's relative dominance. Arrays of strains $U$ and $W$, which possess dominant alleles at both $a$ and $b$ loci, are at the low end of the $V_r, W_r$ regression; arrays of the double recessive strains $V$ and $X$ appear at the high end; and arrays of the strains with an equal proportion of dominant and recessive alleles are in the middle of the regression line.

In Cases II and III involving dominance there are also relationships between parental strain means and the location of each offspring array on the $V_r, W_r$ graph. In Case II, the array of the high-scoring strain $S$ is at the origin at the lower left, the array of strain $Z$ with the lowest phenotypic score is at the upper right, and arrays of intermediate scoring strains are located between $S$ and $Z$. A plot of parental means against either $V_r$ or $W_r$ would have resulted in a straight line of negative slope, indicating a perfect negative correlation between parental means and the proportion of recessive alleles possessed by the parent strain. In actual data in which sampling error was involved, we would have plotted parental means against the sum of $V_r$ and $W_r$; this would have provided a better estimate of the relative dominance of a parental strain than either $V_r$ or $W_r$ alone. In Case III, where dominance favored intermediate rather than either high or low expression of the phenotype, a plot of the relationship between parent means and the sum $V_r + W_r$ is curvilinear.

An example involving true overdominance at individual loci would have produced array variances and covariances similar to those in Figure 10–2, where on the average $V_r$ exceeds $W_r$, resulting in a negative intercept. For reference, the common point found in Case I is also shown. Suppose, however, that the parent strains of the diallel were not completely inbred but were selected lines or other "strains" that possessed some residual genetic heterozygosity. Obviously, some of the within-strain variance would be genetic, which would necessitate correction for inbreeding coefficients when using between- to within-strain mean squares to estimate genetic variance. From the standpoint of genetic architecture, the major effect of residual heterozygosity in parent strains is to shift array $W_r$ and $V_r$ values of each strain toward the single $W_r,V_r$ point representing purely additive genetic variance.

If the inbreeding coefficients of the different parent lines were more or less similar, the shift would be approximately equal for all, and the $W_r,V_r$ regression line would retain a slope of unity but would be moved upward to the left, indicating lower genetic dominance than actually existed for the character, as illustrated in Figure 10–2. If the inbreeding coefficient had differed among parent lines, the amount of shift would have varied in proportion to the degree of inbreeding, resulting in scatter

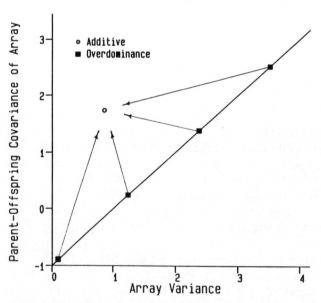

**Figure 10–2** Effect of residual heterozygosity on $V_r,W_r$ relationships in the presence of genetic dominance. The regression line shows the position of strain arrays derived from inbred parents that would be found in the case of overdominance. With increasing residual heterozygosity in parents, $V_r,W_r$ points converge toward the common $V_r,W_r$ point found when only additive genetic dominance is present.

around the $W_r, V_r$ regression line and a slope possibly differing from unity. Although methods exist for incorporating parent inbreeding coefficients into a diallel analysis (e.g., Dickinson & Jinks, 1956), the coefficients are often unknown or are not used, resulting in underestimates of the relative degree of genetic dominance for the character being studied.

## Analyses of Diallel Tables

The analyses of diallel tables take many forms and can be classified in several ways. One distinction can be made between analysis of variance (ANOVA)-based models that partition sums of squares into factors of interest and those based on array characteristics such as those we have been describing. Some ANOVA analyses stop at significance testing; others use expected mean squares to estimate variance components. There are fixed and random models, depending on whether the strains are regarded as a sample drawn from the many possible strains that could be generated from some common ancestral population. Some analyses include the scores of the inbred parent strains; others omit them since inbreds would not exist in the ancestral or base population from which the genetic material was derived. There are also maternal effects and reciprocal effects models, and half diallel designs that exclude reciprocals altogether. Finally, some analyses may be thought of as based on statistical models, where variance components are statistical in nature, and others as based on genetic models, where the analysis attempts more closely to assess gene action.

The merits and limitations of each of the approaches to the analysis of diallel data has been argued and reviewed at length (e.g., Hayman, 1960; Kuehl, Rawlings, & Cockerham, 1968; Singh & Paroda 1984; Wearden, 1964; Wright, 1985). From the standpoint of a behavioral genetic analysis, where the integrity of the data is often corrupted by methodological and measurement problems, the many analytical models available permit a multimethod approach to assess the robustness of genetic interpretations of the data. Assessments of the underlying genetic architecture of a behavioral measure should be consistent across analytical methods if one is to have confidence in conclusions regarding the evolutionary significance of the behavior.

Table 10-3 summarizes the results of two alternative analyses of the diallel data in our three examples. The first analysis, labeled "gene action analysis," is based on the biometrical genetic approach using a fixed model that includes parent strains, as outlined by Hayman (1954a). Summarized below the Hayman analysis are the components derived from a combining abilities ANOVA that excludes data from parent strains, described as Model 3 by Griffing (1956). These two forms of analysis tend to represent the extremes with respect to underlying assumptions and details extracted from a diallel cross. Gene action analysis is demanding with respect to assumptions concerning genetic scaling, linkage equilibrium, and the absence of epistatic interaction, whereas the combining abilities analysis is a relatively robust statistical description of genetic variation expected to exist in the ancestral population. Because ancestral populations would not contain inbred strains, these are omitted from the combining abilities analysis, but at considerable expense with respect to understanding of gene action.

In the gene action analysis, the $D$ term is the variance of the inbred parent strain means and is twice the true additive genetic variance in the genetic material being

**Table 10-3** Estimation of genetic components for Cases I, II, and III[a]

| Genetic component | Case I | Case II | Case III |
|---|---|---|---|
| Gene-action analysis (including parent strains) | | | |
| $D$ | 3.43 | 3.43 | 3.43 |
| $H_1$ | 0 | 3.43 | 2.29 |
| $H_2$ | 0 | 3.43 | 2.29 |
| $h^2$ | 0 | 9.00 | 0 |
| $F$ | 0 | 0 | 0 |
| Dominance ratio $(H_1/D)^{.5}$ | 0 | 1.00 | .82 |
| Proportion of: | | | |
| Increasing alleles | — | .50 | .50 |
| Dominant alleles | — | .50 | .50 |
| Additive variance | 1.72 | 1.72 | 1.72 |
| Dominance variance | 0 | 0.86 | 0.57 |
| Combining-abilities analysis (excluding parent strains) | | | |
| General combining ability | 0.86 | 0.74 | 0.72 |
| Specific combining ability | 0 | 0.69 | 0.80 |
| Additive variance | 1.72 | 1.48 | 1.44 |
| Dominance variance | 0 | 0.69 | 0.80 |

[a]Case I: additive genetic variance, no dominance; Case II: directional dominance favoring high scores; Case III: additive variance plus bidirectional dominance.

studied. $H_1$ and $H_2$ are both estimates of dominance and will be equal if the proportion of dominant (+) and (−) alleles is equal across all loci influencing the trait. In cases of unequal allele frequencies, $H_1$ exceeds $H_2$. The term $h^2$ represents variance associated with mean directional dominance and is simply the square of the difference between inbred parent mean and the overall diallel mean. The final term $F$ reflects the proportion of dominant and recessive alleles influencing the phenotype and takes a positive value if dominant alleles are in excess and a negative value if recessive alleles are in excess. These computations are, of course, simplified by the absence of within-cell environmental variance. The dominance ratio, $(H_1/D)^{.5}$, reflects the proportion of dominant to additive genetic variation, with 0 indicating no dominance, 1 complete dominance, and values greater than 1 indicating overdominance. In Cases II and III, where genetic dominance is involved, estimates of the proportion of increasing (+) and decreasing (−) dominant alleles influencing the trait are given, as are estimates of the proportion of dominant alleles influencing the phenotype. The estimate of additive genetic variance is computed from $D/2$ and dominance variance from $H_1/4$.

It is evident from Table 10-3 that the gene action analysis accurately reflects the underlying gene action built into our diallels for Cases I through III. In Case I only additive genetic variance appears. In Case II dominance variance is half that of additive variance, producing a dominance ratio of 1.0, which indicates complete dominance as constructed in our data. In Case III, mean directional dominance, $h^2$, disappears and the dominance ratio decreases to .82, signifying incomplete dominance, again consistent with the example in which dominance existed only at the $a$ and $b$ loci, with alleles at the $c$ locus acting in an additive manner. In Cases II and III, equal proportions of $(+)$ and $(-)$ dominant alleles are indicated and the proportion of dominant and recessive alleles is shown to be equal, in accord with the data.

The results of the combining abilities analysis and the gene action analysis are directly comparable only in Case I, where additive genetic variance = 2 (general combining ability) = $D/2$. Although providing less information than the gene action analysis, the conclusions from the combining abilities analysis are those one would expect from the three examples—that a substantial portion of genotypic variation in Cases II and III was due to nonadditive sources, while Case I involved only additive genetiv variance. Even with the omission of parental strain data, substantial genetic nonadditivity was detected in Case II. Directional dominance produces more than a simple overall inbred–hybrid mean difference in the diallel—it creates nonadditivity among $F_1$ offspring.

The advantage of a multimethod analytical approach in assessing genetic architecture can be seen in the case of directional dominance by comparing some of the results shown in Tables 10–2 and 10–3 with Figure 10–1. Dominance favoring higher scoring phenotypes produced four distinct outcomes in our analyses: (1) a significant difference in the mean of inbred parent strains and the mean of their $F_1$ hybrid offspring; (2) a regression of offspring array covariances with nonrecurring parents ($W_r$) on array variances ($V_r$) equal to unity; (3) a substantial positive correlation between strain means and the estimated proportion of dominant alleles possessed by the strain, based on array $V_r$s and $W_r$s; (4) evidence of genetic nonadditivity among $F_1$s. Finding less than all four outcomes should reduce confidence in a conclusion of true directional dominance influencing the behavior. A similar concordance of results exists when gene action is purely additive, as in Case I. Combined inbred and combined hybrid means were identical, $W_r$s and $V_r$s were identical in all arrays with $W_r = 2V_r$, and the estimate of additive genetic variance obtained from inbred parents was identical to that obtained from $F_1$ arrays.

## Gene Asymmetry and Correlated Gene Distributions

In Cases I through III, all eight possible allelic combinations were represented in the diallel. There were an equal number of $(+)$ and $(-)$ alleles in each case and the parent strain population was in genetic equilibrium, with no correlation among allelic effects at different loci. Such perfect gene symmetry and gametic phase equilibrium are not often likely to occur in "real life" situations. To begin with, there is sampling error. The chosen inbred strains may not accurately represent the gene pool of the ancestral parent population, thus distorting the proportions of $(+)$ and $(-)$ alleles even when gene symmetry exists. Even more pronounced are spurious gene correlations. The smaller the sample of inbred strains the more likely correlated gene effects across loci will occur as a sampling artifact. But gene asymmetry and correlated gene distributions may also exist in the natural breeding ancestral population as a result of natural selection. Although a polygenic system involving dominance allows much genetic variation to be conserved, natural selection may still decrease the frequency of alleles that push phenotypes away from the population fitness optimum. If natural selection favored higher scores on some character, one might expect to find proportions of $(+)$ alleles to be greater than proportions of $(-)$ alleles at various loci.

Suppose, for example, that in Case II the proportions of (+) to (−) alleles in the ancestral population was actually .6:.4 at each of the three loci, rather than .5:.5. A large number of inbred strains generated from such a population would have produced approximately $.6^3 = 22\%$ of the strains identical to strain S but only $.4^3 = 6\%$ identical to strain Z with respect to loci a, b, and c. In drawing a sample from these strains one would thus be about four times more likely to draw a strain S than a strain Z. Since frequencies of other strains would also vary as a function of the proportion of the increasing and decreasing alleles they possessed, a very large diallel would be needed to have each parent strain represented at its exact expected frequency. For expediency in creating Case IV, it was thus assumed that in the sample of eight strains drawn, strain Z was "missed" and in its place a second inbred strain, identical to strain S at the three loci influencing our hypothetical behavior, was drawn. The resulting diallel outcome is shown in Table 10–4.

With respect to the alleles involved, the effect of dropping strain Z in lieu of an additional strain S is twofold. First, the frequency of (+) to (−) alleles at each locus is now 5:3, or 62.5% (+) and 37.5% (−), rather than 50% of each. Since all (+) alleles are also dominant alleles in Case IV, we also have 62.5% dominant alleles at the three loci. Second, we have created a small correlation of gene effects across loci a and b, since all combinations of (+) and (−) alleles no longer exist with equal frequency at these two loci. By removing the lowest scoring strain from the set we have also increased the inbred mean and reduced the variance among parent strains.

In our last example, also shown in Table 10–4, we mimic a situation in which stabilizing selection has created substantial gametic phase disequilibrium in the ancestral population, causing allelic effects at different loci to become negatively correlated. Since inbred strains generated from such a population are likely to be predominantly like strains U, V, W, and X, with (+) and (−) alleles balancing each other at the a and b loci, we create Case V by replacing extreme scoring strains S and Z with an extra set of V and W strains in our diallel. In doing so we have retained equal proportions of (+) and (−) alleles but have created a substantial negative cor-

**Table 10–4**  Two diallel matrices in which all genotypes are not represented: Directional dominance favoring high scores, strain S replacing strain Z (Case IV) and additive variance plus bidirectional dominance, strains V and W replacing strains S and Z (Case V)

| | Case IV | | | | | Case V | | | | |
|---|---|---|---|---|---|---|---|---|---|---|
| | Parent strain | Array statistics | | | | Parent Strain | Array statistics | | | |
| | S T U V W X Y S | Mean | Var | Covar | | V T U V W X Y W | Mean | Var | Covar |
| S | 6 6 6 6 6 6 6 6 | 6.00 | 0 | 0 | V | 4 5 4 4 3 3 2 3 | 3.50 | 0.86 | 0.86 |
| T | 6 4 6 6 4 4 6 6 | 5.25 | 1.07 | 0.93 | T | 5 4 3 5 2 4 3 2 | 3.50 | 1.43 | 0.86 |
| U | 6 6 4 6 4 6 4 6 | 5.25 | 1.07 | 0.93 | U | 4 3 4 4 3 3 4 3 | 3.50 | 0.29 | 0.29 |
| V | 6 6 6 4 6 4 4 6 | 5.25 | 1.07 | 0.93 | V | 4 5 4 4 3 3 2 3 | 3.50 | 0.86 | 0.86 |
| W | 6 4 4 6 2 4 4 6 | 4.50 | 2.00 | 1.86 | W | 3 2 3 3 2 2 3 2 | 2.50 | 0.29 | 0.29 |
| X | 6 4 6 4 4 2 4 6 | 4.50 | 2.00 | 1.86 | X | 3 4 3 3 2 2 1 2 | 2.50 | 0.86 | 0.86 |
| Y | 6 6 4 4 4 4 2 6 | 4.50 | 2.00 | 1.86 | Y | 2 3 4 2 3 1 2 3 | 2.50 | 0.86 | 0.29 |
| S | 6 6 6 6 6 6 6 6 | 6.00 | 0 | 0 | W | 3 2 3 3 2 2 3 2 | 2.50 | 0.29 | 0.29 |
| Means: | P = 3.75  $F_1$ = 2.36 | 5.16 | 1.15 | 1.04 | | P = 3.00  $F_1$ = 3.00 | 3.00 | 0.71 | 0.57 |
| Var: | P = 2.79  $F_1$ = 0.91 | 0.39 | 0.69 | 0.60 | | P = 1.14  $F_1$ = 0.89 | 0.29 | 0.16 | 0.09 |

erlation or covariance of gene effects at the $a$ and $b$ loci, with balancing (+) and (−) combinations occurring in six of the eight strains of the diallel. A pronounced effect of this change is the substantial reduction in additive genetic variance, even though genetic variability is maintained at all three loci. This reduction in variance is to be expected, since the total contribution to genetic variance from alleles at two loci is $V_a + V_b + 2COV_{ab}$.

The extreme correlated gene distributions of this last example are unlikely to be encountered in the laboratory, even if the natural breeding parent population is in severe gametic phase disequilibrium. Random mating in the laboratory quickly reestablishes equilibrium in a population unless the loci are closely linked (Falconer, 1981). With two unlinked loci, each generation of random mating halves the amount of disequilibrium and the approach to equilibrium is increasing faster with increasing numbers of loci. When many loci influence the trait, equilibrium is approached quickly with random mating even when linkage is fairly strong (Weir & Cockerham, 1979). Thus, for practical purposes, correlated gene distributions are more likely to result from inadequate sampling of parental genotypes in the breeding design than from residual gametic phase disequilibrium in the ancestral population (Nassar, 1965; Kuehl et al., 1968).

The effect of the modest and severe correlated gene distributions of cases IV and V on array variances and parent-offspring covariances can be seen in Figures 10-3 and 10-4, respectively. No longer are array variances and covariances perfectly correlated with unit regression of $W_r$ or $V_r$. In Case IV, where the correlation of gene effects of the $a$ and $b$ loci is slight, the $W_r,V_r$ regression deviates only slightly from

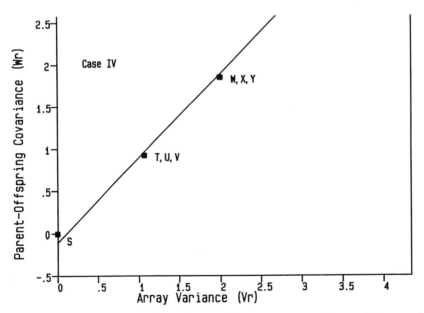

**Figure 10-3** Relationship between $V_r$ and $W_r$ in Case IV, where a slight correlation of gene distributions was created by replacing strain Z with an additional strain S. The slope of the $W_r,V_r$ regression is slightly less than unity, but position along the regression line still corresponds closely to relative dominance of the parent strains.

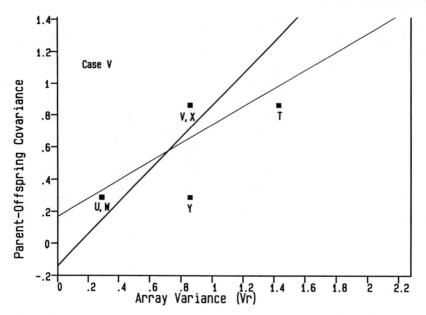

**Figure 10-4** Relationship between $V_r$ and $W_r$ in Case V, which involved a substantial negative correlation of dominant alleles influencing the trait. The slope of the $W_r, V_r$ regression (light line) differs substantially from unity, resulting in an inflated estimate of dominance. The position of a strain array along a unit regression line no longer corresponds closely to the relative proportion of dominant alleles possessed by the parent strain.

unity. In Case V, with a large negative correlation of effects at the $a$ and $b$ loci, the deviation from unit slope is substantial and the relative position of strain arrays along the regression no longer corresponds exactly to the relative dominance of the parent strains. The lack of constancy of $W_r - V_r$ across arrays tests indicates failure of one or both assumptions of an additive-dominance model—independent action of nonallelic genes and the independent distribution of alleles among the parents. In our Cases IV and V, failure was due to gene correlations, but we could as easily have created an example involving epistatic (nonallelic) gene interactions, which would have produced similar distortions, reducing the $W_r, V_r$ correlation and regression to less than unity and in most cases inflating subsequent estimates of dominance.

Summaries of some of the components obtained from the gene action and combining abilities analyses are shown in Table 10-5. In Case IV we replaced the lowest scoring strain Z with another high-scoring strain S. In addition to the reduction in D, reflecting additive genetic variance, we find the two estimates of dominance differing, with $H_2$ falling below the value of $H_1$, indicating unequal proportions of (+) and (−) dominant alleles. The estimate of the proportion of (+) alleles is .62, precisely the 5:3 ratio built in to our example. The results also indicate that dominant alleles influencing the trait outnumber recessives by about 5:3. Because in our example dominant alleles wre synonymous with increasing (+) alleles, the proportions of increasing and dominant alleles are identical. In cases where dominance is not unidirectional, these two estimates would differ.

**Table 10–5** Estimation of genetic components for Cases IV and V[a]

| Genetic component | Case IV | Case V |
|---|---|---|
| Gene-action analysis (including parent strains) | | |
| $D$ | 2.79 | 1.14 |
| $H_1$ | 3.22 | 1.71 |
| $H_2$ | 3.04 | 1.71 |
| $h^2$ | 7.91 | 0 |
| $F$ | 1.40 | 0 |
| Dominance ratio $(H_1/D)^{.5}$ | 1.07 | 1.22 |
| Proportion of | | |
|   Increasing alleles | 0.62 | 0.50 |
|   Dominant alleles | 0.62 | 0.50 |
| Biometrical genetic estimates | | |
|   Additive variance | 1.39 | 0.57 |
|   Dominance variance | 0.80 | 0.43 |
| Random mating estimates | | |
|   Additive variance | 0.78 | 0.57 |
|   Dominance variance | 0.76 | 0.43 |
| Combining abilities analysis (excluding parent strains) | | |
| General combining ability | 0.21 | 0.19 |
| Specific combining ability | 0.57 | 0.60 |
| Additive variance | 0.42 | 0.37 |
| Dominance variance | 0.57 | 0.60 |

[a]Case IV: Directional dominance favoring high score, strain $S$ replacing strain $Z$; Case V: additive variance plus bidirectional dominance, strains $V$ and $W$ replacing strains $S$ and $Z$.

Because components of genetic variance in random mating populations are dependent on gene frequencies of dominant alleles (e.g., Falconer, 1981, p. 118; Mather & Jinks, 1982, p. 216), we must make a distinction between the *random mating* or *statistically additive* forms of these genetic components and the *biometrical* or *genetic* forms of these variance components. Only when the frequency of (+) and (−) dominant alleles is equal across loci are the random mating and biometrical estimates identical, a situation that does not exist in Case IV. Confusion concerning the random mating and biometrical forms of estimates is common, yet the distinction between them typifies the distinction between the attempt to provide a statistical description of genetic variance in a base or ancestral population and the estimation of genetic components more closely related to gene action in a specific sample. Additive genetic variance in a random mating population is a function of additive gene action, nonadditive gene action, asymmetry of (+) and (−) dominant alleles, and the relative proportions of dominant and recessive alleles influencing the character. The random mating form of dominance variance ($H_2$ in Hayman's analysis, sca in Griffing's), therefore, reflects somewhat less than all of the true dominance effect. Hayman also suggested an analysis of variance of diallel tables (Hayman, 1954b) in which his $a$ and $b_3$ terms represent the random mating forms of additive and dominance mean squares, respectively, testing the equivalent of Griffing's gca and sca rather than $D$ and $H_1$.

Although these random mating forms of components are the appropriate estimates for the reference population and thus for computing heritabilities of this population, the ratio of these variances does not measure the degree of dominance in terms of gene action. In Case IV, gene asymmetry leads to a considerable reduction in the random mating form of additive variance and to a small reduction in dominance variance. When dominant alleles are in excess they will contribute substantially less to additive genetic variance and somewhat less to dominance variance than when the frequencies of dominants and recessives are equal. Computation of a dominance ratio (2 dom var/add var)$^{.5}$ from these two random mating estimates, no matter whether from the biometrical analysis including parents or the combining abilities analysis excluding patents, results in the incorrect conclusion of substantial overdominance in Case IV. In contrast, the appropriate dominance ratio, $(H_1/D)^{.5}$, in the gene action analysis realistically reflects the complete dominance built into our data of Case IV.

The effects of the severe distortion created by correlated distributions of alleles in Case V can be seen in the final column of Table 10–5. The negative genetic covariance at the $a$ and $b$ loci substantially reduces additive variance and greatly inflates the dominance ratio, suggesting apparent overdominance rather than the partial dominance built into the example. Nevertheless, from a statistical standpoint, random mating estimates from the combining abilities analysis would appropriately describe the genetic variation in the reference population, indicating that broad heritability would be considerably larger than narrow heritability for the trait examined. From the standpoint of gene action, however, the biometrical estimates are of little use because of the failure of the assumption of independent allele distributions revealed in Figure 10–4.

## DEVELOPMENTAL GENETIC DESIGNS

### Genotype × Age Interactions

Diallel and other breeding designs can be replicated across environments or ages to explore possible genotype × environmental interactions (e.g., Henderson 1966, 1970a,b) or age-related changes in the genetic architecture of a particular phenotype (e.g., Henderson 1978, 1981). The latter are particularly useful when there are reasons for expecting that the optimal level of expression of a phenotype should change with development. A developing organism can be thought of as passing through a series of age-specific ecological niches, which often exert different selection pressures on a particular behavior. The genetic architecture underlying the trait at any single developmental stage should reflect the selection pressures exerted on the phenotype at that time, modified by developmental constraints and subsequent phenotypic demands. With sufficient life history knowledge, one should be able to predict general changes in genetic architecture of a behavioral character during development as well as genetic correlations of character expression across ages.

Infant mice, for example, have limited sensory and locomotor abilities and limited energy resources. This would suggest that, when they are removed or fall from the home nest, a combination of low activity and attempts to elicit maternal retrieval would be an optimal strategy. By the second week of life, however, these juveniles can walk in a straight line and follow odor cues. At this stage an optimal strategy following removal from the nest would be rapid return to the nest. At eye opening, a litter of young mice will tentatively leave the nest, but will not yet have developed rapid escape and nest return responses to predators. At this age they exhibit a "popcorn" response to sudden noise or visual stimuli—the litter suddenly jumps in all directions—again probably an optimal strategy in terms of survival. At each of the three ages there appears to be a different optimal fitness strategy with respect to locomotor behavior, low activity changing to rapid nest return as sensory-motor skills develop and progressing to explosive jumping behavior at the time of eye opening and beginning exploration. In each case, genetic analyses have supported the idea that these are optimal strategies during development. The genetic architecture changes from dominance favoring low out-of-nest activity at 4 days of age to dominance favoring rapid nest return at 11 days of age to dominance favoring vigorous jumping behavior at the time of eye opening (Henderson, 1986). At later ages, intermediate activity levels or situation-specific levels often appear optimal from a fitness standpoint. Again, studies of the genetic architecture of locomotor and exploratory behavior are usually consistent with these expectations (Broadhurst, 1979; Henderson, 1979, 1986).

## Cross-sectional and Longitudinal Designs

Cross-sectional developmental designs incorporating a diallel matrix at each age level have several advantages not shared by designs using genetically heterogeneous subjects. Since the identical genotypes are replicated at each age using different individuals, neither genetic architecture nor genetic correlations of the character across ages are confounded by prior measurement effects or environmental correlations. Conversely, if developmental environmental effects and prior experiences are an important consideration for a research problem and measurement contamination is a minor issue, a repeated-measurement longitudinal design provides a powerful assessment of these influences, independent of genetic factors.

## ISSUES IN INTERPRETING GENETIC RESULTS

Following some specific genetic examples, we return to the issue of the value and limitations of such genetic analyses with respect to understanding evolutionary history. Because the relationship between genetic architecture and fitness is clearest in cases involving dominance, overdominance, or epistasis favoring high (or low) phenotypic expression of the behavior, this example is used to raise some issues regarding genetic studies and evolution. A behavioral phenotype exhibiting hybrid vigor may be one that is directly related to fitness, at least within the range of expression observed in the experiment. The strength of such a conclusion hinges on two broad issues concerning the results. The first is methodological—does the inbred–hybrid difference observed represent a true genetic effect involving directional influences in the ancestral population, or is the difference due to an experimental artifact of measurement drift, scaling, or inadequate genetic samples? The second issue concerns the genetic basis of the effect—does the behavior in itself constitute a component of fitness in the sense that the population mean is below optimum due to factors limiting selection, or is the inbreeding depression due to a few deleterious recessive alleles that affect nervous system function, anatomy, and/or physiology, producing widespread behavior deficits?

## Methodological Issues

The issue of measurement drift becomes important in behavioral studies because of the ease of confounding large-scale genetic experiments with time trend effects such as seasonal factors and changes in laboratory procedures and behavior measurement. Resource constraints often result in design compromises such as testing parent strains earlier on the average than hybrid offspring and thereby confounding effects of measurement drift with inbred–hybrid differences. The analyses of the example diallels demonstrated that a multimethod approach can provide several cross checks on the internal validity of such data. Since artifacts such as measurement drift or observer bias rarely mimic the complete pattern of genetic effects, an inbred–hybrid difference without supporting evidence of dominance from other analyses would implicate a methodological rather than a genetic origin. Other compromises (for example, breeding and testing of closely related genetic groups such as reciprocal matings or offspring of a common parent in close temporal proximity relative to other genotypes) can distort genetic components, including measures of gene symmetry, and environmental components, including maternal effects. Careful analysis can again provide clues of bias, such as unusually small reciprocal differences relative to other sources of random variation. Unfortunately, multimethod crosschecks of diallel or other behavioral genetic data are rare in published articles.

Apparent directional dominance can sometimes appear because of an inappropriate measurement scale in a genetic analysis. Two parental strains with phenotypic scores of 1 and 100 are crossed to produce $F_1$ offspring with a score of 10, for example. On the basis of raw scores the data indicate dominance favoring low scores, but on the basis of a log scale they indicate intermediate inheritance. Numerous scaling tests have been devised to test the adequacy of the underlying metric for analysis when multiple genotypes are available (e.g., Cavalli, 1952; Hayman, 1954a; Mather, 1949). Meeting statistical assumptions of a standard ANOVA often also results in a scale adequate for genetic analyses. The transformations frequently make sense on logical grounds as well, particularly when skewed distributions, rates, and proportions are involved. Occasionally various objectives are in conflict with respect to the appropriate metric for the analysis of a behavioral character, including situations where a transformed scale conflicts with a commonsense notion of an underlying metric for fitness. These situations may reflect a lack of knowledge about the character being studied.

## Genetic Sampling and Ancestral Populations

The greatest methodological problems involving a genetic approach to evolutionary questions center around the adequacy of genotypic sampling. Sampling is, in fact, frequently at the center of many issues concerning the appropriate analytical model and interpretation of the outcome of a breeding experiment. It is helpful in this regard to think of a population of organisms as being able to exist in any one of the following seven levels:

1. An ancestral population under natural selection
2. A reference population descended from the ancestral population
3. A genetic sample drawn from the reference population
4. Domesticated stock descended from the genetic sample
5. Controlled genetic lines derived from the domesticated stock
6. Offspring from a breeding experiment using controlled genetic lines
7. A random mating population descended from level 6

Under ideal conditions these seven levels represent various forms of the same population of organisms with respect to gene frequencies. Quantitative genetic data are usually obtained at level 6 with the assumption that the genetic architecture observed is representative of all other levels. Field biologists or psychologists, making observations at level 3 and not involved with genetics, are concerned only that their samples be representative of the reference population at level 2. In agricultural genetics the primary objective is to make predictions from level 6 to expected outcomes of various breeding manipulations carried out in its descendent population at level 7. Inferences concerning evolutionary history, on the other hand, extend backward across several levels to the ancestral population. Unlike the agricultural case, direct verification of evolutionary conclusions is impossible, and ensuring the fidelity of gene frequencies across levels 1 through 6 is extremely difficult. In agricultural genetics the descendent population derives directly from level 6; thus the two levels are likely to be highly similar in terms of gene frequencies. In contrast, using level 6 genetic data to make evolutionary inferences involves the assumption that selection and genetic drift created by sampling bias, domestication, and genetic bottlenecks between levels 2 and 6 has not seriously distorted the genetic picture of the ancestral population.

If the procedures used to create parental lines at level 5 have been sufficient to maintain close genetic similarity to the reference population and if a random sample of the lines is drawn for the breeding experiment, sampling conditions are met that allow the estimation of genetic parameters in the ancestral population. An extensive breeding design such as a diallel analysis would be based on a random effects model and would omit data from inbred strains, since homozygotes would represent an extremely small proportion of a natural breeding population. In contrast to a model in which a fixed set of parent strains is used in the breeding design and inferences concerning genetic architecture are limited to those genotypes, the precision of genetic estimates in random models depends not only on the accuracy of the estimated means of each of the genotypes studied but also on the extent of the genetic sampling involved. Standard errors of genetic components derived from random model diallels are dependent on within-cell sampling errors, the number of parent strains involved, the size of the component being estimated, and often the size of additional genetic components.

Although the complexity of these standard errors makes it difficult to provide a simple relationship between the extent of genetic sampling and precision for most genetic components, with some simplifications we can get a sense of this relationship. Figure 10-5 shows the proportional size of the standard errors of two components

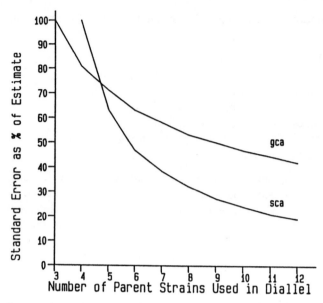

**Figure 10-5** Accuracy of genetic estimates in random model diallel designs as a function of the number of parent strains used. Standard errors of estimates are shown as a percentage of the estimate itself. It is assumed that sample sizes within genotypes are extremely large, resulting in negligible sampling variation of cell means. Standard errors of general combining ability (gca) estimates also assume the absence of specific combining ability (sca). Failure of either assumption results in even larger standard errors for these two components.

estimated from diallels—general and specific combining abilities as a function of the number of parent strains involved. In both cases it is assumed that cell means are estimated without error with infinitely large genotype samples and, in the case of general combining ability (gca), that no specific combining ability (sca) exists so that gca reflects only additive genetic variance. Even under these unrealistically ideal conditions it can be seen that the precision of these genetic estimates is quite low for small breeding designs.

In a 4 × 4 diallel, for example, in the best-case condition of infinite group sizes, standard errors of the gca and sca and related genetic components are at least as large as the estimates themselves. It would require the pooling of data from several replications of a 4 × 4 diallel, each using a different set of inbred parent strains, to obtain even modestly accurate estimates of these two genetic components. An unreplicated 4 × 4 design is thus limited largely to detecting substantial directional dominance or epistasis. For practical purposes, diallels of sizes less than 10 × 10 provide too little precision to be of value in the estimation of the genetic parameters of interest in a reference population. Estimates from moderately sized replicate diallels, such as a pair of 6 × 6 designs each with different parent strains, provide slight gains in efficiency, but in general there is "no free lunch" with respect to obtaining accurate estimates of population parameters using random models. Given the imprecision of estimates of univariate genetic components with limited genetic sampling, it should hardly be surprising that genetic correlations based on pairs of these components have considerable standard errors, making them relatively meaningless estimates of genetic correlations in a reference or an ancestral population.

Under idealized sampling conditions the number of parental genotypes required for accurate estimates of genetic parameters in an ancestral population is considerably larger than that typically found in behavioral research. Furthermore, it is unlikely that sufficient resources will be made available to permit research of necessary scale with mammalian species to obtain worthwhile estimates of genetic components of reference populations. Genetic precision, however, is not the only limiting factor in this endeavor. Another is the myth of the ancestral population. The idealized situation depicted above, in which the genetic material at level 6 is carefully derived from a common naturally breeding ancestral population, is rare in mammalian research, with only a few efforts in this direction (e.g., Connor, 1975; Hewitt & Fulker, 1981). Most behavior genetic research using mice and rats is based on strains or lines from diverse or unknown origins rather than from a common reference population. If selection pressures differed substantially for a trait across the different populations from which parent lines were derived, the genetic architecture obtained from such an experiment would have no meaning with respect to any common natural ancestral population.

Even in cases where lines are generated from a single reference population, evolutionary interpretations can be clouded. If the reference population from which the lines were derived had been under environmental stress as a result of relatively recent colonization of a new habitat, for example, the proportion of additive genetic variance for fitness-related traits could be greater than under optimal conditions (Parsons, 1982). Attempts to estimate genetic parameters of a specific naturally breeding ancestral population are probably best limited to plant and invertebrate species. The success in linking ecological and laboratory behavioral genetic data in *Drosophila*, thereby increasing the understanding of evolutionary patterns in natural populations of this species (Parsons, 1983), is noteworthy in this regard.

## Estimating Population Parmaeters Versus Studying Gene Action

Accurate estimates of genetic parameters of localized mammalian populations under unique selection pressures may be beyond practical reach. Yet, for many behavioral, neurological, and physiological characters, selection pressures may be more or less similar across differing natural populations of a species. To a large extent it is these traits or constructs, whose relationships to fitness transcend specific ecological niches, that are of most interest in psychology and behavioral biology. Genetic drift may cause gene frequencies to differ among isolated natural populations, altering random mating additive and nonadditive genetic variance and heritability estimates derived from these variances. If, however, we acknowledge that most estimates of heritabilities and random mating forms of genetic components are of little use with respect to specific reference populations, this fruitless pursuit can be abandoned in lieu of the strategy of studying gene action using fixed models for clues about evolutionary history. Parenthetically, it should be noted that the use of random factor models has never caught on in the experimental fields of psychology, even though most treatment factors used in these areas fit such models.

The sampling demands of studying gene action in fixed genetic samples are less than those of estimating population parameters in random models. At the same time the analysis is often more informative, as demonstrated in the gene action analyses of the example diallels. True additive genetic variance, $D$, which is independent of the gene frequencies of dominant alleles influencing the trait, and true genetic dominance, $H_1$, which has the same coefficient as $D$, are estimated along with several other genetic components to provide a substantial picture of the underlying genetic architecture of the trait. Even in fixed-model designs, however, reasonable genetic sampling is necessary if there is to be any attempt to generalize beyond the immediate genotypes used.

The smaller the sample of genotypes used the greater the potential for distorting gene action, in part because of the greater sampling variance of small samples. The fewer the number of inbred strains a breeding plan begins with, for example, the more likely that alleles at a given locus will be the same in all strains, preventing genes at this locus from having any influence on genetic variation. Systematic bias is also likely to accompany small samples due to spuriously created gene correlations. Various limited mating combinations in the Case III diallel, for example, produced all types of apparent genetic outcomes, ranging from complete genetic additivity to apparent dominance and overdominance in both directions. In general, the more limited the genetic sampling the more likely distortions due to spuriously correlated gene distributions will occur.

Some breeding experiments begin not with a random set of inbred strains but with a set of partially inbred lines, possibly selected for high and low expression of a genetic character related to the phenotype being studied. As indicated in Figure 10–2, dominance variance is underestimated in the diallel if no adjustment is made for incomplete inbreeding in the parents, with additional complexities occurring when the degree of inbreeding differs among parent strains. Underestimation of dominance among crosses of parent lines with substantial residual homozygosity is often overlooked in data interpretation. There has been a long history of hybridization experiments involving wild and domestic stocks, for example, in which the typical result produced offspring midway between the parent stocks (see Price, 1984, for a summary of these and other data on domestication). The results are not surprising. Even if one of the parent stocks possessed a higher proportion of dominant alleles favoring its phenotype than the other parent stock, genetic heterozygosity would bias the hybrid mean toward a mid-parent value.

Estimates of additive genetic variance reflect the effects of any artificial selection that has occurred in the parental lines that is genetically correlated with the trait being studied. Since artificial selection usually involves producing both high- and low-scoring lines for a given phenotype, inclusion of such lines as parents necessarily increases additive genetic variance. Estimates of additive variation from such experiments thus apply to no reference population, not even one from which selected lines were derived. This situation exists because the variance-increasing effect of the divergence of high and low lines is a function of duration and intensity of selection, as well as of additive genetic variance. The use of uncorrected data from partially inbred artificially selected parental lines thus produces inflated estimates of additive variance, deflated estimates of dominance, exaggerated asymmetry of $(+)$ and $(-)$ dominant alleles, and, sometimes, scaling distortions suggesting epistasis or correlated gene distributions. Although it would be prudent to avoid taking component estimates or dominance ratios too seriously in these situations, useful approximate pictures of gene action sometimes still emerge from such data.

## Determining If Selection Is Acting Directly on a Character

Suppose a trait exhibits a genetic architecture that is consistent across several analyses and the gene action observed suggests a history of prior selection toward a fitness optimum. Genetic sampling has been reasonably wide and there is no reason to believe the results are due to methodological artifacts. The question then remains whether selection has been operating on a behavior closely related to what was measured or on other correlated genetic characters rather different from the investigator's construct. Directional dominance observed for some specific behavior might, for example, be due to low-frequency detrimental recessive alleles affecting the nervous system, to health, or to some other factor that produces widespread behavior decrements. The inbreeding depression observed for a particular behavior may thus be a consequence of selection for a very different character rather than selection favoring high performance on the behavior itself. Similarly, bidirectional dominance may be reflecting opposing directional selection on a pair of characters correlated to the trait being studied, rather than indicating stabilizing selection on the trait itself.

Additive genetic correlations can be a key in determining whether the architecture of a measured behavior is due to selection acting on the behavior itself or to selection acting on correlated or developmental prerequisite characters. In the first case genetic correlations can be expected to be low or negative; in the second case one can expect moderately high positive correlations. When a pair of characters are under simultaneous natural selection, pleiotropic genes affecting both characters in a favorable direction will rapidly be brought to fixation and thus have little influence on genetic variation. In contrast, the pleiotropic genes having opposing fitness effects on the two characters will tend to remain at intermediate frequencies and thus will remain major determinants of genetic variance for the two characters. As a consequence, the additive genetic correlation between the characters is expected to become negative. Since many traits are being acted on simultaneously by natural selection, the average correlation among these characters cannot reach a large negative value. Genetic correlations among fitness-related traits should therefore typically be near zero or be slightly negative. In contrast, if a character under strong selection pressure is a developmental or biological prerequisite for the behavior being studied (e.g., visual acuity required for visual discrimination), the correlation between the studied behavior and its genetic prerequisite will be positive. These correlations can become particularly high when fitness is associated with high trait expression for some environments or ages and with low trait expression for other environments or ages (Henderson, 1986).

The expected low or negative genetic correlations between simultaneously selected characters and expected positive correlations between contingent characters provide a vehicle to help determine the role of natural selection on the behavior of interest. Such correlations require data on fitness-related measures or on characters thought to have widespread influences on other behaviors. Such data are rarely available in a single study, since few studies are truly multivariate. Since genetic correlations can be computed from inbred strain data alone, data from other published studies can sometimes be used for this purpose, particularly with inbred mice, *Mus domesticus*, a species in which substantial inbred strain data have accumulated. As many such data are reported, strain correlations across studies will become an increasingly valuable strategy for studying selection history and for consolidating phenotypic measures. Publishing intermediate statistical analyses of dubious value in lieu of summary descriptive statistics on genotypes tested is to be discouraged.

## SUMMARY

Certain relationships between genetic architecture and evolution seem sufficiently strong to make genetic studies an option for understanding the history of natural selection of behavior. Although powerful breeding designs such as the diallel cross are available, which build in scaling tests and multiple analyses to provide converging evidence for genetic conclusions, the sampling demands required to estimate genetic parameters in ancestral populations are formidable and rarely met in behavioral research. A gene action approach, which avoids dealing with a single ancestral reference population and random mating forms of genetic components, is probably the only viable approach with mammalian populations. Extensive genetic sampling is still a requisite of a gene action approach if sample bias and spurious gene correlations are to be avoided. Even when the genetic architecture observed fits a clear pattern, suggesting that the degree of expression of a behavioral trait may be functionally related to Darwinian fitness, the possibility that selection has operated on a correlated character should be examined. This examination requires a multivariate approach or a substantial inbred strain data base to allow genetic correlations between the behavior of interest and other genetic characters. Despite the failure of most individual behavioral genetic studies concerned with evolutionary history to meet underlying assumptions and methodological requirements, taken as a group the studies in this area and those in ecobehavioral genetics show a reasonably consistent pattern of results.

## REFERENCES

Broadhurst, P. L. (1979). The experimental approach to behavioral evolution. In J. R. Royce & L. P. Mos (Eds.), *Theoretical Advances in Behavioral Genetics* (pp. 43–95). Alphen aan den Rijn, Netherlands: Sjthoff & Noordhoff.

Cavalli, L. L. (1952). An analysis of linkage in quantitative inheritance. In E. C. R. Reeve, & C. H. Waddington (Eds.), *Quantitative Inheritance* (pp. 135–144). London: Her Majesty's Stationery Office.

Connor, J. L. (1975). Genetic mechanisms controlling the domestication of a wild housemouse population *(Mus musculus L). Journal of Comparative and Physiological Psychology, 89*, 118–130.

Crusio, W. E. (1987). A note on the analysis of reciprocal effects in diallel crosses. *Journal of Genetics, 66*, 177–185.

Dickinson, A. G., & Jinks, J. L. (1956). A generalized analysis of diallel crosses. *Genetics, 41*, 65–77.

Ehrman, L., & Parsons, P. A. (1981). *Behavior Genetics and Evolution.* New York: McGraw-Hill.
Falconer, D. S. (1981). *Introduction to Quantitative Genetics* (2nd ed.). London: Longman.
Fuller, J. L., & Thompson, W. R. (1978). *Foundations of Behavior Genetics.* St. Louis, MO: Mosby.
Gould, S. J., & Lewontin, R. C. (1979). The spandrels of San Marco and the Panglossian paradigm: A critique of the adaptionist programme. *Proceedings of the Royal Society of London, 205,* 581-598.
Griffing, B. (1956). Concept of general and specific combining ability in relation to diallel crossing systems. *Australian Journal of Biological Science, 11,* 219-245.
Hayman, B. I. (1954a). The theory and analysis of diallel crosses. *Genetics, 39,* 789-809.
Hayman, B. I. (1954b). The analysis of variance of diallel crosses. *Biometrics, 10,* 235-244.
Hayman, B. I. (1960). The theory and analysis of diallel crosses. III. *Genetics, 45,* 155-172.
Henderson, N. D. (1966). Inheritance of reactivity to experimental manipulation in mice. *Science, 153,* 650-652.
Henderson, N. D. (1970a). Brain weight increases resulting from environmental enrichment: A directional dominance in mice. *Science,* 169, 776-778.
Henderson, N. D. (1970b). Genetic influences on the behavior of mice can be obscured by laboratory rearing. *Journal of Comparative and Physiological Psychology, 72,* 505-511.
Henderson, N. D. (1978). Genetic dominance for low activity in infant mice. *Journal of Comparative and Physiological Psychology, 92,* 118-125.
Henderson, N. D. (1979). Adaptive significance of animal behavior: The role of gene-environment interaction. In J. R. Royce & L. P. Mos (Eds.), *Theoretical Advances in Behavioral Genetics* (pp. 243-287). Alphen aan den Rijn, Netherlands: Sjthoff & Noordhoff.
Henderson, N. D. (1981). Genetic influences on locomotor activity in 11-day-old housemice. *Behavior Genetics, 11,* 109-225.
Henderson, N. D. (1986). Predicting relationships between psychological constructs and genetic characters: An analysis of changing genetic influences on activity in mice. *Behavior Genetics. 16,* 201-220.
Hewitt, J. K., & Fulker, D. W. (1981). Using the triple test cross to investigate the genetics of behavior in wild populations. I. Methodological considerations. *Behavior Genetics, 11,* 23-35.
Kearsey, M. J., & Barnes, B. W. (1970). Variation for metrical characters in *Drosophila* populations. II. Natural selection. *Heredity, 25,* 11-21.
Kuehl, R. O., Rawlings, J. O., & Cockerham, C. C. (1968). Reference populations for diallel experiments. *Biometrics, 24,* 881-901.
Lewontin, R. C. (1964). The interaction of selection and linkage. II. Optimum models. *Genetics, 50,* 757-782.
Mather, K. (1941). Variation and selection of polygenic characters. *Journal of Genetics, 41,* 159-193.
Mather, K. (1943). Polygenic inheritance and natural selection. *Biological Reviews, 18,* 32-64.
Mather, K. (1949). *Biometrical Genetics* (1st ed.). London: Methuen.
Mather, K., & Jinks, J. L. (1982). *Biometrical Genetics* (3rd ed.). Ithaca, NY: Cornell University Press.
Nassar, R. F. (1965). Effect of correlated gene distribution due to sampling on the diallel analysis. *Genetics, 52,* 9-20.
Parsons, P. A. (1982). Adaptive strategies of colonizing animal species. *Biological Review of the Cambridge Philosophical Society, 57,* 117-143.
Parsons, P. A. (1983). Ecobehavioral genetics: Habitats and colonists. *Annual Review of Ecological Systems, 14,* 35-55.

Price, E. O. (1984). Behavioral aspects of animal domestication. *Quarterly Review of Biology, 59*, 1–32.

Singh, O. & Paroda, R. S. (1984). A comparison of different diallel analyses. *Theoretical and Applied Genetics, 67*, 541–545.

Waddington, C. H. (1957). *The Strategy of the Genes.* London: Allen and Unwin.

Walters, D. E., & Gale, J. S. (1977). A note on the Hayman analysis of variance for a full diallel table. *Heredity, 38*, 401–407.

Wearden, S. (1964). Alternative analyses of the diallel cross. *Heredity, 19*, 669–680.

Weir, B. S., & Cockerham, C. C. (1979). Estimation of linkage disequilibrium in randomly mating populations. *Heredity, 42*, 105–111.

Wright, A. J. (1985). Diallel designs, analyses, and reference populations. *Heredity, 54*, 307–311.

# 11

# Changes in Genetic Control during Learning, Development, and Aging

JOHN K. HEWITT

Biometrical genetics has had a major influence on both animal and human behavior genetics. Studies of animal behavior have been given unity through the central idea that the genetic control revealed in crossbreeding experiments can be related to the history of natural selection acting on the trait under study. Biometrical genetics has shown how this genetic control can be analyzed in detail even when the trait is influenced by many genes, each of small effect in relation both to each other and to the effects of environmental influences. Since such polygenic control is typical of most variation that is of interest to students of behavior, biometrical genetics, experimental breeding designs, and related methods of analysis have assumed considerable importance. Much of animal behavior genetics has been directed toward answering evolutionary questions by making inferences from a detailed analysis of the genetic variation for a behavior.

Students of human behavior have been less concerned with evolutionary inferences than with describing contemporary genetic and environmental influences on psychological or biomedical traits and their covariation. Once again, biometrical genetic methods, including path analysis, have been central, because major gene effects on behavior, which can be subject to classic Mendelian pedigree methods, segregation analysis, and linkage studies, make little contribution to phenotypic variation.

In the study of the development of behavior, biometrical genetics will again play an important role alongside other "hardware-oriented" molecular genetic, neurochemical, and physiological methods. In this chapter I illustrate ways in which changes in genetic control during learning, development, and aging might be approached quantitatively. For animals there are good examples of how such changes can be understood in relation to evolution. For humans, where controlled breeding is not an option, we are less able to provide a detailed analysis of the genetics. Nevertheless, progress is being made toward describing developmental processes in ways that are genetically informative. Furthermore, in both animal and human behavior genetics, one outcome of a biometrical analysis is a model for the action of the environment. For the cases I describe here, we will be led to suggest an important difference between how genetic variation and how environmental variation influences development.

## GENETIC ARCHITECTURE OF LEARNING

A general observation in crossbreeding studies of animal learning is that, on average, outbred or hybrid animals outperform their inbred relatives. For example, Broadhurst's (1979) literature compilation showed, as did Whalsten's (1972) earlier review, that the majority of studies report genetic directional dominance or hybrid vigor toward better performance in tasks requiring learning. In this respect, learning to avoid unpleasant stimuli or to find a way through a maze is like rapid and successful mating or rapid and successful escape from a water tank. It is different, however, from the response to a mild stressor like noise, bright lights, or being placed in a novel environment; in other words, from emotional reactivity.

The interpretation of such results relies on the premise that the genetic architecture controlling trait variation reflects the evolutionary pressures that have acted on the trait. This is brought about either by a genuine evolution of the genetic system in Darlington's sense (Darlington, 1939), or by the selective survival of newly arising mutations which de novo display the appropriately adaptive interactions with other alleles or loci. (Hewitt & Broadhurst, 1983; Hewitt, Fulker, & Broadhurst, 1981). Fisher (e.g. 1928, 1958), observing that most mutant alleles are recessive to their wild-type counterparts, argued that the dominance properties of the wild type might themselves have evolved. Thus, although initially a newly arising mutation may in the heterozygous condition result in an intermediate phenotype, any modification of this expression toward that of the fitter homozygote will be favored. If the mutation is recurrent, then eventually modifying genes that give dominance in the direction of the favored wild-type homozygote will become fixed, and directional genetic dominance toward the fitter phenotype will be established. An alternative view (e.g., Wright, 1929) is that only those alleles that produce sufficient gene product as heterozygotes from the outset will be selected and become established as wild-type alleles. Either mechanism results in genetic dominance in the direction favored by natural selection for traits related to fitness. Furthermore, where there is considerable polymorphism, the conditions for the evolution of dominance are optimized (Sheppard, 1967). Mather and his associates have demonstrated that polymorphism is the rule rather than the exception for polygenically controlled traits, that is, traits influenced by many genes each of relatively small effect (Mather, 1941, 1973; Mather & Harrison, 1949; Mather & Jinks, 1983). The theoretical expectation of directional dominance toward higher fitness for such polygenically controlled fitness characters was amply confirmed by experiments on traits such as viability (Breese & Mather, 1960), mating speed and mating success (Fulker, 1966), and litter size (Falconer, 1960). The argument has been extended by Mather (1960, 1973), who shows how, with characters for which intermediate expression is optimal, either little or no dominance variation will be observed or dominance will act at some loci in one direction and at others in the opposite direction. An example of this was provided by Breese and Mather (1960) for bristle numbers in *Drosophila,* which exhibit this kind of ambidirectional dominance; it can be demonstrated in cage populations that maximum fitness in *Drosophila* coincides with intermediate bristle scores (Kearsey & Barnes, 1970).

These arguments suggest that the discovery of directional dominance for a polygenically controlled trait indicates a history of evolutionary pressure toward one extreme of the trait, or that the trait has been linearly correlated with fitness. The absence of dominance or the presence of ambidirectional dominance indicates neutrality or stabilizing selection toward intermediate scores. However, while it may be accepted that known fitness characters are usually controlled by systems displaying dominance in the direction of high fitness, it does not necessarily follow that wherever we find directional dominance we have discovered a phenotype of adaptive importance. There are, of course, numerous major locus conditions that show dominance for the deleterious phenotype. But these are usually very rare or, like Huntingdon's disease, have some special characteristic such as late age of onset that mitigates the effects of natural selection against them. Additionally, were dominance effects occurring haphazardly, then the more genes in the system the less likely is a strong directional effect and the more likely is the production of ambidirectional dominance. Polygenic control is, as Fisher (1958) pointed out, to be expected for characters related to complex adaptations and fitness, because the probability of moving closer to an optimal adaptive point decreases with the size of a random change (mutation) and does so the more rapidly the more dimensions of control or potential change are involved. Hence, many genes of small effect, rather than a few genes of large effect, are likely to be involved in complex adaptations, and this has usually been found to be the case empirically (Mather & Jinks, 1983).

Following these arguments, the inference that has been widely drawn from the kinds of data summarized in Table 11-1 is that in many situations learning well—that is, changing behavior rapidly and appropriately in the face of novel environmental contingencies—is correlated with fitness in rodents (Bruell, 1967) and even in organisms as apparently simple as the fruit fly, *Drosophila melanogaster.* Our own crossbreeding and selection studies of a *Drosophila* learning index, developed by Benzer and his associates (Quinn, Harris, & Benzer, 1974) to screen *Drosophila* for mutants of learning and memory, showed that it had the genetic properties of a fitness-related characteristic: hybrid vigor, asymmetric response to artificial genetic selection, and inbreeding depression (Hewitt, Fulker, & Hewitt, 1983). Subsequently, Hall and his associates (Siegel & Hall, 1979) have identified aspects of the courtship process that may well invoke this learning ability in *Drosophila.*

Table 11-1  Genetic dominance for behavior

| General category of behavior | Number of studies reporting genetic dominance | | |
|---|---|---|---|
| | Toward low performance | Absent or for intermediate performance | Toward high performance |
| Mating | 1 | 2 | 7 |
| Escape from water | 0 | 0 | 6 |
| Emotional reaction to stress | 3 | 7 | 1 |
| Avoidance learning | 0 | 7 | 10 |
| Other learning | 0 | 5 | 9 |

*Source:* Broadhurst, 1979.

## CHANGES IN GENETIC ARCHITECTURE DURING LEARNING AND DEVELOPMENT

Not always is there directional dominance or hybrid vigor toward the response that the psychologist thinks appropriate. The best established case in the behavior genetic literature was first reported by Wilcock and Fulker (1973). Shuttle box escape and avoidance conditioning in rats is studied in a small box with electrifiable grid floors and other stimuli. The rat is placed in the box and at predetermined intervals a warning stimulus—a buzzer or a light—is turned on. This is followed by a foot shock if the rat does not respond to the warning by crossing from one side of the box to the other. Experimental details may vary, but the behavior of the typical rodent will involve several trials on which it does not cross to safety until the foot shock is applied, that is, it escapes, followed by trials in which it avoids the shock by crossing from one side of the box to the other in response to the warning signal. Eventually the animal should be shuttling back and forth under the control of the warning signal or, in psychological parlance, the conditional stimulus.

For this behavior in rats, Wilcock and Fulker (1973) showed that there is an initial significant tendency for genetically homozygous inbred lines to perform better than more genetically heterozygous hybrid crosses; only after 15 to 20 trials does this tendency reverse and genetic dominance for superior performance express itself. The simplest interpretation of this result is that, for a rat, the adaptive defense reaction to being placed in a metal box, hearing a buzzer, and receiving a foot shock is to "freeze." My observations of wild caught rats (Hewitt, 1978) and of wild-laboratory hybrid rats (Hewitt & Fulker, 1984a) suggest that undomesticated rats, on being placed in a brightly illuminated, noisy, open field apparatus, will first engage in a frantic motor discharge, and that this is followed by complete immobility or freezing if no escape has been effected. Once elicited, such a defense reaction would clearly interfere with the active locomotion required by the experimenter for successful avoidance of the foot shock in the shuttle box. After 20 or so trials, the initial defense reaction, which has proved ineffective, is superseded and directional dominance for successful avoidance behavior is observed.

This genetic result has been confirmed numerous times in experiments by the animal behavior genetics group working in Peter Broadhurst's laboratory in Birmingham, England, as well as in those performed by myself and David Fulker at the Institute of Psychiatry, London (Hewitt, 1978; Hewitt & Fulker, 1983, 1984a). Although much of the Birmingham work was not published in detail, a summary of the most important evidence for this phenomenon of directional dominance for high avoidance scores in the shuttle box only in the later trials of conditioning can be found in Hewitt et al. (1981).

Figure 11–1 characterizes the change in directional dominance found in an 8 × 8 diallel cross of inbred and selected lines of rats (Wilcock & Fulker, 1973) and in a triple test cross of wild and laboratory rats (Hewitt & Fulker, 1981, 1983; Hewitt et al., 1981). In the triple test cross, the interpretation of results for the early trials may be complicated by the effects of genotype × environment interaction (Hewitt, 1980),

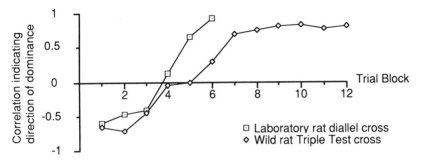

**Figure 11-1** Direction of dominance for avoidance conditioning performance in rats. (From Hewitt, Fulker, & Hewitt, 1981.)

but there is no doubt that for both laboratory and wild samples of rats, directional dominance for high performance in the escape-avoidance shuttle box task emerges only late in the sequence of conditioning. Mice may show a different defense reaction in such situations (Henderson, 1964, 1968).

This example demonstrates one way in which the genetic control of performance may change during the course of a learning sequence, a telescoped episode of development, as different components of the overall behavior pattern are invoked. The changes observed in directional genetic dominance can, in this case at least, be related quite plausibly to the effect of natural selection.

Another example of change in the genetic architecture of phenotypic variation during development was provided by Henderson's (1978) studies of the genetic control of activity in mice. Spontaneous activity in adult male mice typically shows little or no directional dominance or dominance toward higher levels of activity. But if, according to Henderson (1978):

> young [mice] are removed from the nest, a high rate of locomotion should be maladaptive. Activity is as likely to result in the animal's moving farther from the nest as in its moving toward it [since it has extremely limited visual and auditory sensitivity and its locomotor coordination is poor]. [This] increases the probability of attack by predators, including male mice from other nests. More adaptive would be low activity and dependence on the highly efficient maternal retrieval response. [p. 119]

As predicted, an extensive crossbreeding study demonstrated that in 4-day-old mice there is directional dominance for low activity, with hybrid $F_1$ animals (resulting from the intercrossing of inbred lines) moving significantly less than their inbred counterparts. Indeed in this particular series of studies, Henderson was able to show that wild caught mice locomoted even less under the same conditions. This suggests that, as one might expect, the selection pressures on this behavior have been relaxed during domestication. Nevertheless we have another illustration of a change during development in genetic control, and in particular genetic dominance effects, that can be related to natural selection pressures acting differently at different stages of development.

## INCREASES IN HERITABILITY DURING LEARNING

Although directional genetic dominance effects are of considerable theoretical interest in relating phenotypic variation to evolution, in terms of describing the total phenotypic variance they make only a small contribution. [We may note in passing that this is one reason for the relatively elaborate crossbreeding schemes that have to be devised by biometrical geneticists to detect this and other forms of nonadditive genetic variation (Mather & Jinks, 1983).] More important are the additive genetic effects that determine the narrow heritability and, together with dominance effects and other nonadditive genetic influences, the broad heritability (Hewitt, 1987). If we look again at phenotypic variation and its contributions from genes and the environment, what we find in experiments involving learning can once more be instructive.

In a shuttle box escape-avoidance conditioning task, a trial consists of a warning stimulus (e.g., a buzzer) followed by a foot shock if the animal (usually a rat or a mouse) has not avoided it by running from one side of the box to the other following the onset of the warning stimulus. One way to summarize how well the animal is performing is to consider blocks of five trials at a time and score the number in each five in which the animal successfully avoided the foot shock. Typical performance curves for three different genotypes are shown in Figure 11-2.

When considering phenotypic variability we must recognize that with scores bounded by 0 and 5 we expect some scalar interactions such that when average scores are low (usually early in the conditioning sequence) and when they are high there will be an attenuated variance from "floor" and "ceiling" effects. But this does not

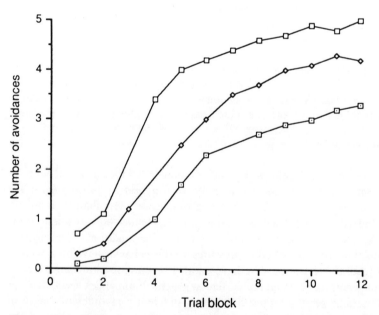

**Figure 11-2** Typical avoidance performance curves for three different genotypes. (From Hewitt, 1978.)

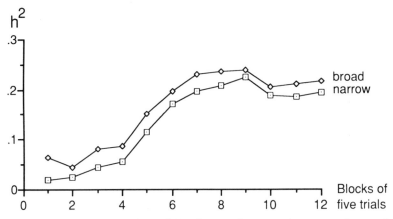

**Figure 11-3** Increasing heritability for avoidance performance observed over 60 trials of conditioning (From Hewitt, 1978; and Hewitt & Fulker, 1983.)

account for the results in Figure 11-3, which shows the change in the estimated heritability from a study of escape-avoidance conditioning using the offspring of a sample of wild caught rats crossed to genotypes of laboratory rats in a triple test cross (Hewitt & Fulker, 1981, 1983, 1984a,b). The heritability ($h^2$) is very low indeed at the start of the sequence and gradually rises to a moderate level around .20 to .25.

In this case individual differences in a behavior, which are in a sense entirely dependent on environmental input, come to depend to an increasing extent on the animal's genotype as learning takes place. Such increases in heritability during a sequence of learning trials were first reported by Vicari (1929), and although they are not always found they are reported sufficiently often to require explanation. It is of course not difficult to think of mechanisms that might give rise to such observations. Let us suppose, for example, that the learning we are studying conforms to one of the very simplest of learning models, either the all-or-none model or the linear model (see Wickens, 1982, for a very useful elementary introduction to these and other models). Under the all-or-none model an organism can be in one of two states, "guessing" or "learned." There are also two responses the organism can make (e.g., to run—to cross from one side of the box to the other in response to the warning stimulus and thereby appropriately to avoid the shock—or to freeze—not to cross from one side to the other during the warning stimulus and, thereby, to fail to avoid the shock). Figure 11-4 summarizes the basic model that involves (1) a mapping from the animal's state to its response, together with (2) a transition matrix from state at trial $t$ to state at trial $t + 1$.

As shown, the model assumes that if the animal is in the learned state it will always respond correctly, and it will respond correctly with probability $g$ before it has learned. Further, once the animal has learned it remains in that state (i.e., learning is an absorbing state) and on any trial there is a probability (alpha) that a guessing animal will learn. This is clearly an oversimplification. An alternative oversimplified learning model is the so-called *linear model*, which predicts the same average learning curves as the all-or-none model but may be distinguished through different predictions for runs of errors or sequences of responses.

States:   Guessing(G)   Learned(L)

Response Mapping:

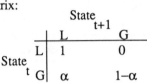

Transition Matrix:

$$\begin{array}{c} & \text{State}_{t+1} \\ \text{State}_t & \begin{array}{c|cc} & L & G \\ \hline L & 1 & 0 \\ G & \alpha & 1-\alpha \end{array} \end{array}$$

**Figure 11-4** Summary of a simple all-or-none learning model. (Adapted from Wickens, 1982.)

Under models of either kind, consistent genotypic differences in the trial-to-trial transition probabilities lead to increasing heritabilities until the point when most genotypes are performing close to their limit. An example of this process is shown in Table 11-2.

Any apparent paradox has been quite simply resolved by assuming, as do simple models for learning, that environmental variation is stochastic, random variation when we are considering responses on a given trial. The only systematic environmental variable here is the *number* of trials. Thus in trial by trial analyses, all our subjects have been exposed to the same environment. What differentiates subjects are the genotypic transition parameters. To be sure, we could have assumed environmental influences on these parameters, and the actual situation will be a matter for experimental determination; but to account for the increasing heritability we will need to invoke at least some genotypic control of the transition probabilities.

**Table 11-2** Illustrative changes in heritability during learning

| Trial | Proportion of correct responses[a] expected for genotypes with: | | | Genotype[b] variance | Environmental[c] variance | $h^2$ |
|---|---|---|---|---|---|---|
| | $\alpha = .015$ | $\alpha = .030$ | $\alpha = .045$ | | | |
| 1  | .10 | .10 | .10 | .000 | .018 | .00 |
| 10 | .21 | .32 | .41 | .005 | .042 | .11 |
| 20 | .32 | .40 | .62 | .011 | .048 | .19 |
| 30 | .42 | .63 | .76 | .015 | .045 | .25 |
| 40 | .50 | .73 | .85 | .016 | .039 | .29 |
| 50 | .57 | .80 | .91 | .015 | .032 | .32 |
| 60 | .63 | .85 | .94 | .013 | .027 | .32 |

[a] A guessing rate of .1 correct responses has been assumed.
[b] Genotypic frequencies of 0.25, 0.50, and 0.75 for $\alpha = .01, .05,$ and .09 respectively, have been assumed.
[c] Based on means of five trials.

The key features of this account of the control of individual variation in performance during learning are the following:

1. The genotype exerts an influence on the expected level of behavior, and it does so through a common mechanism throughout the sequence of trials, that is, no special genotypic effects are switched on in later trials to account for the increased heritability.
2. The effects of genotypic variation on any particular trial may be transmitted forward to the next trial since, for example, moving to the learned state results in a permanent change.
3. The effects of the environment are random or occasion specific and not correlated from trial to trial.
4. Like the effects of the genotype, the consequences of this random or occasion-specific variation may be transmitted from trial to trial.

## CHANGES IN HERITABILITY DURING DEVELOPMENT AND AGING

The key features outlined for the control of performance during a learning sequence are shared by the path model for developmental continuity and change proposed by Eaves, Long and Heath (1986) to account for human cognitive development, a model that is further discussed and elaborated by Eaves, Hewitt, Meyer and Neale in this volume (Chapter 13). Although in many respects the approaches are quite different, there are qualitative similarities in the general predictions that follow from the assumptions of the models. The Eaves et al. (1986) analysis of the pattern of monozygotic (MZ) and dizygotic (DZ) twin correlations and the increasing test-retest reliability for cognitive measures during childhood in the Louisville Twin Study suggested, to a first approximation, a model that allowed that

1. Common genetic influences act consistently throughout development without additional age-specific genetic effects.
2. Environmental influences, whether shared by members of a family or individually idiosyncratic, are occasion specific.
3. There is a persistence of effects on the phenotype. This results in an increase in the relative magnitude of genetic influences because they, and not the environmental influences, are correlated from occasion to occasion.

Two consequences of such a model are

1. A very low initial heritability rises to a higher level during development.
2. The phenotypic variance, in the absence of standardization, rises during development.

These are both consequences similar to those observed for our learning curves during the earlier stages. With the possible exception of traits like handedness or lateral preferences, these consequences are characteristic of most human psychometric phenotypes: MZ correlations diverge from DZ correlations during development while test-retest correlations increase (Hewitt, Eaves, Neale, & Meyer, 1988).

A general developmental model based on that described by Eaves et al. (1986), by Eaves, Hewitt, and Heath (1988), and by Eaves et al. (Chapter 13) is outlined in Figure 11-5.

As we have indicated for cognitive development Eaves et al. (1986) found the paths $g_s$, measuring the age specific action of genes, and $e_c$, measuring the action of environments correlated across ages, (for both shared family and individually unique environments) to be zero, with the transmission or persistence paths $j$ and $z$ equal to 0.99 for genetic and shared family environmental influences and to 0.66 for unique environments. The initial heritability was estimated to be as low as 0.0036, a value that would rise through the accumulation of the genetic influences to 0.7975. Such a pattern may well characterize, in broad outline, the development of a wide range of traits. However, the same general model with different parameter values can accommodate declining heritabilities as well. An illustration of this has been suggested by Hewitt, Carroll, Sims, and Eaves (1987) in formulating a developmental, or perhaps we should say aging, hypothesis for adult blood pressure in humans.

Hewitt et al. (1987) noted that, as shown in Figure 11-6 based on Robert and Maurer's (1977) epidemiological data from 17,854 white adults in the U.S. Health and Nutrition Survey, there is a large increase in the population variance for systolic blood pressure, and to a lesser extent for diastolic blood pressure, from early adulthood (<30 years) to middle age (>50 years). Such an increase is a predictable consequence of the general model of developmental change whenever a transmission process is superimposed on an otherwise stationary process; that is to say, if with the transmission paths $j$ and $z$ in Figure 11-4 equal to zero the population variance and the covariances between occasions are constant, then introducing positive $j$ and/or $z$ developmental transmission paths will lead to an increase in population variance until a new equilibrium value is attained (assuming $j, z < 1$).

In terms of the model, the additive genetic variance will rise from

$$V_{A_0} = h^2(g_c^2 + g_s^2)$$

to

$$V_{A\infty} = h^2 \left[ g_c^2 \left( \frac{1}{1-j} \right)^2 + g_s^2 \left( \frac{1}{1-j^2} \right) \right]$$

while the genetic covariance between the initiation of the process and the new equilibrium will be

$$\text{Cov}_{A_0,\infty} = h^2 g_c^2 \left( \frac{1}{1-j} \right)$$

Similarly, the individual environmental variance will rise from

$$V_{E_0} = e^2 (e_c^2 + e_s^2)$$

initially to

$$V_{E\infty} = e^2 \left[ e_c^2 \left( \frac{1}{1-z} \right)^2 + e_s^2 \left( \frac{1}{1-z^2} \right) \right]$$

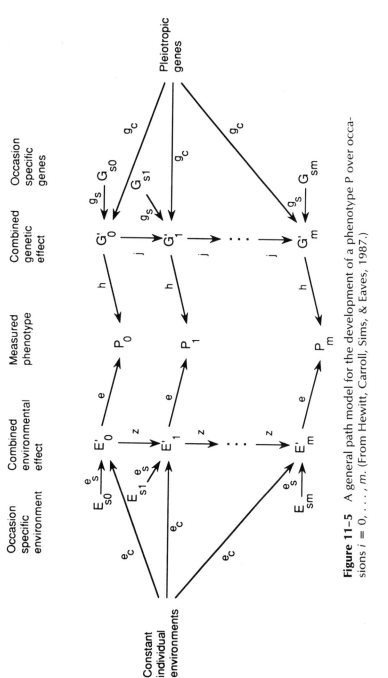

**Figure 11-5** A general path model for the development of a phenotype P over occasions $i = 0, \ldots, m$. (From Hewitt, Carroll, Sims, & Eaves, 1987.)

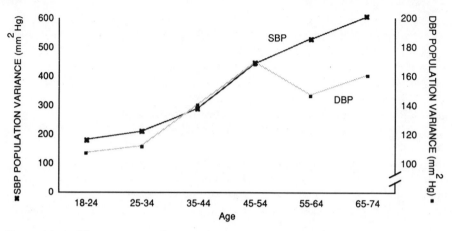

**Figure 11–6** The increase in the population variance for blood pressure (SBP, systolic; DBP, diastolic) with age. (Adapted from Robert & Maurer, 1977.)

at the new equilibrium. Thus even in the absence of longitudinal data, if we can reasonably assume that measurements have been taken at or before the onset of developmental transmission as well as at a point close to the new equilibrium, it is possible to reinterpret traditionally estimated genetic and environmental parameters in terms of the developmental model.

To illustrate this, we can consider systolic blood pressure data from a twin and family study of cardiac reactivity and hypertension (Carroll, Hewitt, Last, Turner, & Sims, 1985; Hewitt, Carroll, Last, Turner, & Sims, 1984; Sims, Carroll, Hewitt, & Turner, 1987). The variance-covariance and correlation matrices shown in Table 11–3 come from 85 balanced pedigrees each comprising a pair of healthy male twins obtained from the population-based Birmingham Family Study Register and both of their parents. The 40 MZ pairs and 45 DZ pairs were between 16 and 24 years of age (mean age 19.1 ± 3.0 years) and their parents were middle aged (mothers' mean age = 49.1 ± 6.0 years; fathers' mean age = 51.5 ± 6.0 years). Details of the blood

**Table 11–3** Covariance and correlation matrices for systolic blood pressure[a]

|  | Twin 1 | Twin 2 | Mother | Father |
|---|---|---|---|---|
| Monozygotic twin families ($N = 40$) | | | | |
| Twin 1 | 89.8 | 0.59 | 0.36 | 0.19 |
| Twin 2 | 54.6 | 94.7 | 0.33 | 0.16 |
| Mother | 57.7 | 54.6 | 284.6 | 0.12 |
| Father | 27.5 | 23.7 | 31.6 | 225.1 |
| Dizygotic twin families ($N = 45$) | | | | |
| Twin 1 | 127.6 | 0.42 | 0.25 | 0.27 |
| Twin 2 | 52.6 | 124.9 | 0.35 | 0.40 |
| Mother | 49.3 | 68.5 | 312.81 | 0.10 |
| Father | 59.3 | 85.4 | 33.1 | 366.1 |

[a]Variances and covariances are given in the lower triangle and correlations in the upper triangle of each matrix.

*Source:* Adapted from Sims, Carroll, Hewitt, & Turner, 1987.

pressure measurement are given in Sims, Hewitt, Kelly, Carroll, and Turner (1986). Traditional genetic and environmental models were fitted by maximum likelihood procedures and tested for adequacy by likelihood ratio chi square. The most parsimonious model that adequately accounted for the data and allowed for the genetic contribution to the familial aggregation for systolic blood pressure is summarized in Table 11–4 along with two simpler but inadequate models for comparison. This model allows individual environmental influences affecting young adult offspring ($E_{1t}$) to differ from those affecting the middle-aged parents ($E_{1p}$), while the additive genetic influences ($V_A$) are constant.

In terms of the developmental model, the constant additive genetic influences across this age span imply that $j = 0$, while the increasing environmental variance implies that $z \geq =0.9$, given that the environmental effects are occasion specific. With these data alone we cannot distinguish between constant and occasion-specific environments, although in the absence of the transmission of genetic effects and the presence of even moderate environmental transmission ($z = 0.4$), anything other than low or zero values for $e_c$ results in increasing test-retest reliabilities with decreasing MZ correlations to an extent that is incompatible with what is known about most phenotypes (Hewitt et al., 1987). Thus we can propose a hypothesis at least, and this is summarized in Figure 11–7. This is our best consistent hypothesis for the developmental changes in systolic blood pressure between young adulthood and middle age and has $j = g_s = e_c = 0$ and $z > 0.9$.

The hypothesis summarized in Figure 11–7 is no more than that, but it serves to draw attention to the kind of formalized quantitative description of developmental change that may be useful in making sense of these kinds of data. Not only have the analyses of Sims et al. (1987) and of Province and Rao (1985) suggested lower heritabilities for older than for younger adults' blood pressures, but declining heritabilities for bilirubin, blood urea nitrogen, glucose, and uric acid measures over a ten-year period have been recently reported in the National Heart Lung and Blood Institute Twin Study (Kalousdian, Fabsitz, Havlik, Christian, & Rosenman, 1987). For such measures as plasma glucose concentration there is a marked increase in population variance from younger to older adults (Hadden & Harris, 1987). These kinds of observations may be explained within the framework of the model discussed here, although longitudinal, genetically informative data are necessary for a proper resolution of the model (Hewitt et al., 1987).

Suppose that such a hypothesis were valid. What mechanisms might underlie the

**Table 11–4** Outcome of traditional genetic model fitting to twin-family data for systolic blood pressure

| Model[a] | Goodness of fit | | | Parameter estimates for adequate model | |
|---|---|---|---|---|---|
| | df | $\chi^2$ | $p$ | | |
| $E_1, V_A$ | 18 | 54.2 | <.001 | | |
| $E_{1(twins)}, E_{1(parents)}$ | 18 | 52.8 | <.001 | | |
| $E_{1(twins)}$ | | | | $E_{1(twins)}$ | = 35.8 ± 7.5 |
| $E_{1(parents)}$ | 17 | 8.2 | .963 | $E_{1(parents)}$ | = 222.8 ± 33.0 |
| $V_A$ | | | | $V_A$ | = 77.0 ± 14.7 |

[a] $E_1$, individual environment effects; $V_A$, additive genetic effects.

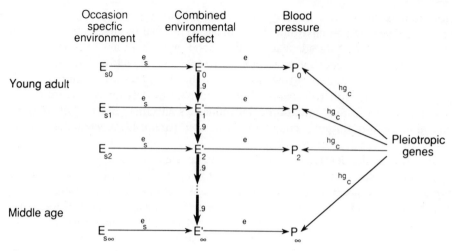

**Figure 11-7** Hypothesis for the development of adult systolic blood pressure. (From Hewitt, Carroll, Sims, & Eaves, 1987.)

phenomenon of occasion-specific environmental variation that is transmitted forward from occasion to occasion together with genetic influences that are consistent throughout adulthood but whose effects do not accumulate? Again, as with the data on learning we discussed earlier, a simple stochastic model suggests itself. Systolic blood pressure can be thought of as an index of functioning of the multicomponent cardiovascular system. During young adulthood we observe a relatively constant mean level for the population with a normal distribution of individual differences that is set by the genotypically determined properties of this system. Sometime between young adulthood and middle age, components of the system begin to fail, randomly we will assume. If each component failure raises the observed blood pressure, the mean level will increase, as will the population variance. For example, if the failures approximated a Poisson process, then the probability of $i$ component failures in time interval $T$ between young adulthood and middle age would be

$$P(i \text{ failures}) = \frac{(\lambda T)^{-i}}{i!} e^{-\lambda T}$$

with mean and variance equal to $\lambda T$. Thus if each component failure increments blood pressure by an amount $I$, the mean increment and additional variance will be $\lambda TI$ and $\lambda TI^2$, respectively. The mean increase in systolic blood pressure during this period is around 10 mm Hg, whereas the additional population variance is about 200 mm² Hg. With these values the process would require increments of 20 mm Hg per component failure with a mean of 0.5 component failures. The percentage of the population experiencing zero, one, two, three, or four failures would be 60.7, 30.3, 7.6, 1.2, and 0.2, respectively.

To illustrate how such a process would match the kinds of twin and family data reported by Sims et al. (1987) we can use their values of genetic and environmental variation to simulate data under the hypothesis. Their initial heritability is .68, their initial population variance is 112.8, and the increment in variance for the parents, which we assume results from the Poisson process, is 187.01. Taking a mean increment in blood pressure of 10 mm Hg gives an increment magnitude of 18.7 mm Hg per component failure and an average of 0.535 components failed. Using these values, 10,000 numerically simulated families give the covariance and correlation matrices shown in Table 11–5. It is clear that this is a reasonable approximation to the pattern found empirically.

Another feature of the hypothesis that fits with empirical observations is illustrated in Figure 11–8. The distribution of systolic blood pressure is initially normal (ex hypothesi) and gradually becomes skewed to the right as components fail. A pattern of such skewness increasing with age is precisely that observed for blood pressure distributions in large-scale epidemiological studies of aging populations (see, e.g., Robert & Maurer, 1977).

With this example we have another illustration of how the parameters of genetic and environmental control may change during development, or in this case aging, in ways for which we can offer more than just a mere description of the change. And once again we have suggested that a considerable portion, if not all, of the environmental variation behaves as if its primary influence is occasion specific, haphazard, or even entirely random, though its effects may well persist beyond the occasion on which they acted. Genetic variation, on the other hand, whether the effects persist beyond their occasion of action (as for learning and developing intelligence) or not (as in our hypothesis for the aging of adult blood pressure), behaves as if its influence is largely consistent across the period of learning, development, or aging.

The practical difference between consistent genetic effects that are persistent ($j > 0$) and those that are not ($j = 0$) is that preventing the expression of genetic influences on a given occasion, for example, by drug treatment or school intervention, will have consequences for the magnitude of genetic influences on subsequent occasions only if genetic effects are persistent. If the genetic effects are merely consistent, then with the withdrawal of treatment the full genetic effect is restored.

**Table 11–5**  Covariances and correlations (above the diagonal) between twins' and their parents' blood pressures based on 10,000 simulated families.

|  | Twin 1 | Twin 2 | Mother | Father |
|---|---|---|---|---|
| Monozygotic twin families ($N = 10{,}000$) | | | | |
| Twin 1 | 113.0 | 0.68 | 0.19 | 0.21 |
| Twin 2 | 77.4 | 113.0 | 0.20 | 0.21 |
| Mother | 35.4 | 37.1 | 300.5 | −0.00 |
| Father | 38.6 | 38.5 | −1.2 | 302.6 |
| Dizygotic twin families ($N = 10{,}000$) | | | | |
| Twin 1 | 115.6 | 0.36 | 0.21 | 0.22 |
| Twin 2 | 41.3 | 114.9 | 0.21 | 0.21 |
| Mother | 40.3 | 39.7 | 306.81 | −0.02 |
| Father | 40.8 | 39.7 | −5.5 | 301.3 |

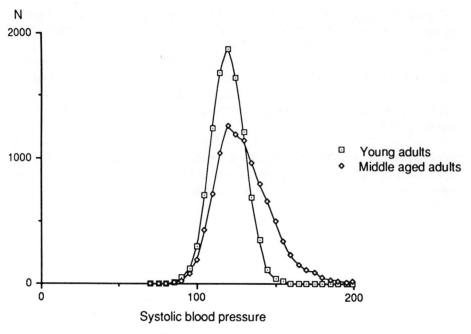

**Figure 11-8** Simulated distributions of systolic blood pressure for young and middle-aged adults based on a component failure model.

## CONCLUSION

With these examples from the animal and human literature we can recognize developmental changes in genetic control of two different kinds. First, we have the kind of systematic shifts in the pattern and direction of nonadditive genetic variance, principally genetic dominance effects, that were illustrated for escape and avoidance conditioning in laboratory and wild rats and for activity in mice. Such changes are predictable from evolutionary considerations, in particular Matherian arguments about the relationship between genetic architecture and natural selection.

Second, we have systematic changes in the proportional importance of genetic variation at different points in a developmental sequence. We have seen two examples in which increasing heritability is predicted by simple models of the developmental process that, effectively, allow genetic influences to determine the parameters of a growth process and to do this consistently over time, while environmental influences are occasion specific. In Eaves' model for cognitive development, this is expressed as occasion-to-occasion transmission of the effects of consistently acting genetic individual differences. In our discussion of performance differences during learning we explicitly suggested genetic control of parameters determining transition to the learned state, and that all environmental variance was stochastic and hence occasion specific.

In our last example of this second kind of change we illustrated a simple, and in some respects quite plausible, mechanism for declining heritability during an aging sequence. This again has the environment as an essentially haphazard or random influence whose action is occasion specific, but the effects of this action persist. Random failures of the components of a system are the obvious example. Clearly the data presented are inadequate to determine whether this hypothesis for adult blood pressure is an approximation to reality. However, as illustrations, the kinds of models discussed here and in other chapters in this section provide ways to think about the quantitative aspects of behavioral development that go beyond traditional cross-sectional biometrical genetics. They suggest a variety of simple but important generalizations that now need to be tested: Environmental influences are either occasion specific and persistent (learned), or act consistently throughout development but are not conserved. Genetic influences tend to act consistently throughout development, and hence, if the effects on the phenotype persist, give rise to increasing heritabilities during development (growth). Once development is accomplished there is a plateau in mean, variance, and variance composition at young adult values. Sometime between young adulthood and old age a new process begins that is characterized by failures of the components of behavioral or physical systems, which failures are unrelated to the index values attained during development. If these component failures are essentially random we will observe declining heritabilities; if they are influenced by the onset, in the postreproductive period, of maladaptive genetic influences that were not subject to the rigors of selection we will observe a declining genetic correlation between younger and later adulthood. The possibilities for the analysis of the genetic control of aging are taken up by Eaves et al. in Chapter 13.

## REFERENCES

Breese, E. L., & Mather, K. (1960). The organization of polygenic activity within a chromosome in *Drosophilia*. II. Viability. *Heredity, 14,* 375-399.

Broadhurst, P. L. (1979). The experimental approach to behavioral evolution. In J. R. Royce & L. P. Mos (Eds.), *Theoretical Advances in Behavior Genetics.* Germantown, MD: Sijthoff and Noordhoff.

Bruell, J. (1967). Behavioral heterosis. In J. Hirsch (Ed.), *Behavior Genetic Analysis.* New York: McGraw Hill.

Carroll, D., Hewitt, J. K., Last, K. A., Turner, J. R., & Sims, J. (1985). A twin study of cardiac reactivity and its relationship to parental blood pressure. *Physiology and Behavior, 34* 103-106.

Darlington, C. D. (1939). *The Evolution of Genetic Systems.* Cambridge: Cambridge University press.

Eaves, L. J., Hewitt, J. K., & Heath, A. C. (1988). The quantitative study of human developmental change: A model and its limitations. In B. S. Weir, E. J. Eisen, M. M. Goodman, & G. Namroony (Eds.) *2nd International Conference on Quantitative Genetics.* Sunderland, MA: Sinauer Associates.

Eaves, L. J., Long, J., & Heath, A. C. (1986). A theory of developmental change in quantitative phenotypes applied to cognitive development. *Behavior Genetics, 16,* 143-162.

Falconer, D. S. (1960). *Quantitative Genetics.* Edinburgh: Oliver and Boyd.

Fisher, R. A. (1928). The possible modification of the response of the wild type to recurrent mutations. *American Naturalist, 62,* 115-126.

Fisher, R. A. (1958). *The Genetical Theory of Natural Selection* (2nd ed.). New York: Dover.

Fulker, D. W. (1966). Mating speed in *Drosophila melanogaster:* A psychogenetic analysis. *Science, 153,* 203-205.

Hadden, W. C., & Harris, M. I. (1987). *Prevalance of Diagnosed Diabetes and Impaired Glucose Tolerance in Adults 20–74 Years of Age.* Bethesda, MD: U.S. Department of Health and Human Services, (PHS) 87-1687.
Henderson, N. D. (1964). A species difference in conditioned emotional response. *Psychological Reports, 15,* 579–585.
Henderson, N. D. (1968). Genetic analysis of a conditoned fear in mice. *Journal of Comparative and Physiological Psychology, 65,* 325–330.
Henderson, N. D. (1978). Genetic dominance for low activity in infant mice. *Journal of Comparative and Physiological Psychology, 92,* 118–125.
Hewitt, J. K. (1978). The genetic control of activity, reactivity and learning in a population of wild rats *(Rattus norvegicus).* Doctoral dissertation, University of London.
Hewitt, J. K. (1980). A note on the test for direction of dominance in the triple test cross in the presence of genotype × environment interaction. *Heredity, 45,* 293–295.
Hewitt, J. K. (1987). Heritability. *Science Progress, Oxford, 71,* 37–49.
Hewitt, J. K., & Broadhurst, P. L. (1983). Genetic architecture and the evolution of aggressive behavior. In E. C. Simmel, M. E. Hahn, & J. K. Walters (Eds.), *Aggressive Behavior: Genetic and Neural Aspects.* London: Lawrence Erlbaum.
Hewitt, J. K., Carroll, D., Last, K., Turner, J. R., & Sims, J. (1984). A twin study of cardiac reactivity and parental hypertension. *Behavior Genetics, 14,* 604.
Hewitt, J. K., Carroll, D., Sims, J., & Eaves, L. J. (1987). A developmental hypothesis for adult blood pressure. *Acta Geneticae Medicae et Gemellologiae, 36,* 475–483.
Hewitt, J. K., Eaves, L. J., Neale, M., & Meyer, J. (1988). Resolving causes of longitudinal continuity or "tracking": I. Twin studies during growth. *Behavior Genetics, 18,* 133–151.
Hewitt, J. K., & Fulker, D. W. (1981). Using the triple test cross to investigate the genetics of behavior in wild populations. I. Methodological considerations. *Behavior Genetics, 11,* 23–35.
Hewitt, J. K., & Fulker, D. W. (1983). Using the triple test cross to investigate the genetics of behavior in wild populations. II. Escape-avoidance conditioning in *Rattus norvegicus. Behavior Genetics, 13,* 1–15.
Hewitt, J. K., & Fulker, D. W. (1984a). Using the triple test cross to investigate the genetics of behavior in wild populations. III. Activity and reactivity. *Behavior Genetics, 14,* 125–135.
Hewitt, J. K., & Fulker, D. W. (1984b). Genetic differentiation of Roman selection lines did not disappear with inbreeding. *Behavior Genetics, 14,* 389–395.
Hewitt, J. K., Fulker, D. W., & Broadhurst, P. L. (1981). Genetics of escape-avoidance conditioning in laboratory and wild populations of rats: A biometrical approach. *Behavior Genetics, 11,* 533–544.
Hewitt, J. K., Fulker, D. W., & Hewitt, C. A. (1983). Genetic architecture of olfactory discriminative avoidance conditioning in *Drosophila melanogaster. Journal of Comparative Psychology, 97,* 52–55.
Kalousdian, S., Fabsitz, R., Havlik, R., Christian, J., & Rosenman, R. (1987). Heritability of clinical chemistries in an older twin cohort: The NHLBI Twin Study. *Genetic Epidemiology, 4,* 1–11.
Kearsey, M. J., & Barnes, B. W. (1970). Variation for metrical characters in *Drosophila* populations. II. Natural selection. *Heredity, 25,* 11–21.
Kearsey, M. J., & Jinks, J. L. (1968). A general method for detecting additive, dominance and epistatic variation for metrical traits. I. Theory. *Heredity, 23,* 403–409.
Mather, K. (1941). Variation and selection of polygenic characters. *Journal of Genetics, 41,* 159–193.

Mather, K. (1960). Evolution in polygenic systems. *Academia Mazionale dei Lincei, 47,* 131–152.

Mather, K. (1973). *Genetical Structure of Populations.* London: Chapman and Hall.

Mather, K., & Harrison, B. J. (1949). The manifold effect of selection. *Heredity, 3,* 1–52.

Mather, K., & Jinks, J. L. (1983). *Biometrical Genetics* (3rd ed.). London: Chapman and Hall.

Province, M. A., & Rao, D. C. (1985). A new model for the resolution of cultural and biological inheritance in the presence of temporal trends: Application to systolic blood pressure. *Genetic Epidemiology, 2,* 363–374.

Quinn, W. G., Harris, W. A., & Benzer, S. (1974). Conditioned behavior in *Drosophila melanogaster. Proceedings of the National Academy of Sciences USA, 71,* 708–712.

Robert, J., & Maurer, K. (1977). *Blood Pressure Levels of Persons 6–74 Years.* Bethesda, MD: U.S. Department of Health, Education and Welfare, (HRA) 78-1648.

Sheppard, P. M. (1967). *Natural Selection and Heredity.* London: Hutchinson.

Siegel, R. W., & Hall, J. C. (1979). Conditioned responses in courtship of normal and mutant *Drosophila. Proceedings of the Natural Academy of Sciences USA, 76,* 3430–3434.

Sims, J., Carroll, D., Hewitt, J. K., & Turner, J. R. (1987). Developmental effects upon blood pressure variation: A family study. *Acta Geneticae Medicae et Gemellologiae, 36,* 467–473.

Sims, J., Hewitt, J. K., Kelly, K. A., Carroll, D., & Turner, J. R. (1986). Familial and individual influences on blood pressure. *Acta Geneticae Medicae et Gemellologiae, 35,* 7–21.

Vicari, E. M. (1929). Mode of inheritance of reaction-time and degrees of learning in mice. *Journal of Experimental Zoology, 54,* 31–88.

Whalsten, D. (1972). Genetic experiments with animal learning: A critical review. *Behavioral Biology, 7,* 143–182.

Wickens T. D. (1982). *Models for Behavior.* San Francisco: W. H. Freeman.

Wilcock, J., & Fulker, D. W. (1973). Avoidance learning in rats: Genetic evidence for two distinct behavioral processes in the shuttle-box. *Journal of Comparative and Physiological Psychology, 82,* 247–253.

Wright, S. (1929). Fisher's theory of dominance. *American Naturalist, 63,* 274–279.

# 12
# What Can Adoption Studies Tell Us about Cognitive Development?

ROBIN P. CORLEY AND DAVID W. FULKER

Most British and American children born since the Second World War have spent what are commonly known as their "formative" years growing up in a household headed by their biological parent or parents. Although a great deal of effort by developmental psychologists has gone into tracing how differences in these children's rearing situations might have been responsible for observed differences in the children, these studies suffer from a serious confound, since genetic similarity between parents and their biological children may be contributing to the observed correlation between aspects of the rearing home and characteristics of the child. Adoption studies use a natural experiment, the rearing of children by biologically unrelated parents, to estimate the extent to which resemblance between parents and their children in intact families is due to cultural transmission rather than genetic similarity.

## LOGIC OF THE ADOPTION DESIGN

To the extent that a child given up for adoption is placed at random in the home of rearing parents, this experiment approaches the ideal degree of control of a two-way analysis of variance design. The two orthogonal effects consist of the rearing environment of the child and the child's genetic background, as inferred from the performance of the child's biological parents. The more placement in the rearing home is selective, based on characteristics of either the child or the biological parents, the less compelling is the experiment. And if either the biological parents or the adoptive parents represent a narrow range of potential influence on the child, the ability to extrapolate the findings of the natural experiment to society as a whole is reduced. The inclusion in an adoption study of a control group of children reared by their biological parents strengthens the logic of the design. The family environments of the two groups can be compared to detect restricted variation in presumably important characteristics of the homes in which the children are reared. The ideal adoption design might actually include three control groups of parents rearing their natural children: (1) parents matched to the biological parents of the adopted children, (2) parents matched to the adopting parents of the adopted children, and (3) parents representative of the population as a whole. None of the adoption studies conducted to date has been able to approach such a level of control.

Both critics (e.g., Lewontin, 1975) and advocates (e.g., DeFries & Plomin, 1978) of behavioral genetic methodology have agreed on the theoretically compelling nature of adoption designs. It is the persuasiveness of the data arising from actual studies that has been most questioned. Lewontin (1975) has argued that extrapolating findings from any of the existing studies to a more general population is unwarranted, because the study samples cannot be shown to be representative on the basis of factors, such as social class, education, urban versus rural upbringing, and religion, that have been shown to be related to differences in performance. Whether adoption studies can ever hope to assess the complete range of rearing environments has been questioned by Goldberger (1976), since prospective adoptive families typically undergo a screening process that may eliminate from consideration families with what are considered to be deviant beliefs or practices. In contrast, DeFries and Plomin (1978) have argued that adoption studies can be informative even in the presence of sampling and placement inadequacies, because many of these potential biases can be eliminated through statistical adjustments.

DeFries and Plomin have also argued for the importance of longitudinal adoption studies. Most existing studies have included children at only one time point, with varying ages in the sample. By comparing results from several studies with subjects of different mean ages, it is possible to derive a somewhat fuzzy picture of what developmental changes might be occurring in a particular domain of interest, but the image can be sharpened by studying the same children across several time points. Prospective adoption studies represent one of the most powerful and convincing methods of isolating purportedly environmental influences on the development of children. Whether performance is measured on an absolute scale that can reveal changes in average level (height, for example), or whether performance is measured relative to average level and degree of variation (IQ), developmental change is revealed in adoption studies by age changes in the degree of covariation between performance of family members or between environmental assessments in the rearing situation and child performance. Some major aspects of genetic influences on development can be inferred from changing relationships between biological parental scores and those of their children. When such parent-offspring adoption data are augmented by the study of siblings and/or twins, a more complete picture of genetic and environmental influences on development is revealed.

This definition of development as change in the relative standing of children across ages is only one aspect of what development encompasses. It ignores possible buffering influences acting to keep relative position constant, as well as changes in mean levels of performance at various ages. It does, however, appear to be an appropriate definition with which to study potential sources of individual variation in performance at particular ages.

## WHY DOES THIS CHAPTER FOCUS ON GENERAL INTELLIGENCE?

This chapter will focus on one domain of development, that of cognition, and specifically on those measures of cognitive functioning derived from the psychometric tradition (Kail & Pellegrino, 1985). The concept of general intelligence has decreased in research importance as alternative research traditions in cognition have risen in prominence. Sternberg (1985) has categorized psychologists' thinking about intelligence into six models of intelligence that have distinctive motivating questions. But as McCall (1981) and Kail and Pellegrino have discussed, students of individual differences in cognition have primarily used measures in the psychometric tradition. Although the number of studies of individual differences that use experimentally based measures is growing (Bornstein & Sigman, 1986; Hunt, 1983), most familial studies of intelligence and most long-term longitudinal studies of cognition have used measures of general cognitive ability based on paper-and-pencil group tests or individualized IQ tests. For this reason, the primary rationale for discussing cognitive development on the basis of tests of general intelligence is to achieve comparability with other studies. The second argument for using measures of general intelligence is that tests at different ages purportedly measure the same underlying ability domain, so that developmental change detected with these tests may be reflecting changes in the unobserved ability.

Some inight into whether measures of general intelligence do, in fact, measure the same underlying ability dimension can be obtained from strictly longitudinal research such as that reported by Bayley (1949). Wohlwill (1980) has listed three general propositions derived from the longitudinal data. First, measures of cognitive performance in infancy, to the age of 12 to 18 months, are nonpredictive of later intelligence test scores. Second, test-retest correlations decrease in direct proportion to the interval between tests (a simplex pattern). Third, for a constant test-retest interval, test-retest correlations increase with age.

Failure of the infant intelligence tests to predict later IQ led to a reassessment of the notion that intelligence is a stable trait over the entire life span (Brooks-Gunn & Weinraub, 1983). As Wohlwill (1980) has noted, the lack of stability from infancy to older ages can be interpreted in at least two ways: as a discrepancy between the construct of intelligence as measured at the two ages or as an indication that individual development of the same construct over time varies considerably. Developmentalists have differed in the extent they stress these two explanations. McCall (1979, 1981) and Scarr (1983) have emphasized the possibility of highly canalized development in infant intelligence, with the general specieswide developmental pattern under strong genetic control. Each normal human infant passes through the substages of infant development and achieves mastery of such concepts as object permanence within the first two years. McCall (1979) argued that only around 18 months of age do behaviors such as verbal fluency emerge that will show differential performances throughout life. The individual differences in infancy about the specieswide developmental function are simply not important for future development. As the developmental function becomes less canalized, variability in environments and genetic circumstances produces more extreme deflections from the species norm, and there is less self-righting or catchup (McCall, 1981). In contrast, Fagan (1984) has emphasized that some infant behaviors are predictive of subsequent performance on childhood intelligence tests, and that most infant intelligence tests may lack predictive power because they measure the wrong things. Evidence from a recent twin study (DiLalla et al., 1989) shows high correlations between Fagan's measure and parental IQ, as well as moderate heritability in the infant, suggesting that genetic continuity may contribute to stability for some measures of infant intelligence.

Plomin (1986) has suggested how behavior genetic analysis of measures of general cognitive ability from infancy and childhood can contribute to the understanding of what intelligence tests measure at different ages. To the extent that the genetic correlation between measures of general cognitive ability at different ages approaches 1.0, we conclude that the same genetic systems are affecting performance on the measures, given that both measures show nonzero heritability. Conversely, a genetic correlation that approaches 0 is a strong indication that different genetic systems are affecting performance, and that the two measures are clearly nonisomorphic. The genetic contribution to the phenotypic correlation between two measures is the product of three things: the genetic correlation parameter, $r_G$; the square root of the heritability of measure 1, $h_1$; and the square root of the heritability of measure 2, $h_2$. When the measures are scores at different ages, all three parameters must be nonzero for there to be a genetic contribution to phenotypic stability. If all are substantial, they may account for a large proportion of the observed phenotypic stability. Plomin (1986) argued that the application of multivariate designs to longitudinal data produces results that have considerable psychological interest because parameters such as $r_G$ have theoretical as well as empirical importance. Major developmental transitions are hypothesized to occur in the second year, between ages 5 and 7, and between ages 11 and 13. These potential genetic reorganizations might be reflected in lower age-to-age genetic correlations across the transitional ages (Plomin, 1986).

## PREVIOUS ADOPTION STUDIES

### Classic Adoption Studies

Scarr and Carter (1983) identified certain older adoption studies as classic on the basis of thoroughness, size, and influence. Burks (1928) studied 214 adoptive and 105 control families in California. The children in the study were between 5 and 14 years of age, with a mean age of 9 years. The level of selective placement was low, judged by paternal occupational status. The children in adoptive homes were placed before they reached 12 months of age. Besides evaluating the mental levels of the child and both rearing parents in the adoptive and control homes, Burks and her assistants assessed the home environments using several standardized measures.

Leahy (1935) studied 194 adoptive and 194 control families in Minnesota; the children were also between 5 and 14 when tested. Leahy and Burks both used correlations between child and adoptive parents versus correlations between control child and natural parents as their primary means of assessing the relative importance of additive genetic variation and residual sources of variation. In Burks' study, the correlation between adoptive father's mental age and the child's IQ was .07 ($N = 178$), with a correlation of .19 ($N = 204$) between adoptive mother's mental age and child's IQ. The corresponding correlations in nonadoptive homes were .45 ($N = 100$) for father's mental age and .46 ($N = 105$) for mother's mental age. In Leahy's study, adoptive fathers' Otis IQ scores correlated .15 ($N = 178$) with their adoptive children's Stanford-Binet IQ scores, whereas the correlation for adoptive mothers' scores was .20 ($N = 186$). The correlations in the control families were .51 ($N = 175$) for fathers and .51 ($N = 191$) for mothers. In both Burks' and Leahy's studies, the major measure of the rearing environment correlated discrepantly with children's IQ scores in adoptive and control families, suggesting that these measures might be so heavily confounded with parental abilities as to necessitate adjustment before environmental relationships can be inferred.

Skodak and Skeels (1949) reported on 100 adopted children who had been tested at four points during their childhood and most (63) of whose biological mothers had been given IQ tests, typically when the decision was made to relinquish the child. Their most striking result was that the correlation between maternal and child IQ scores continued to increase as the children aged, even though the biological mothers were no longer in contact with their children. At the first test session, when the average age of the children was 2 years and 2 months, the correlation with maternal IQ of the children's IQ scores was 0. At the fourth test session, when the children had achieved the average age of 13 years and 6 months, the correlation with maternal IQ was .38 for one child test and .44 for a second test. Although Skodak and Skeels did not report adoptive parent test data, they did obtain parental education levels. No systematic trends were found in the correlations between parental education and children's IQ scores as the children aged, the highest correlation being reported as .10 between maternal education and the adopted children's IQ scores at an average age of 7 years.

The data from older studies were extensively critiqued by Kamin (1974). Each of the three adoption studies just mentioned as classic came in for criticism. The Burks (1928) study was criticized for inadequate matching between adoptive and control families on the basis of parental age and presence or absence of sibs and for the inclusion of some adopted children with extremely low scores, probably due to organic damage. The Leahy (1935) study was criticized because the matching between adoptive and control families did not produce comparable scores for environmental variables other than parental occupation and educational status, with the adoptive homes being better off in terms of "environmental status." Both the Burks and the Leahy studies used IQ measures that did not successfully control for age effects. Because adoptive parents are older, inadequate age adjustments could bias results. The Skodak and Skeels (1949) study was criticized because of the extent of selective attrition over the course of the extended study, the 100 final subjects and their families being considered unrepresentative of the group with which the project started. Evidence for selective placement on the basis of maternal education was also noted. The extent to which these potential biases might actually alter the findings from these studies was not examined in detail by Kamin.

## Recent Adoption Studies

Two adoption studies carried out in Minnesota by Scarr and Weinberg (1983) were designed to investigate specific research questions, but together suggest an interesting developmental trend in cognitive resemblance among siblings. The Transracial Adoption Study (Scarr & Weinberg, 1976) investigated 101 adoptive families with 130 adopted children socially defined as black. The adopted children in the study ranged in age from 4 to 18 years, with an average age of 7 years. Because the biological parents were not tested for IQ, years of education was used as a proxy measure of general intelligence. The major findings were that the adopted children were performing at a higher mean level than comparable children raised in the black community, that individual differences among the children were more closely related to the educational level of the biological parents than to the IQ scores of the adoptive parents, and that the correlation among 140 adoptive siblings for IQ (.44) was actually higher than the correlation among the 107 natural siblings in the study (.42).

The second of the Minnesota projects, the Adolescent Adoption Study (Scarr & Weinberg, 1978), focused on older adoptees in 115 families and on nonadopted children from 120 different families. The children were from 16 to 22 years of age at the time of testing. In this study, education was once again used as a proxy for intelligence in the biological mothers. Adopted children's IQ scores correlated as highly with their biological mothers' and fathers' educational levels as the nonadopted children's scores did with those of their parents. But in contrast to the sibling data from the Transracial Adoption Study (and also from the Texas Adoption Project, see below), the correlation among 84 adoptive siblings' IQ scores ($-.03$) was markedly lower than the correlation among the IQ scores of 168 nonadoptive siblings (.35). Scarr and Weinberg (1978) concluded that older adolescents are largely independent from their family influence and have pursued courses increasingly in line with their own talents and interests. To the extent that adoptive siblings follow talents and interests more differentiated than those of biological siblings, such a loosening of parental influence will make them less alike than siblings in nonadoptive families.

The Texas Adoption Project (Horn, 1983; Horn, Loehlin, & Willerman, 1979) investigated resemblances among biological mothers, adoptive parents, and the adopted and biological offspring of the adoptive parents. Data were obtained from 364 of the biological mothers of 469 adopted children in 300 adoptive families. At the time of testing, 86 percent of the adopted children were between 5 and 15 years of age. Some selective placement was observed, with correlations of .14 and .11 between the performance IQs of the adoptive mothers and fathers and the biological mothers.

A number of interesting comparisons are possible with this design. The correlation between the IQ scores of the biological mothers and their adopted-away children was .28, whereas the correlations between the scores of the adoptive mothers and fathers and the adopted children were .15 and .12, respectively. The scores of the adoptive mothers and fathers correlated .21 and .29 with the scores of their biological children. When full-scale IQ scores were used to compare biologically related and unrelated siblings, the correlations were .35 and .26 for the two sibling types, suggesting an unexpectedly high degree of familial environmental influences shared by the adoptive siblings. The additional results reported by Horn (1983) when the sample was subdivided by age of the child suggest that age was an important factor affecting the size of the correlations among the scores of family members. The correlation of adopted children's scores with scores of their biological mother was .36 for 235 children between the ages of 5 and 9, but was $-.02$ for 62 children between the ages of 10 and 14, suggesting the possibility of a decrease in the importance of the genetic contribution to individual differences in cognitive performance as children start to reach adolescence. The environmental contribution may increase over the same age range, since the correlation between adoptive parental scores and scores of adopted children in the younger age range was .14 ($N = 280$) for mothers and .08 ($N = 280$) for fathers, going up to .19 ($N = 121$) and .23 ($N = 125$) for the older group of adopted children.

Although the data from these adoption projects suggest that developmental trends may be occurring in general intelligence, none of the reported studies has drawn sufficiently precise age bands to allow a detailed investigation of age-to-age trends. It is unclear whether apparently contradictory findings, such as an increasing trend in biological mother-adopted child resemblance in the Skodak and Skeels study and a decrease in the biological mother-adopted child correlation across two age groups in the Texas Adoption Project, might be due to differing age distributions.

## THE COLORADO ADOPTION PROJECT

The most ambitious of the contemporary adoption projects studying cognition is the Colorado Adoption Project (CAP), with which we are associated. The CAP (Plomin & DeFries, 1985; Plomin, DeFries, & Fulker, 1989) is an ongoing longitudinal, prospective adoption study that is designed to collect a vast amount of information about the development of adopted and nonadopted children during their infancy and childhood. The core CAP sample consists of 245 adoptive families and 245 nonadoptive families that are matched to the adoptive families on the basis of the gender of the adopted proband, the number of children in the family, the occupation and education level of the father, and paternal age. The biological and adoptive parents of the proband adoptees are typically tested before the adopted child reaches age 1, as are the parents of the nonadopted probands. Approximately 20 percent of the biological fathers of the adoptees have been tested, yielding a much richer data base on this group than that obtained in previous behavior genetic studies. Data are collected from the children and their families at yearly intervals. Home visits are made to assess the children and their rearing environments near the time of their first, second, third, and fourth birthdays. Additional information is collected by phone and questionnaire at 18 and 30 months of age and again at 5 and 6 years of age. Children are again tested by home or laboratory visit after completing first grade (approximately age 7). As of this year the youngest proband children have reached age 6 and the oldest have reached age 14.

In addition to collecting information from the adopted and nonadopted probands and their parents, an effort is made to collect data on the first younger child added to the adoptive and nonadoptive families through birth or adoption, following the same testing schedule and using the same test battery as with the proband sibling. Younger siblings continue to be recruited as they are born. Sibling data thus continue to be added for each age for as long as six years after the proband sample has been completely assessed.

Although the design of CAP is an augmented version of earlier adoption projects such as those of Burks and of Skodak and Skeels, its implementation owes much to two other large-scale projects in behavior genetics. The Hawaii Family Study of Cognition (HFSC; DeFries et al., 1979) examined resemblances among parents and their biological offspring, when all children had reached adolescence and were capable of being tested on the same cognitive battery. The collected data represent a rich source of information about familial resemblance on a wide variety of cognitive and noncognitive measures. Many of these were incorporated into the testing regimen for CAP adults. The Louisville Twin Study (LTS; Wilson, 1983; Chapter 2 in this volume) is an ongoing longitudinal study that has as its goal the testing in a laboratory setting of twins at 3, 6, 9, 12, 18, 24, 30, and 36 months and 4, 5, 6, 7, 8, 9, and 15 years of age. This large number of test sessions allows a detailed picture of cognitive development to emerge. Many of the CAP measures used in assessments of children at different ages have been used in the LTS.

The primary measure of general cognitive ability in the CAP adults is their first principal component score (GFAC) from a battery of 16 specific cognitive abilities tests administered as part of a longer test session lasting approximately three hours. The 16 tests from the domains of spatial visualization, verbal abilities, perceptual speed, and visual memory are rescored into 13 measures. Individual subjects are allowed to have missing data on up to two measures and still receive a principal component score. The 13 cognitive measures are separately age- and gender-corrected within the three major subject groupings of biological, adoptive, and non-adoptive parents before a component score is generated. For a subsample of 183 adoptive and nonadoptive parents, the first principal component score correlated .73 with a full-scale Wechsler Adult Intelligence Scale (WAIS) IQ score obtained six years later.

The primary measure of general cognitive ability in CAP children at years 1 and 2 is the Bayley Mental Development Index (MDI) from the Bayley Scales of Mental Development (Bayley, 1969). In the CAP the MDI is administered in the child's home by a trained tester as part of a longer test session of approximately two hours. The primary measure of general cognitive ability at ages 3 and 4 is the Stanford-Binet IQ test (SBIQ). At year 7, CAP children are given the Wechsler Intelligence Scale for Children–Revised (WISC–R) battery and their full-scale IQ scores are used as the primary measure of general cognitive ability at this age.

Table 12-1 presents the correlations between the parental cognitive scores and proband children's IQ scores at five ages in adoptive and nonadoptive families for all data collected through the summer of 1987. In addition, the table lists the correlations between scores of siblings at the corresponding ages and scores of probands. The first part of Table 12-1 gives the correlations for the entire sample, while the second part of the table includes only data from probands who have already been tested at year 7. Some families are no longer participating in the study by year 7 testing, whereas some proband children have not yet completed first grade. The general similarity of the correlations in the two parts of the table suggests that possible selective attrition and cohort differences are not strongly affecting the degree of resemblance between family members in adoptive and nonadoptive homes.

The increasing correlation between the cognitive score of the biological mother and the IQ scores of her adopted-away offspring as the child grows up is similar to the pattern reported by Skodak and Skeels (1949). On the basis of pattern of correlations for adoptive parents, it appears that the strongest effect of rearing parents is felt at ages 3 and 4.

**Table 12-1** Correlations of proband IQ scores for five ages with first principal cognitive component score of their parents and IQ scores of their siblings at the same age.

| Family member | Year of IQ data | | | | |
|---|---|---|---|---|---|
| | 1 | 2 | 3 | 4 | 7 |
| A. All proband data collected through September 30, 1987 | | | | | |
| | | Adopted proband | | | |
| Biol. Father | .31 ( 50) | .31 ( 47) | .20 ( 46) | .43 ( 43) | .11 ( 38) |
| Biol. Mother | .10 (240) | .07 (216) | .15 (207) | .18 (197) | .33 (160) |
| Adop. Father | .07 (234) | .05 (213) | .21 (204) | .13 (195) | .13 (158) |
| Adop. Mother | .10 (239) | .09 (216) | .15 (206) | .17 (197) | .06 (160) |
| Sibling[a] | .07 ( 84) | .02 ( 78) | .30 ( 68) | .11 ( 59) | .14 ( 17) |
| | | Nonadopted proband | | | |
| Father | .08 (242) | .14 (228) | .13 (211) | .12 (211) | .17 (139) |
| Mother | .01 (241) | .14 (227) | .13 (210) | .17 (210) | .16 (138) |
| Sibling[b] | .36 (102) | .41 ( 86) | .36 ( 79) | .21 ( 71) | .51 ( 21) |
| B. Data only for probands tested at age 7 before September 30, 1987 | | | | | |
| | | Adopted proband | | | |
| Biol. Father | .35 ( 38) | .27 ( 36) | .17 ( 36) | .42 ( 35) | .11 ( 38) |
| Biol. Mother | .10 (157) | .15 (147) | .17 (146) | .24 (141) | .33 (160) |
| Adop. Father | .07 (155) | .17 (145) | .23 (143) | .16 (139) | .13 (158) |
| Adop. Mother | .07 (157) | .03 (147) | .08 (145) | .14 (141) | .06 (160) |
| Sibling[a] | .07 ( 61) | .08 ( 61) | .32 ( 59) | .09 ( 54) | .14 ( 17) |
| | | Nonadopted proband | | | |
| Father | .08 (139) | .23 (134) | .13 (128) | .12 (131) | .17 (139) |
| Mother | −.01 (138) | .20 (133) | .15 (127) | .18 (130) | .16 (138) |
| Sibling[b] | .44 ( 71) | .42 ( 64) | .35 ( 63) | .20 ( 61) | .51 ( 21) |

*Note:* Numbers in parentheses are the pairwise number of observations.
[a]Sibling data for the adopted probands include both adopted siblings and biological children of the adoptive parents.
[b]Sibling data for the nonadopted probands include only full siblings.

One troubling aspect of Table 12-1 is the relatively low level of the correlations between the scores of the control parents and the scores of their children at each age. Children raised by their natural parents should resemble them due to both genetic and cultural transmission, yet the correlations for these nonadoptive families do not consistently exceed the correlations for either the biological or adoptive parents of the adopted children. Several possible explanations for lowered correlations in the nonadoptive families have been investigated, such as inadequate age correction or reduced reliabilities for the parental test battery for the control parents, but none of the investigated potential causes has yielded a satisfying explanation. In the absence of an alternative explanation, the general practice in the CAP has been to attribute the lower-than-expected correlations to chance and to combine the nonadoptive sample with the adoptive families in model-fitting exercises. It should be noted, however, that all three of the contemporary adoption studies cited here have reported nonadoptive correlations that appear lower than expected if both genetic and cultural transmission act to increase resemblance between parents and offspring. Horn (1983) reported in the Texas Adoption Project a correlation of 0.36 between biological mother IQ and the IQ scores of adopted-away offspring in the 5- to 9-year age range, but the correlations between the IQ scores of the adoptive mother and father and their natural children are only 0.24 and 0.16, respectively. In the Adolescent Adoption Study (Scarr & Weinberg, 1983), correlations of .28 and .43 are reported for educational level of the biological mother and father with IQ of the adopted children, but the corresponding correlations for nonadopted children are .17 and .26. In the Transracial Adoption Study (Scarr & Weinberg, 1976), the correlations between the educational levels of the biological mother and father and the IQ scores of their adopted-away children are .33 and .43, but the corresponding correlations for the adoptive mother and father with their own natural children are no higher at .34 and .39.

## THE UNIVARIATE MODEL OF PHENOTYPIC DEVELOPMENT

Cognitive data collected from the CAP have been fit to a number of variants of a model of development that allows for assortative mating and selective placement. The basic model for influences affecting a single phenotype in an individual is given in Figure 12-1. Two latent variables $G$ and $E$ are causally related to the observed phenotypic variable $P$. All three variables are conceived to be standardized variables with a mean of zero and a standard deviation of one. The latent variable $G$, in the context of adoption data, represents the effect of the additive genotype. The latent variable $E$ represents all other sources of variation, including nonadditive genetic sources of variation such as dominance and epistasis, measurement error, idiosyncratic experiences, and shared familial experiences. The proportion of environmental variance is calculated as a residual, as seen in the following. The symbol $s$ corresponds to the correlation between the genotype and the environment contributing to the phenotype of interest. Using the path analytical rules of Wright (1934), it is pos-

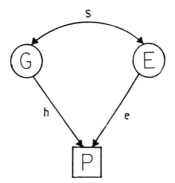

**Figure 12-1** Univariate path model of the genetic ($G$) and environmental ($E$) components of the phenotypic value ($P$). The path coefficients $h$ and $e$ represent the square roots of heritability and environmentality, respectively. $s$ corresponds to the correlation between genotype and environment.

sible to calculate the following expected correlations between the three variables in the diagram:

$$R_{GE} = s \tag{12-1}$$

$$R_{GP} = h + se \tag{12-2}$$

$$R_{EP} = e + sh \tag{12-3}$$

The variance of the observed standardized variable $P$ is

$$V_P = 1 = h^2 + e^2 + 2hse \tag{12-4}$$

This formulation allows us to calculate $e$, the square root of environmentality, using the quadratic formula, as follows:

$$e = (s^2h^2 - h^2 + 1)^{1/2} - sh \tag{12-5}$$

This basic model for an individual can be extended to include data collected from several family members, allowing for the estimation of the model's parameters. Fulker (1982) suggested a model and method for the analysis of data from twins and their parents that allows for assortative mating, cultural transmission, and genotype–environment covariance due to parental influence. Fulker and DeFries (1983) presented a similar model for adoption designs that include data from both adoptive and nonadoptive families. When multiple relationships from the same family are assessed, correlations between family members are no longer independent, and covariance matrices, rather than correlations, are compared with their expected values. Fulker (1982) and Fulker and DeFries (1983) used the likelihood ratio statistic employed by Jöreskog (1969) to obtain maximum-likelihood estimates of the model's parameters.

Figure 12-2 represents the CAP's working model for resemblance between phenotypes of parents and child in nonadoptive families. Parental genotypes of mother and father are connected to child genotype by a path marked 1/2, reflecting the average additive effect of meiotic partitioning of parental chromosomes and subsequent recombination in the offspring. Phenotypic assortative mating for the trait of interest is represented by a conditional path $p$ (Carey, 1986). The phenotypes of the parents are also allowed to impact the environment in which the child is being raised through cultural transmission parameters $m$ and $f$. In this univariate model, only the direct influence of the parental phenotype of interest is included in the cultural transmission parameters. Other parental influences that might be related to the child's phenotype, such as child-rearing practices, are not modeled explicitly, but rather are included among the residual contributors to the child's environment.

Figure 12-3 represents the CAP's working model for resemblance between parental phenotypes and child phenotype in adoptive families. The adoptive family represents a situation in which genotypic transmission occurs from the biological parents and cultural transmission from the adoptive parents. The two sources of influence on the child are not independent when selective placement occurs. Only selective placement based on resemblance between biological and adoptive parental

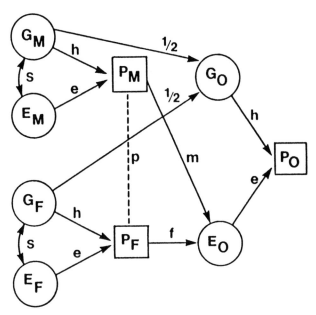

**Figure 12-2** Path model of cultural and genetic transmission in nonadoptive families. The subscripts on the latent and manifest variables are given as M, F, and O for mother, father, and offspring. See text for explanation of variables and discussion.

phenotypes is considered a potential confound when children are placed in the first year of life, as in the CAP. Four selective placement parameters $x_1$ to $x_4$ are included in the model to represent correlations between the phenotypes of the biological and adoptive parents. The working model also allows for a potential discrepancy between the degree of assortment in wed (adoptive and nonadoptive) couples, designated by parameter $p$, and assortment in the largely unwed biological couples, parameter $q$.

The IQ data collected from CAP children at ages 1, 2, 3, 4, and 7 and the cognitive data collected from their parents have recently been systematically explored using the same series of models for each year of childhood data (Fulker, DeFries, & Plomin, 1988). Three covariance matrices were calculated at each year, two for subsets of the adoptive families and one for the nonadoptive families. The adoptive families split into two groups on the basis of the presence or absence of biological father scores. Table 12-2 includes parameter estimates for each year from a simplified model in which the selective placement parameters have been set to zero, cultural transmission parameters for mothers and fathers have been equated, and the two assortative mating parameters for wed and unwed couples have been combined into one. This particular simplified model represents an acceptable alternative to the full model at each year based on differences between their likelihood ratio statistics.

The results from both the full and reduced univariate models for each year of cognitive data suggest two major developmental conclusions: that the importance of additive genetic influences for cognitive test performance increases as children reach middle childhood, and that the importance of cultural transmission based on parental phenotype does not systematically increase with length of exposure to the rearing

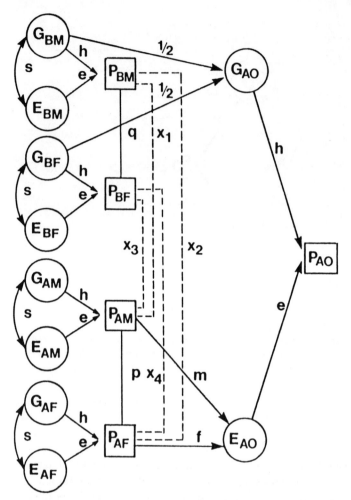

**Figure 12-3** Path model of cultural and genetic transmission in adoptive families. The subscripts on the latent and manifest variables are given as BM, BF, AM, AF, and AO for biological mother, biological father, adoptive mother, adoptive father, and adopted offspring. See text for explanation of variables and discussion.

parents. Tests of alternative reduced models in which the cultural transmission parameters are set to zero indeed suggest that the influence of exposure to parental phenotypes may peak around age 3 and subsequently decline as children enter the school system.

## THE DEVELOPMENT OF MULTIVARIATE MODELS OF CAP DATA

The univariate working model employed in the CAP has been extended in several ways. Rice and colleagues (Rice, Carey, Fulker, & DeFries, 1989; Rice, Fulker, & DeFries, 1986) applied a multivariate version to the simultaneous analysis of specific

**Table 12-2** Parameter estimates resulting from the fit of a parsimonious model of genetic and cultural transmission to general cognitive ability data in the Colorado Adoption Project

| Parameter | Age (yr) | | | | |
|---|---|---|---|---|---|
| | 1 | 2 | 3 | 4 | 7.7 |
| $h^2$ | .09 | .14 | .10 | .20 | .36 |
| $m = f$ | .04 | .05 | .10 | .07 | .01 |
| $p = q$ | .27 | .25 | .27 | .27 | .21 |
| $s$ | .02 | .03 | .05 | .04 | .01 |
| $x^2$ | 33.52 | 32.86 | 29.84 | 30.13 | 25.48 |
| $df$ | 26 | 26 | 26 | 26 | 26 |
| $p >$ | .10 | .10 | .20 | .20 | .40 |

*Source.* Table 3 from Fulker, DeFries, & Plomin (1988). Reprinted by permission from *Nature*, 336, 769. Copyright © 1988, Macmillan Magazines Ltd.

cognitive abilities in CAP children at age 4, concluding that the specific abilities were highly correlated genetically. Environmental indices have been included to assess the importance of influences other than parental phenotypes on the child's environment (Rice, 1987; Rice, Fulker, DeFries, & Plomin, 1988). Several purportedly environmental factors related to cognitive performance at ages 3 and 4 appear to be most plausibly related to the child's phenotype through their correlation with the parental genotype, rather than through their direct contribution to the child's environment. However, in infancy, an assessment of the quality of the child's home-rearing environment did appear to have a direct effect on the child's phenotype, indicating that at this age the factor of quality of home environment could be truly considered an environmental effect (Rice et al., 1988).

The derivation of expectations for multivariate extensions of the CAP working model has been simplified by the use of some additional path rules codified by Vogler (1985ab). The expectations are derived from path diagrams where the symbols for latent and observed variables refer to vectors of variables of a particular type. In form, the path diagram may be identical to its univariate representation, but the actual number of variables indicated by each symbol can range from 1 to $k$, depending on the application. The symbol **h** now corresponds to a diagonal matrix in which the square roots of the heritabilities for each trait are placed on the diagonal. The symbol **e** represents a diagonal matrix with the square roots of the environmentalities on the diagonal. The symbol **s** is a full square matrix of correlations between the vector of genotypes and the vector of environments. The following rules for deriving multivariate expected correlations using such a diagram are given by Vogler (1985a):

> Rule 1: A matrix of path coefficients is untransposed when a path is traced backward against the direction of the arrow; the path coefficient matrix is transposed when the direction of traversal is forward in the direction of the arrow.
>
> Rule 2: A matrix of correlations is incorporated into the equation for expected correlations at the point where a change of direction occurs in tracing a path from backward to forward. This can occur in two ways. First, when the direction change occurs by tracing backward along one single-headed arrow and then forward along a different causal path, the matrix is that defining the intercorrelations of the set of variables from which the two

causal paths lead. Second, when a direction change occurs through a double-headed arrow, the matrix is that defining the intercorrelations among the two sets of variables at opposite ends of the path.

Rule 3: A correlation matrix that defines a double-headed arrow is untransposed when it is traced in one direction and transposed when traced in the opposite direction. The definition of which direction involves the untransposed form of the matrix and which direction involves the transposed form is arbitrary, depending on how the user defines the matrix.

Rule 4: Matrices are multiplied together in the same order in which they are encountered when tracing a path.

Rule 5: When a path is traced in a single direction only, the equation for the expected correlations is premultiplied (when tracing forward) or postmultiplied (when tracing backward) by the correlation matrix of the set of variables at the causal end of the path. [pp. 39–40]

Using these rules the following expectation, the analog of equation (12–4) is derived for the correlation matrix between phenotypes:

$$\mathbf{R_P = h\ R_G\ h' + e\ R_E\ e' + h s e' + e s' h'} \qquad (12\text{–}6)$$

The model of assortative mating used in the multivariate extension of the basic CAP model has been modified since its original formulation (Rice, Fulker, & DeFries, 1986). Carey (1986) has shown that the expectation for this matrix can be written as

$$\mathbf{M = R_P\ D\ R'_P} \qquad (12\text{–}7)$$

where the phenotypic correlation matrix $\mathbf{R_P}$ is assumed to be equivalent in both genders and $\mathbf{D}$ is the nonsymmetric matrix of conditional paths that index the extent of direct matching between traits in the mother and traits in the father. The process of assortative mating on the basis of phenotypic variables induces correlations between the latent variable vectors $\mathbf{G}$ and $\mathbf{E}$ of the spouses. The induced correlational paths sum to yield the same phenotypic correlations, but are more easily dealt with using the multivariate path rules of Vogler.

One other recent extension of the CAP model adds additional sources of data. LaBuda, DeFries, Plomin, and Fulker (1986) and DeFries, Plomin, and LaBuda (1987) have combined CAP data with results from the Louisville Twin Study for comparable measures at ages 1, 2, 3, and 4. In the univariate model, the estimate of heritability, $h^2$, is actually composed of the product of the square roots of the heritabilities for the childhood and adult measures, and the genetic correlation between them. Using heritability estimates derived from the LTS twin data, it is possible to arrive at estimates of the genetic correlations between adult general cognitive ability and childhood IQ scores. At ages 1, 2, 3, and 4, those estimates, based on a parsimonious model that set the levels of the cultural transmission parameters to zero, were .67, .85, .79, and .90, respectively (DeFries et al., 1987). This suggests that similar genetic influences affect cognitive performance across the age span from infancy to adulthood.

## COMBINING SIBLING AND PARENT-OFFSPRING CAP COGNITIVE DATA

One other extension to the working model (Corley, 1987) will be discussed in greater detail. A variant of the multivariate model that dropped selective placement and equated assortative mating in wed and unwed couples was extended to include younger siblings present in the adoptive and nonadoptive families. Measures of general cognitive ability in infancy, childhood, and adulthood are considered separate phenotypes. The implemented version of the model is bivariate, one variable being the adult principal cognitive component score and the other the childhood IQ score at a particular age. Published data from complementary studies were added to help resolve the model, and certain parameters were fixed to zero because of assumptions about the critical phenotypes for assortative mating and cultural transmission. Despite the rather precarious nature of an enterprise that applies a bivariate model to a sample in which the two measures of interest have not been measured in the same individual, we feel that such a parameterization is more appealing than assuming that scores from early childhood cognitive tests and adult cognitive tests necessarily measure the same underlying ability domain.

### Description of the Multivariate Model with Siblings

Figure 12-4 represents the version of the multivariate model applied to nonadoptive families in which two offspring are present. This figure includes both the conditional path representation for phenotypic assortment in the parental generation and the induced correlations among the genotypes and environments of the spouses. The induced links allow for easier treatment by the path rules of Vogler, although by collection and substitution of terms the Greek symbols can be replaced in the expectations for familial covariances. To be noted is that the phenotypes of the two offspring are connected not only through paths traced back through their parents, but

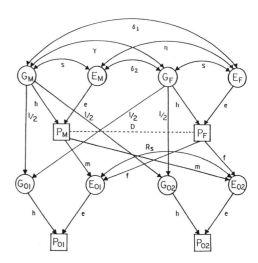

**Figure 12-4** Multivariate path model of cultural and genetic transmission in nonadoptive families with two offspring. See text for explanation of variables and discussion.

also by an additional matrix, $R_S$. $R_S$ is a measure of the strength of effects of parental influences not directly assessed by the parental phenotypic vectors, such as child-rearing practices. It also is a measure of those shared sibling effects that are not directly linked to the parents, such as community or cohort effects (although not born in the same year, adjacent siblings in the same family may be made more similar by growing up at nearly the same time), and more specific nonparental shared environmental influences such as schooling and playmates. The expectations for covariances among nonadoptive family members are given in Table 12-3.

Figure 12-5 represents one type of CAP adoptive family with two offspring. In this figure, one of the offspring, whose phenotypic vector of scores is labeled $P_{NO}$, is the biological child of the adopting parents. The other child, $P_{AO}$, is adopted. Figure 12-6 represents a second type of adoptive family in which both of the children are adopted. There are now three sets of parents, two biological and one adoptive. Because selective placement is assumed to be absent, the resemblance between offspring in these two sorts of families is viewed as due to cultural transmission through the phenotypes of the adoptive parents, and due to the same matrix of residual sib influences, $R_S$, seen in nonadoptive families with two offspring. The complete adoptive family expectations that have been derived from Figures 12-5 and 12-6 are given in Table 12-4.

## Additional Data Used in the Analyses

The combined parent-offspring and sibling adoption design is theoretically capable of resolving the genetic and environmental parameters at all tested ages, except for the assortative mating matrix $D$. This statement assumes that the degree of covariance between genotype and environment in the parental generation, $s$, is equivalent to the G-E covariance in the child generation produced in nonadoptive families in which parents are rearing their own biological children. Because information was unavailable on the level of resemblance between parents on the childhood measures, and because we assumed that parents choose mates on the basis of their adult phe-

**Table 12-3** Expected covariances between nonadoptive family members derived from Figure 12-4

| Covariance | Expectation for covariance |
|---|---|
| $P_M, P_F$ | $V_P^{1/2} (R_P D R_P) V_P^{1/2}$ |
| $P_M, P_{01} = P_M, P_{02}$ | $V_P^{1/2} [\frac{1}{2} (\tau + R_P D \tau) h + (R_p m' + Mf') e] V_P^{1/2}$ |
| $P_F, P_{01} = P_F, P_{02}$ | $V_P^{1/2} [\frac{1}{2} (\tau + R_P D' \tau) h + (R_P f' + M' m') e] V_P^{1/2}$ |
| $P_{01}, P_{02}$ | $V_P^{1/2} \{\frac{1}{2} h [R_G + \frac{1}{2} (\gamma + \gamma')] h + e [m(R_p m' + Mf') + f(R_p f' + M' m') + R_S] e + \frac{1}{2} h [(\tau' + \tau' D R_P) f' + (\tau' + \tau' D' R_P) m'] e + e [m(\tau + R_P D \tau) + f(\tau + R_P D' \tau)] \frac{1}{2} h\} V_P^{1/2}$ |
| $P_M, P_M = P_F, P_F = P_{01}, P_{01} = P_{02}, P_{02}$ | $V_P^{1/2} (R_P) V_P^{1/2}$ |

*Note:* Expectations use the composite matrices $\tau$ ($hR_G + e\, s'$), $\mu$ ($R_{PM}DR_{PF}'$), and $\gamma$ ($\tau'D\tau$). $R_P$ is assumed to be identical in all family members, so subscripting is dropped in expectations. $R_P$, $R_G$, $R_S$, $h$, and $e$ are symmetrical, and thus transposition is ignored in these expectations.

# WHAT CAN ADOPTION STUDIES TELL US?

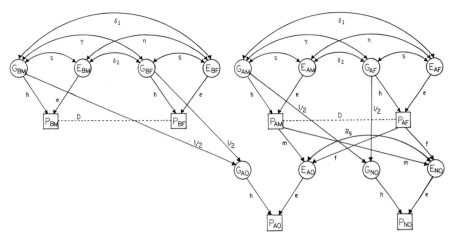

**Figure 12-5** Multivariate path model of cultural and genetic transmission in adoptive families with one adopted child (AO) and one natural child (NO) of the adoptive parents. See text for explanation of variables and discussion.

notypes, the matrix **D** was constrained to have only one nonzero element, measuring the level of assortment on the adult cognitive phenotype. There is, however, a large discrepancy between the theoretical ability of the combined design to resolve the parameters of the model and the current situation in the CAP. Since the oldest children in the study have been tested only to age 13, and the number of siblings in the study, particularly those who have been tested at the older ages, remains small, we have chosen to add data from complementary studies to these analyses.

Table 12-5 presents the additional correlations taken from the literature, which have been incorporated into the estimation procedure. The data from the Louisville Twin Study (Wilson, 1983) represent the only reported large-scale twin study of both

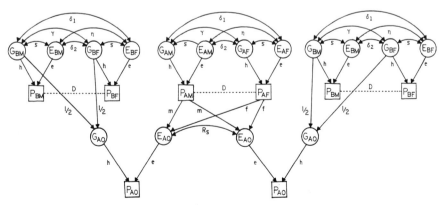

**Figure 12-6** Multivariate path model of cultural and genetic transmission in adoptive families with two adopted children from different sets of biological parents. See text for explanation of variables and discussion.

**Table 12-4**  Expected covariances between adoptive family members derived from Figures 12.5 and 12.6

| Covariance | Expectation for covariance |
|---|---|
| $P_{BM1},P_{BF1} = P_{AM},P_{AF} = P_{BM2},P_{BF2}$ | $V_P^{1/2}[R_P\ D\ R_P]V_P^{1/2}$ |
| $P_{BM1},P_{AM} = P_{BM1},P_{AF} =$ | 0 |
| $P_{BM1},P_{BM2} = P_{BM1},P_{BF2} =$ | |
| $P_{BF1},P_{AM} = P_{BF1},P_{AF} =$ | |
| $P_{BF1},P_{BM2} = P_{BF1},P_{BF2} =$ | |
| $P_{BM2},P_{AM} = P_{BM2},P_{AF} =$ | |
| $P_{BF2},P_{AM} = P_{BF2},P_{AF}$ | |
| $P_{BM1},P_{AO1} = P_{BM2},P_{AO2}$ | $V_P^{1/2}[\frac{1}{2}(\tau + R_P\ D\ \tau)\ h]\ V_P^{1/2}$ |
| $P_{BF1},P_{AO1} = P_{BF2},P_{AO2}$ | $V_P^{1/2}[\frac{1}{2}(\tau + R_P\ D'\ \tau)h]\ V_P^{1/2}$ |
| $P_{BM1},P_{AO2} = P_{BM2},P_{AO1} =$ | 0 |
| $P_{BF1},P_{AO2} = P_{BF2},P_{AO1} =$ | |
| $P_{BM1},P_{NO} = P_{BM2},P_{NO} =$ | |
| $P_{BF1},P_{NO} = P_{BF2},P_{NO}$ | |
| $P_{AM},P_{NO}$ | $V_P^{1/2}[\frac{1}{2}(\tau + R_P\ D\ \tau)\ h + (R_P\ m' + Mf')\ e]\ V_P^{1/2}$ |
| $P_{AF},P_{NO}$ | $V_P^{1/2}[\frac{1}{2}(\tau + R_P\ D'\ \tau)\ h + (R_P\ f' + M'\ m')\ e]\ V_P^{1/2}$ |
| $P_{AM},P_{AO1} = P_{AM},P_{AO2}$ | $V_P^{1/2}[R_p\ m' + M'\ f')\ e]\ V_P^{1/2}$ |
| $P_{AF},P_{AO1} = P_{AF},P_{AO2}$ | $V_P^{1/2}[R_p\ f' + M'\ m')\ e]\ V_P^{1/2}$ |
| $P_{AO1},P_{AO2}$ | $V_P^{1/2}[e(m(R_p\ m' + M\ f') + f(R_p f' + M'\ m') + R_s)\ e]\ V_P^{1/2}$ |
| $P_{AO1},P_{NO} = P_{AO2},P_{NO}$ | $V_P^{1/2}[e(m(R_p\ m' + M\ f') + f(R_P\ f' + M'\ m') + R_s)\ e]\ V_P^{1/2}$ |
| $P_{BM1},P_{BM1} = P_{BF1},P_{BF1} =$ | $V_P^{1/2}(R_P)V_P^{1/2}$ |
| $P_{AM},P_{AM} = P_{AF},P_{AF} =$ | |
| $P_{BM2},P_{BM2} = P_{BF2},P_{BF2} =$ | |
| $P_{NO},P_{NO}$ | |
| $P_{AO1},P_{AO1} = P_{AO2},P_{AO2}$ | $V_P^{1/2}(h\ R_G\ h + e\ R_E\ e)\ V_P^{1/2}$ |

*Note:* See note to Table 12-3.

infant and child development. The CAP was designed to complement the LTS, and used identical IQ measures at ages 1, 2, 3, and 7. The advantage of adding these data is that they provide more power for estimating heritability for the childhood measures. No large-scale adult twin studies have yet been reported that used a cognitive measure like the first principal component score of the CAP adult test battery or of the closely related Hawaii Family Study of Cognition test battery. However, the Loehlin and Nichols (1976) twin study of National Merit Scholarship Qualifying Test results obtained a total score based on five cognitive subtests for a very large set of monozygotic (MZ) and dizygotic (DZ) twins. Their estimates of .86 and .62 for MZs and DZs are in close agreement with the median correlations for general intelligence of .85 and .58 based on 34 and 41 studies, respectively, reported by Bouchard and McGue (1981) in a review of familial studies of intelligence. The use of adult twin data allows for estimation of the genetic transmission parameter $h$ for the adult measure.

DeFries et al. (1979) reported HFSC parent-offspring correlations as adults for the first principal component score of a cognitive battery from which the CAP battery was derived. The HFSC is the largest study of parent-offspring cognitive resemblance

# WHAT CAN ADOPTION STUDIES TELL US?

**Table 12-5** Supplementary correlations ($r$) from the literature used in parameter estimation of the model

| Age of subjects | Correlation type | $r$ | $N$ of pairs | Source |
|---|---|---|---|---|
| Year 1 | MZ Twins | .68 | (89) | Wilson (1983) |
| Year 1 | DZ Twins | .63 | (92) | Wilson (1983) |
| Year 2 | MZ Twins | .81 | (88) | Wilson (1983) |
| Year 2 | DZ Twins | .73 | (115) | Wilson (1983) |
| Year 3 | MZ Twins | .88 | (104) | Wilson (1983) |
| Year 3 | DZ Twins | .79 | (125) | Wilson (1983) |
| Year 4 | MZ Twins | .83 | (105) | Wilson (1983) |
| Year 4 | DZ Twins | .71 | (120) | Wilson (1983) |
| Year 7 | MZ Twins | .84 | (116) | Wilson (1983) |
| Year 7 | DZ Twins | .59 | (119) | Wilson (1983) |
| Adult | MZ Twins | .86 | (1300) | Loehlin & Nichols (1976) |
| Adult | DZ Twins | .62 | (864) | Loehlin & Nichols (1976) |
| Adult | Father-offspring | .33 | (1357)[a] | DeFries et al. (1979) |
| Adult | Mother-offspring | .38 | (1358)[b] | DeFries et al. (1979) |
| Adult | Nonadoptive siblings | .31 | (455)[c] | DeFries et al. (1979) |

MZ, monozygotic; DZ, dizygotic.
[a] Pooled father-son and father-daughter correlations for Americans of European ancestry (AEA).
[b] Pooled mother-son and mother-daughter AEA correlations.
[c] Pooled from three categories of AEA sibling correlations.

yet reported; it has the advantage of having tested both parents and offspring with the same battery. Although the HFSC reported results separately by offspring gender, the correlations were pooled before model fitting was attempted, and were weighted with the inverse of the sampling variance for the correlations. The use of parent-offspring adult data is necessary to resolve the cultural transmission parameters $m$ and $f$ for the adult measure.

The expectations for the parent-offspring and sibling data from the HFSC come directly from the expectations for adult measures in nonadoptive families listed in Table 12-3. The extension of the expectations for the basic sibling model in nonadoptive families to the twin data was straightforward. For DZ twins, the only difference in expectations is the substitution of a residual twin ($\mathbf{R}_T$) matrix for the corresponding residual sibling ($\mathbf{R}_S$) matrix. The expectations for MZ twins are identical to those for DZ twins except for the additive genetic component, which reflects the identity of their genotypes at each age. Table 12-6 presents the multivariate expectations for covariances of measures between MZ and DZ twins.

## Additional Assumptions Made When Fitting the Model

A number of more or less defensible assumptions were made in fitting the model to the data. The assumption that there is no direct contribution of the childhood cognitive phenotypes to marital assortment appears largely justified because most dating and tentative assortment occurs after childhood. Because children are exposed only to their parents as adults, only those elements of **m** and **f** that correspond to cultural

**Table 12-6** Expected covariances for monozygotic (MZ) and dizygotic (DZ) twins

| Covariance | Expectation for covariance |
|---|---|
| $P_{MZ1}, P_{MZ2}$ | $V_P^{1/2} \{h\, R_G\, h + e[m(R_p\, m' + M\, f') + f(R_P\, f' + M'\, m') + R_T]\, e + \frac{1}{2} h[(\tau' + \tau'\, D\, R_P)\, f' + (\tau' + \tau'\, D'\, R_P)\, m']\, e + e\, [m(\tau + R_P\, D\, \tau) + f(\tau + R_p\, D'\tau)] \frac{1}{2} h\} V_P^{1/2}$ |
| $P_{DZ1}, P_{DZ2}$ | $V_P^{1/2} \{\frac{1}{2} h\, [R_G + \frac{1}{2}(\gamma + \gamma')]\, h + e\, [m\, (R_p\, m' + M\, f') + f(R_p f' + M'\, m') + R_T]\, e + \frac{1}{2} h\, [(\tau' + \tau'\, D\, R_P)\, f' + (\tau' + \tau'\, D'\, R_p)\, m']\, e + e\, [m\, (\tau + R_p\, D\, \tau) + f(\tau + R_p\, D'\, \tau)] \frac{1}{2} h\} V_P^{1/2}$ |

*Note:* Expectations use the composite matrices $\tau$ ($h\, R_G + e\, s'$), M ($R_{PM}\, D\, R_{PF}'$), and $\gamma$ ($\tau'\, D\, \tau$). $R_p$ is assumed to be identical in all family members, so subscripting is dropped in expectations. $R_p$, $R_G$, $R_T$, h, and e are symmetrical, and thus transposition is ignored in these expectations.

transmission through the adult phenotype are left free. The other elements are set to zero.

The major assumption that selective placement for cognitive variables is negligible in the CAP sample is supported by several lines of evidence. The adoption agencies through which the CAP sample was recruited report that they do not attempt matching between biological parents and adoptive parents on the basis of perceived cognitive level. The univariate model fitting at each year found that the selective placement parameters could be set to zero without significant loss of fit, since the four observed correlations between the measures of general cognitive ability in the biological and adoptive parents ranged from 0 to $-.03$ in the complete CAP sample. The model's assumption that the level of assortment in wed and unwed couples is comparable was similarly supported by the results from univariate model fitting and by the very similar correlations of .29, .31, and .25 found for biological, adoptive, and nonadoptive couples in the full sample.

In the full matrix specification of the model, off-diagonal elements of $R_S$ and $R_T$ appear. However, in the current data set there are no data points with expectations involving these off-diagonal elements, and thus they are not estimated. These parameters are of considerable theoretical importance in measuring the persistence of shared environmental variance across time, but their exploration in the CAP will require patient waiting until both probands and siblings reach cognitive maturity.

The most problematic aspect of simplifying the model's parameters to fit the available data is obtaining reasonable estimates of phenotypic correlations between childhood measures and the adult measure. Phenotypic correlations among the infancy and childhood measures are directly estimable from the CAP offspring data, but only a fraction of the sample has usable IQ data as late as age 7. To estimate $R_E$ it is necessary to employ reasonable, nonzero point estimates of elements of $R_P$. Two plausible solutions meet this requirement. One is to use "best estimates" of child-adult phenotypic correlations based on the several reported longitudinal studies that tested their subjects' general cognitive ability at different ages. The second is to estimate what the observed correlations should be between childhood and adult measures on the basis of some sort of predictive formula. Jensen (1969) developed a simple formula for estimating cross-age correlations for which he claimed the virtues of simplicity and close fit to the observed results through age 10. Accepting age 10 as a suitable substitute for adulthood, the appropriate elements of $R_P$ can be derived.

Table 12-7 presents pooled estimates and derived estimates of the phenotypic correlations. In practice, estimates from the prediction formula were used to estimate $R_E$. It should be stressed, however, that although the estimates of $R_E$ are dependent on the assumed value for $R_P$, the other parameters of the model are estimated directly from the data and are not affected by the choice of an estimate for $R_P$.

## How the Data Were Fit to the Model

Data from the adoptive and nonadoptive CAP families were organized into two sets of pedigrees. Lange, Westlake, and Spence (1976) have demonstrated how a maximum-likelihood approach to parameter estimation can be extended to such unbalanced pedigrees. Their approach assumes that the distribution of scores in a pedigree is multivariate normal, an assumption that should be approximately true for polygenic systems in which there is no directional dominance or genotype–environment interaction (Eaves, Last, Young, & Martin, 1978). For a given pedigree, the log-likelihood of obtaining the vector of scores x is:

$$L = -\frac{1}{2} \ln |\Sigma| - \frac{1}{2} (\mathbf{x} - \mathbf{\mu})' \Sigma^{-1} (\mathbf{x} - \mathbf{\mu}) - \frac{1}{2} n \ln 2\pi \quad (12\text{-}8)$$

where $\Sigma$ is the expected covariance matrix of the vector of scores, x is the vector of scores in the pedigree, $\mu$ is the vector of expected means for the pedigree, and $n$ is

Table 12-7 Estimates of phenotypic correlations between early IQ tests and adult general cognitive ability

| Two ages correlated | Pooled estimate | Formula estimate[a] |
|---|---|---|
| Year 1, adult | .25[b] | .24 |
| Year 2, adult | .49[c] | .34 |
| Year 3, adult | .36[d] | .42 |
| Year 4, adult | .53[e] | .50 |
| Year 7, adult | .78[f] | .77 |

[a] Calculated using Jensen's (1969) formula for estimating the correlation between tests given at times 1 and 2:

$$\text{Est. } r_{12} = r_{tt} * \text{sqrt} \left(\frac{CA_1}{CA_2}\right)$$

where $r_{tt}$ is the reliability of the test, $CA_1$ is the chronological age at first testing, $CA_2$ is the chronological age at second testing, and $CA_2$ is truncated to age 10 for older ages.
[b] Median correlation of 7- to 12-month tests and 8- to 18-year follow-ups from McCall (1979).
[c] Median correlation of 19- to 30-month tests and 8- to 18-year follow-ups from McCall (1979).
[d] Median correlation of three-year test and 14- to 18-year follow-ups from Bayley (1949) and Honzik, MacFarlane, and Allen (1948).
[e] Median correlation of 4-year test and 14- and 18-year follow-ups from Bayley (1949) and from Honzik, MacFarlane, and Allen (1948).
[f] Median correlation of seven-year test and 14- and 18-year follow-ups from Bayley (1949) and from Honzik, MacFarlane, and Allen (1948).

the number of observations in the pedigree. The joint log-likelihood of obtaining a set of pedigrees is simply the sum of the log-likelihoods of the individual pedigrees. Maximum likelihood parameter estimates are obtained by minimizing twice the negative of the joint log-likelihood of the set of pedigrees, using the generalized numerical optimization package MINUIT (CERN, 1977). Alternative submodels with fixed parameters can be evaluated from pedigree data, since twice the difference between the joint log-likelihood of the reduced model and the joint log-likelihood of the full model yields a log-likelihood ratio that is distributed asymptotically as chi square (Lange et al., 1976).

Two other elements are added to the function value before minimization. One concerns overparameterization of the model. The model is parameterized in terms of the following matrices: $R_P$, $R_G$, $R_E$, $R_S$, $R_T$, $D$, $h$, $s$, $m$, and $f$. One parameter is also allotted for the expected phenotypic standard deviation of each measure. Certain constraints exist among these parameters:

$$\exp R_P = h R_G h + e R_E e + h s e + e s' h \qquad (12\text{-}9)$$

$$\exp s = \frac{1}{2}[ R_G h + s e + (R_G h + s e) D' R_p ] m'$$

$$+ \frac{1}{2}[ R_G h + s e + (R_G h + s e) D R_P ] f' \qquad (12\text{-}10)$$

The expected matrices are subtraced from the parameterized matrices on an element-by-element basis. The discrepancies are multiplied by a penalty weight and summed, and the total discrepancy is added to the function value, which is minimized (James, 1972). A series of convergence runs is utilized in which the penalty weight is increased each run until the discrepancies have been eliminated and no longer contribute to the function value. At that point the number of free parameters has been reduced by the number of parameters in $R_P$ and s.

Second, likelihoods for the non-CAP correlations used in the estimation are added to the function. Each correlation is treated as the off-diagonal element of a two-by-two covariance matrix that has ones on the diagonal. Each covariance matrix is compared with an expected covariance matrix derived from the expectations for that correlation. The function calculated for each such pair is:

$$F = (n - 1)[ \ln | \Sigma | - \ln | S | + \text{tr}(S \Sigma^{-1}) - p ] \qquad (12\text{-}11)$$

where $n$ is the degrees of freedom of the matrix, $p$ is the order of the matrix, $S$ is the observed covariance matrix, and $\Sigma$ is the expected covariance matrix. The sum of these $F$s, when minimized, yields a log-likelihood ratio statistic that is asymptotically distributed as chi square in large samples under the assumption of multivariate normality of the observed covariance.

The adequacy of alternative submodels that fix various parameters to hypothesized values can be tested by comparing the likelihood of the full model and the likelihood of the reduced model. The difference between log-likelihood ratios of alternative models is also distributed as chi square, with degrees of freedom equal to the

## Results from Fitting the Data to the Models

Table 12-8 lists parameter estimates when the full model was fit to data collected through March, 1987, at years 1, 2, 3, and 4 (Corley, 1987) and to data collected through September, 1987, at year 7. In addition to the parameters previously described, a few parameters having to do with the means and standard deviations of the measures are listed. The IQ scores for children were rescored before the derivation of pedigrees. A common mean for children with the same expectations in a family, for example, nonadoptive probands and their siblings, was calculated and subtracted from each child's actual score, and the difference was then divided by the theoretical standard deviation of 15. Both parental and child scores should thus be centered around 0 with a standard deviation near 1. This table reveals that many of the parameters are functions primarily of the extra correlations that were added to the CAP data set and not of the particular CAP pedigrees used at a given year. Thus the estimates of adult $h$, maternal and paternal phenotypic transmission effects on the adult, and adult residual sibling and twin correlations are nearly constant across the five sets of analyses. In addition, the estimates for the childhood residual twin correlations are largely dependent on the correlations for MZ and DZ twins that were taken from the Louisville Twin Study results. The interesting parameters in this table, which reflect changes in the CAP data over the five years of pedigree data, are the genetic and environmental correlations, the maternal and paternal effects on the

**Table 12-8** Estimates of parameters for full models at all five years

| Parameter description | Year 1 | Year 2 | Year 3 | Year 4 | Year 7 |
|---|---|---|---|---|---|
| Phenotypic $r$ | .24 | .34 | .42 | .50 | .77 |
| Genetic $r$ | .27 | .46 | .44 | .73 | .71 |
| Environmental $r$ | .21 | .26 | .36 | .30 | .85 |
| Adult $h$ | .75 | .75 | .75 | .75 | .74 |
| Child $h$ | .45 | .44 | .43 | .50 | .75 |
| Maternal effect on adult | .04 | .05 | .05 | .05 | .07 |
| Maternal effect on child | .03 | .06 | .09 | .14 | −.09 |
| Paternal effect on adult | −.06 | −.05 | −.05 | −.05 | −.03 |
| Paternal effect on child | .04 | .05 | .13 | .07 | .09 |
| Adult residual sibling $r$ | −.03 | −.03 | −.03 | −.03 | −.04 |
| Adult residual twin $r$ | .68 | .68 | .68 | .68 | .68 |
| Child residual sibling $r$ | .23 | .27 | .24 | .14 | .02 |
| Child residual twin $r$ | .61 | .76 | .82 | .72 | .62 |
| Adult assortative mating $d$ | .28 | .25 | .26 | .25 | .22 |
| Standard deviation adult measure | 1.00 | .99 | 1.00 | .97 | .96 |
| Standard deviation child measure | .80 | 1.03 | 1.00 | .84 | .72 |
| Mean of adult measure | −.00 | .02 | .01 | .03 | .07 |
| Mean child measure, adoptive | −.02 | −.02 | −.02 | −.02 | −.02 |
| Mean child measure, nonadoptive | −.01 | .00 | .00 | −.00 | .00 |

child, $h$ at the different years, and the residual sibling correlations at the different years.

The genetic correlations estimated from the full models are low at year 1, moderate at years 2 and 3, and relatively high at years 4 and 7. Until year 7, the environmental correlations are lower than the genetic correlations. The high level of both genetic and environmental correlations for year 7 reflects the high degree of stability in IQ between age 7 and adulthood. The estimates for childhood $h$ based on the combination of pedigree data involving siblings and the twin correlations remain relatively constant until year 7, in contrast to the picture presented by the twin data alone. By year 7, the estimate for $h$ is concordant with that for adult $h$. The maternal and paternal phenotypic transmission effects on children also do not show any strong systematic trend. There is slightly more evidence for these cultural transmission parameters at years 3 and 4 than at years 1 and 2, although the parent with the strongest effect varies between father and mother at the two ages. The apparent decrease in the estimate for the residual sibling correlation at years 4 and 7 from the level seen in the first three years may be simply the product of sampling variation due to smaller sibling samples at later ages, but it may also reflect a decrease in the importance of miscellaneous familial factors as children begin their school years.

Six alternative submodels were systematically tested with each year of data. These models were not exhaustive, but were of substantive interest. If the full model is designated Model 1, then Model 2 is a test of complete genetic continuity between childhood and adulthood, with $R_G$ set to one. In contrast, Model 3 is a test of genetic continuity, with $R_G$ fixed at zero. Model 4 tests whether in addition to lack of genetic continuity, childhood $h$ can be set to zero. Model 5 is a test of the importance of cultural transmission for both children and adults, with all elements of **m** and **f** set to zero. Model 6 tests whether there is any source of cross-generation transmission, with **m**, **f**, $R_G$, and childhood $h$ all set to zero. Model 7 adds to Model 6 one additional restriction, that the residual sibling correlation $R_S$ in childhood be set to zero. These models form a rough hierarchy of constraints, with Model 4, for example, a subset of both Model 1, the full model, and Model 3. Each submodel was tested against the full model and against any other submodel of which it could be considered a subset. If the model produced a function value difference that indicated a significant degradation in fit under the assumption that the differences were distributed as chi square, it was considered an unacceptable alternative model. Table 12–9 indicates which submodels were considered acceptable at each year.

In general, the model-testing procedures support the picture presented by the full models. The hypothesis of no genetic continuity cannot be rejected at year 1, is marginally rejected at year 2, is marginally acceptable at year 3, and is decisively rejected at years 4 and 7. The hypothesis of no cultural transmission can be rejected only at year 3. Only at year 1 can Model 4, which adds the constraint of zero childhood heritability to the constraint of a zero genetic correlation, not be rejected. The hypothesis that there is no source of transmission between adults and their children in childhood is not acceptable at any age. The failure of Model 6 at all ages means that Model 7 is also rejected, since the test whether the residual sibling correlation can be set to zero assumes that other sources of sibling resemblance, such as childhood $h$ and parental transmission, can be set to zero.

Table 12-9  Are particular submodels acceptable alternatives to the full model?

| Model description | Tested against: | Year 1 | Year 2 | Year 3 | Year 4 | Year 7 |
|---|---|---|---|---|---|---|
| Model 2: Complete genetic continuity ($R_G = 1.0$) | Full Model | Yes | Yes | No | Yes | Yes |
| Model 3: No genetic continuity ($R_G = 0$) | Full Model | Yes | No | Yes | No | No |
| Model 4: No genetic continuity ($R_G = 0.0$) and $h_C = 0$ | Full Model, Model 3 | Yes | No | No | No | No |
| Model 5: No cultural transmission ($m_{CA} = m_{AA} = m_{CA} = f_{AA} = 0$) | Full Model | Yes | Yes | No | Yes | Yes |
| Model 6: No transmission across generations (models 4 and 5) | Full Model, Models 4 and 5 | No | No | No | No | No |
| Model 7: No sources of familial resemblance in childhood (Model 6 and $R_S = 0$) | Full Model, Model 6 | No | No | No | No | No |

## SUMMARY

The results presented in the preceding section yield a relatively coherent picture of the development of genetic continuity between measures of general cognitive ability in infancy and early childhood and general intelligence in adulthood. They support the univariate model-fitting results, which suggest that genetic factors consistently contribute more to parent-offspring resemblance than does environmental exposure to the parental cognitive phenotype. They also illustrate the practicality and usefulness of including sibling data in Colorado Adoption Project (CAP) analyses in addition to the primary parent-offspring data. The combination of sibling data from the CAP and twin data from the Louisville Twin Study (LTS) tends to moderate the picture of increasing heritability during the years from age 1 to 7 for standardized IQ tests presented by the LTS results alone. As discussed by DeFries et al. (1987), combining twin correlations with parent-single offspring results from the CAP suggests that both heritability and genetic continuity increase over the four ages they investigated. The results of the analyses in the preceding section suggest that the dramatic increase is in the genetic correlation, whereas childhood $h$ (the square root of the heritability) stays relatively constant over the first four years of life and then jumps to its adult level by age 7. The increase in parent-offspring resemblance during the four years in which $h$ is relatively constant is attributed to an increasing genetic correlation, which goes from .27 at year 1 to .73 at year 4.

Another advantage of including sibling data in behavior–genetic analyses is that their scores reveal something about the persistence of environmental effects across time. For example, a comparison of the childhood residual twin correlation estimates with the corresponding childhood residual sibling correlations listed in Table 12-7 for all five years of IQ data strongly suggests that twins share environmental influences to a much greater extent than do nontwin siblings. A number of reasons are possible, including the fact that twins are typically tested on the same day and may be affected by similar short-term influences such as disturbances in the home or mutual aggravation. Siblings in the CAP are typically tested years apart and yet show a substantial residual sibling correlation on all measures except the year 7 WISC-R IQ scores. Dunn (1987) has reported that mothers of CAP siblings behave similarly to their children when the children are at the same age, on the basis of videotaped dyadic interactions. A persistent parenting style may be one source of similar performances of siblings at the same age. One other interesting finding that Dunn has reported is the relatively inconsistent nature of maternal behavior to the same child videotaped at different ages. Thus maternal behavior, unlike theoretically more stable environmental characteristics such as socioeconomic status, may contribute to sibling resemblance for tests given at the same age, but not to phenotypic stability.

Whether the trend observed for the genetic correlation between years 1 and 7 is due to age changes or to changes in the nature of the tests given at each age remains problematic. The model we have discussed in the preceding section is not a growth model, such as those presented by Eaves, Long, and Heath (1986) and Hewitt, Eaves, Neale, and Meyer (1988). Attempts to apply a similar model to cognitive growth using CAP data are currently being undertaken (Phillips, 1988; Phillips & Fulker, 1988), but it seems that measurement of individual differences in sensorimotor development may not be the best way to get at infant variation that is strongly related to cognitive performance at later ages. We hope that several large-scale collaborative twin projects currently being carried out at the Institute for Behavioral Genetics will yield information about potential alternative measures of general cognitive ability that are more predictive of subsequent cognitive performance and that may show a higher genetic correlation with later IQ measures. Such results would support the notion that increasing genetic correlations between adult cognitive measures and childhood measures over the first four years of life are not necessarily the result of age changes but may be attributable to the types of tests employed.

When compared with the less comprehensive earlier adoption studies, results obtained in the Colorado Adoption Project to date illustrate the power of a full-scale longitudinal prospective adoption study to explore developmental trends in cognitive development. We have focused primarily on intelligence in this chapter because these measures of cognitive development for infants and young children are widely used and because the changing nature of the tests over the years reflects the rapid developmental pace of cognitive growth. As the CAP children move into the period of early adolescence, we believe that developmental trends in other aspects of behavior will emerge with greater clarity. We believe that the longitudinal prospective adoption study is the optimal design for investigating the biological and cultural factors influencing development within the context of the nuclear family. We expect influences from schools and peers to grow in importance relative to parental influence as the children spend an increasing amount of time in other social contexts. Data are collected from the teachers of CAP children from the first grade onward to explore how children behave in a social contest other than the home. We hope that the potential of the Colorado Adoption Project for studying development in a host of domains will become increasingly realized as the sample moves toward adulthood.

## ACKNOWLEDGMENTS

We are grateful to the families participating in the Colorado Adoption Project, who have so generously contributed their time and interest, and to the adoption agencies who make the study possible—Lutheran Social Services of Colorado and Denver Catholic Community Services. This research was supported in part by grants HD-10333 and HD-18426 from the National Institute of Child Health and Human Development (NICHD), by grants BNS-7826204 and BNS-8200310 from the National Science Foundation, and by a grant from the Spencer Foundation. Preparation of the manuscript was facilitated by grant RR-07013-20 awarded to the University of Colorado by the Biomedical Research Support Grant Program, Division of Research Resources, National Institutes of Health. We thank Rebecca G. Miles for expert editorial assistance.

## REFERENCES

Bayley, N. (1949). Consistency and variability in the growth of intelligence from birth to eighteen years. *Journal of Genetic Psychology, 75,* 165–196.

Bayley, N. (1969). *Manual for the Bayley Scales of Infant Development.* New York: Psychological Corporation.

Bornstein, M. H., & Sigman, M. D. (1986). Continuity in mental development from infancy. *Child Development, 57,* 251–274.

Bouchard, T. J. Jr., & McGue, M. (1981). Familial studies of intelligence: A review. *Science, 212,* 1055–1059.

Brooks-Gunn, J., & Weinraub, M. (1983). Origins of infant intelligence testing. In M. Lewis (Ed.), *Origins of intelligence: Infancy and Early Childhood* (2nd ed., pp. 25–66). New York: Plenum.

Burks, B. (1928). The relative influence of nature and nurture upon mental development: A comparative study of foster parent-foster child resemblance and true parent-true child resemblance. *Twenty-Seventh Yearbook of the National Society for the Study of Education, 27*(1), 219–316.

Carey, G. (1986). A general multivariate approach to linear modeling in human genetics. *American Journal of Human Genetics, 39,* 775–786.

CERN (1977). *MINUIT: A System for Function Minimization and Analysis of Parameter Errors and Correlations.* Geneva: Center for European Nuclear Research.

Corley, R. P. (1987). *Genetic and environmental continuity among measures of general cognitive ability in infancy, early childhood, and adulthood using combined parent-offspring and sibling data from the Colorado Adoption Project.* Unpublished doctoral dissertation, University of Colorado, Boulder.

DeFries, J. C., Johnson, R. C., Kuse, A. R., McClearn, G. E., Polovina, J., Vandenberg, S. G., & Wilson, J. R. (1979). Familial resemblance for specific cognitive abilities. *Behavior Genetics, 9,* 23–43.

DeFries, J. C., & Plomin, R. (1978). Behavioral genetics. *Annual Review of Psychology, 29,* 473–515.

DeFries, J. C., Plomin, R., & LaBuda, M. C. (1987). Genetic stability of cognitive development from childhood to adulthood. *Developmental Psychology, 23,* 4–12.

DiLalla, L. J., Thompson, L. A., Plomin, R., Phillips, K., Fagan III, J. F., Haith, M. M., Cyphers, L. H., & Fulker, D. W. (1989). Infant predictors of preschool and adult IQ: A study of infant twins and their parents. (Manuscript submitted for publication).

Dunn, J. (1987, April). *Sibling differences and differential maternal behavior.* Paper presented at Biennial Meeting of Society for Research in Child Development, Baltimore.
Eaves, L. J., Last, K. A., Young, P. A., & Martin, N. G. (1978). Model-fitting approaches to the analysis of human behavior. *Heredity, 41,* 249–320.
Eaves, L. J., Long, J., & Heath, A. C. (1986). A theory of developmental change in quantitative phenotypes applied to cognitive development. *Behavior Genetics, 16,* 143–162.
Fagan, J. F. III (1984). The intelligent infant: Theoretical implications. *Intelligence, 8,* 1–9.
Fulker, D. W. (1982). Extensions of the classical twin method. In B. Bonné-Tamir, T. Cohen, & R. M. Goodman (Eds.), *Human Genetics. Part A: The Unfolding Genome* (pp. 395–406). New York: Alan R. Liss.
Fulker, D. W., & DeFries, J. C. (1983). Genetic and environmental transmission in the Colorado Adoption Project: Path analysis. *British Journal of Mathematical and Statistical Psychology, 36,* 175–188.
Fulker, D. W., DeFries, J. C., & Plomin, R. (1988). Genetic influence on general mental ability increases between infancy and middle childhood, *Nature, 336,* 767–769.
Goldberger, A. S. (1976). Mysteries of the meritocracy. In N. J. Block & G. Dworkin (Eds.), *The IQ Controversy: Critical Readings* (pp. 265–279). New York: Pantheon.
Hewitt, J. K., Eaves, L. J., Neale, M. C., & Meyer, J. M. (1988). Resolving causes of developmental continuity or "tracking." *Behavior Genetics, 18,* 133–151.
Honzik, M. P., MacFarlane, J. W., & Allen, L. (1948). Stability of mental test performance between 2 and 18 years. *Journal of Experimental Education, 17,* 309–322.
Horn, J. M. (1983). The Texas Adoption Project: Adopted children and their intellectual resemblance to biological and adoptive parents. *Child Development, 54,* 268–275.
Horn, J. M., Loehlin, J. C., & Willerman, L. (1979). Intellectual resemblance among adoptive and biological relatives: The Texas Adoption Project. *Behavior Genetics, 9,* 177–207.
Hunt, E. B. (1983). On the nature of intelligence. *Science, 219,* 141–146.
James, F. (1972). Function minimization (report no. 72-21). Geneva: Center for European Nuclear Research.
Jensen, A. R. (1969). How much can we boost IQ and scholastic achievement? *Harvard Educational Review, 39,* 1–123.
Jöreskog, K. G. (1969). A general approach to confirmatory factor analysis. *Psychometrika, 34,* 183–202.
Kail, R., & Pellegrino, J. W. (1985). *Human Intelligence: Perspectives and Prospects.* New York: W. H. Freeman.
Kamin, L. J. (1974). *The Science and Politics of IQ.* Potomac, MD: Lawrence Erlbaum.
LaBuda, M. C., DeFries, J. C., Plomin, R., & Fulker, D. W. (1986). Longitudinal stability of cognitive ability from infancy to early childhood: Genetic and environmental etiologies. *Child Development, 57,* 1142–1150.
Lange, K., Westlake, J., & Spence, M. A. (1976). Extensions to pedigree analysis. III. Variance components by the scoring method. *Annals of Human Genetics, 39,* 485–491.
Leahy, A. M. (1935). Nature–nurture and intelligence. *Genetic Psychology Monographs, 17,* 236–308.
Lewontin, R. C. (1975). Genetic aspects of intelligence. *Annual Review of Genetics, 9,* 387–405.
Loehlin, J. C., & Nichols, R. C. (1976). *Heredity, Environment, and Personality.* Austin, Tex.: University of Texas Press.
McCall, R. B. (1979). The development of intellectual functioning in infancy and the prediction of later IQ. In J. D. Osofsky (Ed.), *Handbook of Infant Development* (pp. 707–741). New York: Wiley-Interscience.
McCall, R. B. (1981). Nature–nurture and the two realms of development: A proposed integration with respect to mental development. *Child Development, 52,* 1–12.

Phillips, K. (1988). *Quantitative genetic analysis of longitudinal trends in IQ in the Colorado Adoption Project.* Unpublished doctoral dissertation, University of Colorado, Boulder.

Phillips, K., & Fulker, D. W. (1988). Quantitative genetic analysis of longitudinal trends in IQ in the Colorado Adoption Project. *Behavior Genetics, 18,* 729.

Plomin, R. (1986). Multivariate analysis and developmental behavioral genetics: Developmental change as well as continuity. *Behavior Genetics, 16,* 25–43.

Plomin, R. & DeFries, J. C. (1985). *Origins of Individual Differences in Infancy: The Colorado Adoption Project.* Orlando, FL: Academic Press.

Plomin, R., DeFries, J. C., & Fulker, D. W. (1989). *Nature and Nurture in Infancy and Early Childhood.* New York: Cambridge University Press.

Rice, T. (1987). *Multivariate path analysis of cognitive and environmental measures in the Colorado Adoption Project.* Unpublished doctoral dissertation, University of Colorado, Boulder.

Rice, T., Carey, G., Fulker, D. W., & DeFries, J. C. (1989). Multivariate path analysis of specific cognitive abilities in the Colorado Adoption Project: Conditional path model of assortative mating. *Behavior Genetics, 19,* 195–207.

Rice, T., Fulker, D. W., & DeFries, J. C. (1986). Multivariate path analysis of specific cognitive abilities in the Colorado Adoption Project. *Behavior Genetics, 16,* 107–125.

Rice, T., Fulker, D. W., DeFries, J. C., & Plomin, R. (1988). Path analysis of IQ during infancy and early childhood and an index of the home environment in the Colorado Adoption Project. *Intelligence, 12,* 27–45.

Scarr, S. (1983). An evolutionary perspective on infant intelligence: Species patterns and individual variations. In M. Lewis (Ed.), *Origins of Intelligence: Infancy and Early Childhood* (2nd ed.) (pp. 191–223). New York: Plenum.

Scarr, S., & Carter, L. (1983). Genetics and intelligence. In J. L. Fuller & E. C. Simmel (Eds.), *Behavior Genetics: Principles and Applications* (pp. 217–335). Hillsdale, NJ: Lawrence Erlbaum.

Scarr, S., & Weinberg, R. A. (1976). IQ test performance of black children adopted by white families. *American Psychologist, 31,* 726–739.

Scarr, S., & Weinberg, R. A. (1978). The influence of "family background" on intellectual attainment. *American Sociological Review, 43,* 674–692.

Scarr, S., & Weinberg, R. A. (1983). The Minnesota adoption studies: Genetic differences and malleability. *Child Development, 54,* 260–267.

Skodak, M., & Skeels, H. M. (1949). A final follow-up of one hundred adopted children. *Journal of Genetic Psychology, 75,* 85–125.

Sternberg, R. J. (1985). Human intelligence: The model is the message. *Science, 230,* 1111–1118.

Vogler, G. P. (1985a). *Multivariate path analysis of cognitive abilities in reading-disabled and control nuclear families and twins.* Unpublished doctoral dissertation, University of Colorado, Boulder.

Vogler, G. P. (1985b). Multivariate path analysis of familial resemblance. *Genetic Epidemiology, 2,* 35–53.

Wilson, R. S. (1983). The Louisville Twin Study: Developmental synchronies in behavior. *Child Development, 54,* 298–316.

Wohlwill, J. (1980). Cognitive development in childhood. In O. G. Brim & J. Kagan (Eds.), *Constancy and Change in Human Development* (pp. 359–444). Cambridge, Mass.: Harvard University Press.

Wright, S. (1934). The method of path coefficients. *Annals of Mathematics and Statistics, 5,* 161–215.

# 13

# Approaches to the Quantitative Genetic Modeling of Development and Age-related Changes

LINDON EAVES, JOHN K. HEWITT, JOANNE MEYER, AND MICHAEL NEALE

Competing theories about the aging process (e.g., Burch, 1968; Burnet, 1974; Holliday, 1975; and many others) often reduce to different ideas about how the effects of genes and environment change and accumulate with age. The main purpose of this chapter is to outline some of the components that might go into a quantitative genetic analysis of developmental and age-related changes in humans.

In spite of the obvious importance of age-related changes for our understanding of human biology and disease, geneticists have given scant attention to how such changes might be modeled. There are models for almost every aspect of family resemblance, including individual genes of large effect (Elston & Stewart, 1971), linkage (Ott, 1985), the effects of the multifactorial background on the expression of single major loci (Lalouel, Rao, Morton, & Elston, 1983), cultural inheritance and social interaction (Cavalli-Sforza & Feldman, 1973; Eaves, 1976a,b; Heath, Kendler, Eaves, & Markell, 1985; Rao, Morton & Yee, 1976; Rice, Clononger, & Reich, 1978), genotype × environment interaction (Eaves, 1984, 1987; Jinks & Fulker, 1970; Kendler & Eaves, 1986), and other processes involving genetic and environmental effects. The effects of age, however, have usually been treated as nuisance effects that have to be removed from family resemblance for continuous traits by more-or-less arbitrary regression techniques, or by corrections for penetrance of major loci based on age-of-onset distributions in the case of disease. None of these methods begins to address the basic question of why age changes occur in the way they do or whether the changes are under genetic control and themselves variable between individuals.

One class of theories of aging (e.g., Curtis, 1966) proposes that aging is associated with the progressive accumulation of random errors (e.g., somatic mutations). Another (e.g., Kirkwood, 1977) proposes that aging represents the progressive and possibly programmed deterioration of repair processes and homeostatic control. Several strands of evidence already suggest that age-related changes are not due just to the impact and accumulation of purely random environmental effects of the kind discussed by Hewitt in Chapter 11. Young, Eaves, and Eysenck (1980) showed that the genetic and environmental components of personality change with age—different genetic effects are expressed at different ages. Longitudinal twin studies, such as that of Wilson (e.g., 1983) show that the "spurts and lags" of cognitive development are themselves more highly correlated in monozygotic than in dizygotic twins—pointing to the genetic control of developmental changes. Province and Rao (1985) showed that the familial correlations for blood pressure changed significantly during adult life. Even the age of onset of such a well-characterized genetic disorder as Huntington's disease shows familial aggregation (Farrar & Conneally, 1985). The same is true of schizophrenia in siblings (Kendler, Tsuang, & Hays, 1988). Large twin studies have shown that health-related habits such as smoking and alcohol consumption show age-of-onset correlations in families (Eaves & Eysenck, 1980; Heath & Martin, 1988). Crucial milestones in people's lives, including age at marriage (Heath, 1983) and age of first intercourse (Martin, Eaves, & Eysenck, 1977), show familial correlations and even evidence for a genetic component in timing. The list could continue. The important point is that age is more than a nuisance when we try to understand the impact of genetic and environmental effects on human behavior, disease, and development. Differences in the timing and duration of events are themselves familial and may hold the key to many mechanisms responsible for human differences and disease.

Currently there is no single theory or model that might be used to analyze and interpret every kind of age-related change in humans. However, several approaches are tailored to particular kinds of questions and data. We shall attempt to describe these by reference to studies of our own that are currently in progress. In each case we try to ask first what the genetic (or environmental) determinants of the developmental (or aging) process might be and only then try to find ways of committing those ideas to a mathematical and, we hope, testable form.

We consider four basic issues, each beginning from a slightly different conception of the developmental process and each leading to a slightly different approach to modeling and data analysis. They are (1) linear time series models for the covariance structure of repeated measures, (2) models for age-related changes in homeostatic regulation of gene action and sensitivity to the environment, (3) multifactorial threshold models for the joint control of disease liability and age of onset, and (4) survival analysis models that incorporate genetic and environmental heterogeneity between members of a population.

## LINEAR TIME SERIES MODELS FOR COVARIANCE STRUCTURE OF KINSHIP DATA

It has been recognized for a long time that correlations between family members measured at different ages hold the key to the genetic analysis of age-related changes (e.g., Eaves, 1978; Eaves & Eysenck, 1980; Eaves, Last, Young, & Martin, 1978; Plomin & DeFries, 1985; Young, Eaves, & Eysenck, 1980). Initially, however, there was no theory to guide the analysis and interpretation of such correlations beyond the bland observation that correlations between relatives might decay as the age difference decreased because different genes might be expressed at different ages.

More recently, several researchers have begun to explore the application of linear time series models (Box & Jenkins, 1970) for the covariance structure of longitudinal measurements of related individuals (Eaves, Hewitt, & Heath, 1988; Eaves, Long, & Heath, 1986), especially twins and parents and their children. Other applications of time series in genetic epidemiology have been suggested for the analysis of sibling interactions (Carey, 1986) and repeated measures (Boomsma & Molenaar, 1987). Such models provide greater insight and promise greater parsimony in predicting the patterns of developmental change in family data. Studies using these kinds of methods are recognized as "state of the art" for the genetic analysis of developmental changes in humans.

The questions addressed by the theory of time series applied to age-related changes in kinship data are (1) Can we detect differences in the accumulation of genetic and environmental effects over time? (2) Do the same genes and environmental effects operate throughout development, or do the sources of genetic and environmental influence change with age? Many longitudinal and cross-sectional family data have been collected without any testable developmental theory to guide their interpretation. The necessary theory is now available in the form of a model for development that makes quantitative predictions about the patterns of variances and covariances that will be observed in family data. Technical aspects of the model are given in Eaves, Long, & Heath (1986) and Eaves et al. (1988). The ideas are simple enough, however. Ignoring nonadditive effects, there are three main sources of individual differences: genes, the environment shared by family members, and the environment unique to the individual. As far as development is concerned, there is no reason why these effects should all work in the same way. For example, the effects of a given set of genes may be exerted continuously throughout development but the quality of the unique environment may fluctuate from time to time. The effects of genes may be stable and uniform over time, but the effects of the environment may gradually accumulate. Many of the consequences of these basic ideas can be traced mathematically using a linear time series model that allows separate terms for the contribution and persistence of genetic and environmental effects. The basic mathematical model assumes that a set of behavioral measures, $\mathbf{X}_t$, made on members of a family at time $t$, can be expressed as a linear combination of previous phenotypic values, $\mathbf{X}_{t-1}$, and currently expressed genetic and environmental effects, $\mathbf{R}_t$. The linear model is thus

$$\mathbf{X}_t = \mathbf{F}\mathbf{X}_{t-1} + \mathbf{G}\mathbf{R}_t$$

The development of the covariance structure with age can then be traced as a function of the coefficients in $\mathbf{F}$ and $\mathbf{G}$. The matrix $\mathbf{F}$ defines the impact of previous values of the phenotype on current values and thus captures some elements of the developmental process. $\mathbf{F}$ may also specify interaction between family members (Carey, 1986). The phenotypic covariance structure also depends on the covariances between the latent genetic and environmental effects ($\mathbf{R}$) at different times. This, in turn, will depend on whether the same genes and environments are expressed at different ages or whether there are also age-specific genetic and environmental effects. Figure 13–1 illustrates one possible model for developmental change in graphic form. In this model we allow not only phenotype-to-phenotype transmission but also the

# APPROACHES TO QUANTITATIVE GENETIC MODELING

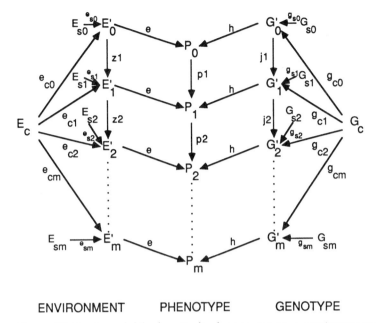

ENVIRONMENT  PHENOTYPE  GENOTYPE

**Figure 13-1** A model for longitudinal measures on a continuous trait.

possibility that genetic effects, $G'_i$, may be transmitted differently than environmental effects, $E'_i$.

Eaves, Long, & Heath (1986) showed that the variances and family correlations of continuous variables are expected to change in different ways depending on the stability and accumulation rates of genetic and environmental effects. These theoretical findings provide the basis for inferring some elementary features of the developmental process from longitudinal and cross-sectional data from genetically informative studies.

A crucial step in theory development in genetic epidemiology is showing that theoretical distinctions can be translated into manageable and informative empirical studies. The strategy evolved to address this issue involves generating, either deterministically or by simulation, hypothetical data sets that reflect one form of the theory and then analyzing them "as if" they were actual data to see how reliably and accurately the original model can be discriminated from equally plausible alternatives. This method has been employed widely in the past (e.g., Eaves, 1972, 1984; Eaves & Jinks, 1972; Eaves, Kendler, & Schulz, 1986; Heath & Eaves, 1985; Heath et al., 1986; Martin, Eaves, Kearsey, & Davies, 1978) for studying power and bias for testing relatively simple hypotheses, involving relatively few degrees of freedom ($df$) for error and small numbers of parameters. A problem in analyzing longitudinal data is that the data rapidly generate a large number of summary statistics and the hypotheses to be compared are sometimes complex. Hewitt and Heath (1988) extended the argument to more complex cases and provided a graph and algorithm for computing the crucial noncentrality parameter of the noncentral chi-square distribution of large $df$. Subsequently, Hewitt, Eaves, Neale, and Meyer (1988) explored the general effects on resolving power of different heritabilities, different degrees of fidelity in transmission of genetic and environmental effects over time, and different numbers of repeated measures. A major question to be answered by such studies was whether it was possible, in practice, to distinguish between developmental models of the kind described here and the less restrictive, but theoretically less potent, multiple factor models for genetic and environmental covariance structure (e.g., Martin & Eaves, 1977). They showed that "providing we measure on at least four occasions, it is easy to detect developmental transmission with workable sample sizes." This finding is significant for the design of future studies and is already the basis for the design of our own longitudinal twin study of cardiovascular function (Schieken, Eaves, Hewitt, Mosteller, Bodurtha, Moskowitz, & Nance, 1989).

Over the past three years we have applied the developmental model to longitudinal studies and cross-sectional data. Although longitudinal studies are more informative, we can still test some developmental hypotheses with cross-sectional kinship data because a genetic or social relationship between individuals measured at different ages reflects, in attenuated form, the relationship to be expected between repeated measures of the same person.

Eaves, Long, & Heath (1986) reanalyzed Wilson's (1983) longitudinal twin data on cognitive development. The best fitting model required genes, shared family environment, and unique environment. Furthermore: (1) the same genes were expressed throughout development with highly persistent and cumulative effects, (2) the effects of the family environment were also highly persistent but were ameliorated in the long term because the quality of the family environment was not consistent over time (different types of environment matter at different times), and (3) the effects of the unique environment were occasion specific and short lived. The analysis illustrates the powerful alliance between genetically informative studies (e.g., twins), longitudinal follow-up, and time series models that differentiate between latent genetic and environmental sources of continuity and change.

Hewitt, Carroll, Sims, and Eaves (1987; and Chapter 11 in this volume) have analyzed a new British data set on systolic and diastolic blood pressure in young adults and their parents. The same genes operate all the time, but their effects do not get bigger with age. The increase in the variance of both systolic and diastolic blood pressure with age is most probably due to increases in the impact of unique, random environmental effects. This finding shows how statistical genetic methods can resolve one theory of the aging process, that is, accumulation of random stochastic error, from others that involve genetic change.

Corey, Eaves, Mellen, and Nance (1986) applied the model, in a "continuous time" derivation, to the analysis by maximum likelihood of cross-sectional height, weight, and blood-pressure data on a group of families of MZ twins in which the individuals were measured at a wide range of different ages. No systematic age changes in the genetic control of blood pressure were found, but the genetic control of body weight differences changed with age. As Hewitt et al. (1987) have pointed out, the negative result for blood pressure of Corey and co-workers may have resulted from the attempt to model data across the entire life span in terms of a single developmental process.

Jardine (1985) tried to use our developmental models to predict age-related changes in twin correlations for alcohol consumption from a study of over 4000 Australian twin pairs. No form of the developmental model could account for her results.

This shows that our model cannot explain everything and stimulates our exploration of alternatives.

Our studies so far have focused on the analysis of single variables and have ignored the developmental consequences of social interactions. However, the model extends very easily to the multivariate case (Eaves et al., 1988) and can incorporate the social interaction between family members as long as the initial conditions underlying the interaction can be specified (Carey, 1986). Thoughtful multivariate studies are likely to become important because genetic epidemiologists are concerned not just about disease end points and individual physiological variables but with some of the intervening risk factors and processes. Thus, for example, the genetic study of cancer and cardiovascular disease cannot be fully effective without measures of salient environmental variables. However, these environmental indices have a life of their own at the same time that they contribute to risk. They change with time, they are correlated in families, and they may be affected by the behavior of family members and peers. Thus, to analyze in detail age-related changes in health and behavior we have also to characterize, jointly, the developmental changes in salient covariates.

Any model is an approximation. Among the limitations of the time series models are the assumptions of genetic additivity and additivity of genetic and environmental effects; the problem of handling assortative mating and cultural inheritance when the timing of milestones such as marriage and birth affect the contribution of genetic and cultural factors; the fact that our model for continuous variation does not map very well onto the analysis of age of onset for disease states or the timing of developmental milestones. Many of these issues lead to the pursuit of extensions of the time series model and to the alternative approaches that we now discuss.

## AGE CHANGES IN GENE REGULATION AND SENSITIVITY TO THE ENVIRONMENT

It has been argued that homeostatic mechanisms are "tuned" by natural selection to be most effective when selection is strongest, that is, during the typical reproductive years. Afterward there will be little selection pressure to remove deleterious alleles (e.g., Haldane, 1941; Williams, 1957; and others). Normal homeostatic mechanisms established by selection in the reproductive years are expected to break down with advancing age. This theory may be testable with appropriate kinship data.

Homeostatic mechanisms can be of two main kinds—they can involve either regulation of genes by other genes ("genetic homeostasis," Lerner, 1954; or "epistasis," Mather, 1967, 1974) or the control of sensitivity to the environment ("genotype $\times$ environment interaction, $G \times E$," e.g., Mather & Jinks, 1982). If aging reflects the progressive failure of homeostasis, then there should be changes in the pattern of means, variances, and correlations between relatives as individuals age. Turning the argument on its head, it may be possible to *infer* some basic properties of the homeostatic mechanisms involved in aging by analyzing age changes in statistics derived from family data. This requires further theoretical study of the changes in the pattern of age-dependent family resemblance caused by breakdown in epistatic gene regulation and the genetic control of sensitivity to the environment.

The basic argument may be stated as follows:

1. Homeostatic control is tuned by selection during the reproductive years.
2. Selection is relatively weak on phenotypic effects in later life, so control is likely to be poorer in older people.
3. Aging thus may reflect progressive deterioration in homeostatic control.
4. Homeostatic control is due to either (a) regulation of gene expression by other genes (epistasis) and/or (b) genetic regulation of sensitivity to the environment ("genotype × environment interaction").
5. Different kinds of epistasis and G × E result from different types of selection (Mather, 1966, 1974; Mather & Jinks, 1982).
6. Different kinds of gene action and G × E have different effects on means, variances, and correlations in family data (e.g., Eaves, 1982; Mather & Jinks, 1982).
7. *It follows that* (a) age-related changes in homeostasis will produce age-related changes in kinship data (e.g., twins) and (b) analysis of the patterns of change may allow identification of the kinds of regulation involved.

Eaves (1988) used Mather's basic model for epistasis to show that (1) with duplicate gene interactions, a characteristic of traits under strong directional selection, there is a marked reduction in genetic variance and in the correlation between siblings (or dizygotic twins) relative to that for monozygotic twins and (2) complementary gene action, characteristic of strong stabilizing selection, leads to increases in genetic variance and does not reduce the DZ correlation relative to that for MZ's. Thus, if the regulation of gene action deteriorates with age under either set of conditions, we should expect the pattern of variances and covariances between relatives to change as a function of age so that we might infer the underlying mechanisms of change from statistical analysis of family resemblance. Table 13–1 gives an example of how means, variances, and sibling correlations might change when the effects of directional dominance and duplicate gene interactions are removed by alteration of gene regulation. The figures reflect the kinds of changes which might occur in these statistics if such effects are gradually "switched off" with advancing age and make us relatively optimistic that the method may be informative about the relationship between homeostasis and aging, but appropriate power studies are still needed.

**Table 13–1** Example of effects of genetic nonadditivity on population mean ($\mu$), genetic variance ($\sigma^2$), and sibling correlation ($\rho$)

| | | Epistasis ($\theta$) | | | | | | | | |
|---|---|---|---|---|---|---|---|---|---|---|
| | | 0 | | | −0.5 | | | −1.0 | | |
| | Intensity | $\mu$ | $\sigma^2$ | $\rho$ | $\mu$ | $\sigma^2$ | $\rho$ | $\mu$ | $\sigma^2$ | $\rho$ |
| | 0 | 1.20 | 0.64 | 0.50 | 1.02 | 0.34 | 0.48 | 0.84 | 0.20 | 0.37 |
| Dominance ($h$) | 0.5 | 1.52 | 0.36 | 0.47 | 1.23 | 0.15 | 0.45 | 0.94 | 0.05 | 0.31 |
| | 1.0 | 1.84 | 0.31 | 0.33 | 1.42 | 0.10 | 0.32 | 0.99 | 0.02 | 0.13 |

*Notes:* Interaction in a digenic system is assumed with the frequency of the increasing allele being 0.8 at both loci. Positive values of $h$ imply directional dominance for increasing trait value. Negative values of $\theta$ imply duplicate gene interaction when $h$ is positive.

With a constant gene frequency and constant additive deviations, alterations in the mechanism of gene regulation can produce a 30-fold change in genetic variance and a fourfold change in the genetic correlation of siblings.

## RELATIONSHIP BETWEEN LIABILITY AND ONSET IN TIME-DEPENDENT MEASURES

The models outlined previously may extend to the treatment of discontinuous phenotypic *states* (e.g., achievement of a developmental milestone or presence of a disease) and may certainly yield insight about the way in which the changing expression of genes and environment may lead to phenotypic changes over the individual life span. They are, however, most appropriate for the interpretation of continuous variables that can be measured at arbitrarily chosen ages, preferably in a longitudinal study. Often the data that face the genetic epidemiologist do not come in this form. Rather we have data on the *timing* (age of onset) of a particular event. The distinction is somewhat arbitrary, since many diseases such as depression are not persistent and the patient is asymptomatic for longer or shorter periods. Typically, however, the genetic study of disease treats the disease as a phenotype that, once expressed, indicates a latent genetic vulnerability.

Most genetic models for age-related traits treat age of onset as a random variable that is uncorrelated with liability for the disease and uncorrelated between family members. The second of these assumptions is manifestly false for Huntington's disease and schizophrenia. The risk that we compute for a relative of a patient with an age-related disease is a function not just of the biological relationship between proband and relative and the age of the relative seeking advice, but also of the age of onset of the proband and the mechanism relating the causes of the disease to variation in age of onset. Being wrong about either can lead to quite large errors in predicted risk ratios under some circumstances. Furthermore, a common genetic basis to disease liability and age of onset implies that understanding of the etiology of the disease could shed some light on the processes of development and aging. Thus, the ability to analyze simultaneously the control of onset and disease liability for a variety of age-related diseases may have heuristic value in the study of aging itself.

Analysis of the relationship between causes of disease liability and causes of variation in age of onset may be set against a more general theme of the effects of selection, truncation, and ascertainment on the analysis of continuous variables (Neale, Eaves, Hewitt, Maclean, Meyer, & Kendler, 1989; Neale, Eaves, Kendler, & Hewitt, 1989).

We start by considering only genetic effects on family resemblance, and assume that the joint effects of genes and the unique environment are normally distributed (i.e., there are no individual genes of large effect—"major genes"). Even within this simple framework, however, a number of possible causal relationships can hold between liability to disease and the control of age of onset. These carry different implications for the analysis of the disease process and the relationship between aging and disease.

Figure 13-2 shows a path diagram that summarizes some of the possible relationships between the genetic control of liability (L) and age of onset (A). The standard rules of path analysis allow us to derive algebraic or numerical expectations for the correlations between onset and liability in any set of kinship data *sampled at random from the population* (see, e.g., Neale, Eaves, Hewitt, Maclean, Meyer, & Kendler, 1989). So far we have considered only pairs of relatives (e.g., twins, siblings,

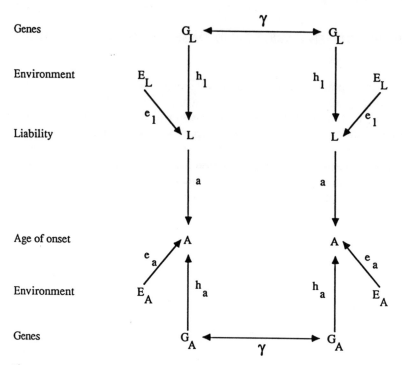

**Figure 13-2** One possible model for the relationship between the genetic control of liability for a disease (L) and age of its onset (A).

parent-child pairs). In practice, a pair will only be ascertained if one or both members have a sufficiently high liability that they develop the disorder. It has usually been assumed that there is a single sharp threshold above which people become affected. This need not be the case, and our approach can deal with either (see Neale, Eaves, Kendler, & Hewitt, 1989). The kind of threshold model *does* affect the correlation between relatives in ages of onset. The expected moments of the truncated samples may be evaluated by numerical integration using Gaussian quadrature. We can thus determine the expected correlation between ages of onset in pairs of individuals concordant for being affected. Typical results are given in Tables 13-2 and 13-3.

For a relatively rare disorder under the threshold model, large correlations in ages of onset can arise only when there is specific genetic control of onset that is independent of liability. This is apparently the case for schizophrenia (Kendler et al., 1988).

In addition to the straightforward correlation in age of onset, we can trace the expected risk to relatives of an affected proband as a function of the age of onset of the proband and the age of follow-up of the relative. Again we find (see Table 13-2) that the predicted risk to the relative of an affected proband as a function of age of onset and follow-up depends very strongly on the heritability of age of onset and the genetic correlation between onset and liability. This is true even when we hold constant the genetic contribution to liability. Thus, the advice we might give to the relative will depend on knowledge not just of the inheritance of the disease, but also on the genetic and environmental control of onset and its relationship to the disease itself.

**Table 13-2** Examples of correlations in ages of onset expected in monozygotic twin pairs concordant for disease of different heritabilities of liability and different correlations between liability and age of onset

|  |  | Heritability of liability | | | |
|---|---|---|---|---|---|
|  |  | .9 | .7 | .5 | .3 |
| Incidence 1% | .9 | 0.178 | 0.084 | 0.040 | 0.016 |
| Liability— |  |  |  |  |  |
| onset | .8 | 0.093 | 0.045 | 0.021 | 0.009 |
| Correlation | .7 | 0.055 | 0.027 | 0.013 | 0.005 |
|  | .5 | 0.020 | 0.010 | 0.005 | 0.002 |
| Incidence 10% |  |  |  |  |  |
| Liability— |  |  |  |  |  |
| onset | .9 | 0.284 | 0.149 | 0.077 | 0.034 |
| Correlation | .8 | 0.160 | 0.087 | 0.045 | 0.020 |
|  | .7 | 0.098 | 0.054 | 0.028 | 0.012 |
|  | .5 | 0.038 | 0.021 | 0.011 | 0.005 |

## GENETIC AND ENVIRONMENTAL VARIATION IN SURVIVAL MODELS

Survival analysis models (e.g., Elandt-Johnson & Johnson, 1979) have been used to analyze the aging process (see, e.g., Burch, 1968). The premise of such methods is that the distribution of survival and failure times (ages of onset) reflects the complexity of the process underlying aging and disease. Such approaches tend to treat all individuals as homogeneous, or regard any heterogeneity as explicable by measurable covariates (e.g., Cox, 1972). However, most of the variation in human traits is not accounted for by measurable covariates but by latent genetic and environmental influences. Thus, survival models that ignore genetic and environmental heterogeneity are not expected to fit the data on the timing of disease and death.

Some attempts have been made to extend survival models to allow for latent classes in the distribution of the survival parameters (e.g., Holt and Prentice, 1974;

**Table 13-3** Example of effect of genetic correlation between onset and liability for threshold disorder with 1 percent frequency in population on the risk to co-twin as a function of proband's standardized age of onset and standardized age of co-twin

| Age of co-twin | Standardized age of onset of proband | | | | |
|---|---|---|---|---|---|
|  | −2.0 | −1.0 | 0.0 | 1.0 | 2.0 |
| −2.0 | .37 | .29 | .25 | .22 | .21 |
| −1.0 | .32 | .25 | .22 | .20 | .18 |
| 0.0 | .20 | .16 | .14 | .12 | .11 |
| 1.0 | .07 | .05 | .04 | .04 | .04 |
| 2.0 | .01 | .01 | .01 | .01 | .00 |

*Note:* Computation assumed following values for paths in Figure 13-2: $h_l = 0.837$; $h_a = 0.0$ (i.e., no specific genetic control of onset); $a = -0.8$.

Hougaard, 1986). However, these either have estimated nuisance parameters to account for population heterogeneity or have assumed stable distributions that do not reflect adequately the well-tried and flexible models of genetic epidemiology for the latent causes of variation. Murphy and his colleagues (e.g., Murphy & Trojak, 1987; Murphy, Trojak, Berger, & Foster, 1987; Murphy, Trojak, Hou, & Rohde, 1981) have developed an elegant model of survivorship—the "bingo model"—that assumes that the ultimate failure of a system is due to the failure of one of a number of competing independent processes. The treatment of genetic effects in this model has shown the folly of applying classic genetic methods (such as the conventional multifactorial model) uncritically to failure time data. However, the practical problem of analyzing the kinds of genetic effects underlying differences in survival has still to be tackled. Our studies have taken a somewhat different approach. We are now at the stage of needing to know how much "analysis" can actually be achieved with collectible data.

Any genetic model of survival must combine three features: (1) the "genetic model" specifying the latent genetic and environmental sources of variation; (2) the "survival model," which accounts for the process that generates failure and survival times among individuals with comparable latent trait values; and (3) the "link model," which relates the latent variables to the parameters of the survival model (Eaves, 1987b).

Meyer and Eaves (1988) have begun to examine the properties of one model. Their genetic model assumed additive polygenic inheritance with residual random environmental effects uncorrelated between family members. The survival model assumed that failure times for individuals with the same survival parameter, $\lambda$, follow the gamma distribution for a given (integer) number of "hits" required for failure. The link model relating genetic and environmental effects to $\lambda$ is assumed to be: $\lambda = \exp(b + dx)$, $b$ and $d$ being constants and $x$ the (standardized) latent variable to which genes and environment are assumed to contribute. This function preserves additivity in the components of the latent variable and retains the necessary constraint that the $\lambda$ be positive for all $x$. The same link model is used in other treatments of heterogeneity between groups in recent models for survival (e.g., Hougaard, 1986). The survival model used by Meyer and Eaves assumes that all the heterogeneity in survival and failure time is caused by variation in the failure rate of components in the organism. Clearly this is not the only possibility. Some individuals may begin life with a greater "load" of disadvantageous alleles, that is, they have fewer components to fail during life because more have failed at the start. The feasibility of resolving this aspect of the survival model from variation in failure rate reflecting a background of latent genetic and environmental variation remains to be explored.

Meyer and Eaves have developed a FORTRAN computer program SURGEN that computes and maximizes the likelihood of a set of paired observations (survival or failure times) with respect to the parameters of a (currently linear) model for the correlations between latent traits and the constants $b$ and $d$ of the link function for a given number of "hits" ($N$) in the survival model. Preliminary results from application of the algorithm to simulated data on MZ and DZ twins show that we can discriminate between alternative models for the number of hits within the framework of a genetic study and estimate with reasonable precision the parameters of the genetic and link model. Questions still to be addressed include analysis of major gene effects, the effects of censoring, and competing causes of failure.

## CONCLUSION

In the past, age has been seen as an inconvenience in the analysis of genetic and environmental effects rather than as a problem of intrinsic appeal. Increasingly, it is clear that many fundamental problems in the study of human behavioral differences and the prediction of disease outcomes require that we begin to specify more exactly the relationship between the expression of genes and the course and timing of developmental processes. Throughout this chapter we have charted the progress that has been made so far with several different approaches to this problem. We have in each case noted the most obvious gaps in our understanding and what we see as the next steps in development of the appropriate theory. The approaches we have outlined do not yet cohere in an elegant whole, but serve to represent some of the basic problems and principles out of which a more coherent strategy might be forged.

## REFERENCES

Boomsma, D. T., & Molenaar, C. M. (1987). Genetic analysis of repeated measures. I. Simplex models. *Behavior Genetics, 17,* 111–124.

Box, G. E. P., & Jenkins, G. M. (1970). *Time Series Analysis: Forecasting and Control.* San Francisco: Holden Day.

Burch, P. R. J. (1968). *An Inquiry Concerning Growth, Disease and Aging.* Edinburgh: Oliver and Boyd.

Burnet, S. M. (1974). *Intrinsic Mutagenesis: A Genetic Approach to Ageing.* New York: John Wiley.

Carey, G. (1986). Sibling imitation and contrast effects. *Behavior Genetics, 16,* 319–342.

Cavalli-Sforza, L. L., & Feldman, M. W. (1973). Cultural versus biological inheritance: Phenotypic transmission from parent to children (a theory of the effect of parental phenotypes on children's phenotypes). *American Journal of Human Genetics, 25,* 618–637.

Corey, L. A., Eaves, L. J., Mellen, B. G., & Nance, W. E. (1986). Testing for developmental changes in gene expression for quantitative traits in kinships of twins: Application to height, weight and blood pressure. *Genetic Epidemiology, 3,* 73–83.

Cox, D. R. (1972). Regression models and life tables (with discussion). *Journal of the Royal Statistical Society, Series B, 34,* 187–220.

Curtis, H. J. (1966). *Biological mechanisms of aging.* Springfield, IL: C. C. Thomas.

Eaves, L. J. (1972). Computer simulation of sample size and experimental design in human psychogenetics. *Psychological Bulletin, 77,* 144–152.

Eaves, L. J. (1976a). A model for sibling effects in man. *Heredity, 36,* 205–214.

Eaves, L. J. (1976b). The effects of cultural transmission on continuous variation. *Heredity 37,* 69–81.

Eaves, L. J. (1978). Twins as a basis for the causal analysis of personality. In W. E. Nance (Ed.), *Twin Research: Psychology and Methodology* (pp. 151–174). New York: Alan Liss.

Eaves, L. J. (1984). The resolution of genotype × environment interaction in segregation analysis of nuclear families. *Genetic Epidemiology, 1,* 215–228.

Eaves, L. J. (1987a). Including the environment in models for genetic segregation. *Journal of Psychiatric Research, 21*, 639–647.

Eaves, L. J. (1987b). Survival analysis: Lessons from quantitative genetics. In A. D. Woodhead, & K. H. Thompson (Eds.) *Evolution of Longevity in Animals,* Plenum Publishing Corporation.

Eaves, L. J. (1988). Dominance alone is not enough. *Behavior Genetics, 18,* 27–33.

Eaves, L. J., & Eysenck, H. J. (1980). The genetics of smoking. In H. J. Eysenck (Ed.), *The Causes and Effects of Smoking.* London: Temple-Smith.

Eaves, L. J., Hewitt, J. K., & Heath, A. C. (1988). The quantitative genetic study of developmental change: A model and its limitations. In B. S. Weir, E. J. Eisen, M. M. Goodman, & G. Nomkoong, (Eds.), *The Second International Conference on Quantitative Genetics.* Sunderland, MA: Sinauer Associates.

Eaves, L. J., & Jinks, J. L. (1972). Insignificance of evidence for differences in heritability of IQ scores between races and social classes. *Nature, 240,* 84–88.

Eaves, L. J., Kendler, K. S., & Schulz, S. C. (1986). The familial versus sporadic classification: Its power for the resolution of genetic and environmental etiologic factors. *Journal of Psychiatric Research, 20,* 115–130.

Eaves, L. J., Last, K. A., Young, P. A., & Martin, N. G. (1978). Model-fitting approaches to the analysis of human behaviour. *Heredity, 41,* 249–320.

Eaves, L. J., Long, J., & Heath, A. C. (1986). A theory of developmental change in quantitative phenotypes applied to cognitive development. *Behavior Genetics, 16,* 143–162.

Elandt-Johnson, R. C., & Johnson, N. L. (1979). *Survival Models and Data Analysis.* New York: John Wiley.

Elston, R. C., and Stewart, J. (1971). A general model for the genetic analysis of pedigree data. *Human Heredity, 21,* 523–542.

Farrar, L. A., & Conneally, P. M. (1985). A genetic model for age at onset in Huntington's disease. *American Journal of Human Genetics, 37,* 350–357.

Haldane, J. B. S. (1941). The relative importance of principal and modifying genes in determining some human diseases. *Journal of Genetics, 41,* 149–157.

Heath, A. C. (1983). Human quantitative genetics: Some issues and applications. Unpublished doctoral thesis, University of Oxford, Oxford.

Heath, A. C., & Eaves, L. J. (1985). Resolving the effects of phenotype and social background on mate selection. *Behavior Genetics, 15,* 15–30.

Heath, A. C., Kendler, K. S., Eaves, L. J., & Markell, D. (1985). The resolution of cultural and biological inheritance: Informativeness of different relationships. *Behavior Genetics, 15,* 439–465.

Heath, A. C., & Martin, N. G. (1988). Teenage alcohol use in the Australian twin register: Genetic and social determinants of starting to drink. *Alcoholism: Clinical and Experimental Research, 12,* 735–741.

Hewitt, J. K., Carroll, D., Sims, J., & Eaves, L. J. (1987). A developmental hypothesis for adult blood pressure. *Acta Geneticae Medicae et Gemellelogiae, 36,* 475–483.

Hewitt, J. K., Eaves, L. J., Neale, M. C., & Meyer, J. (1988). Resolving causes of developmental continuity or "tracking." I. Longitudinal twin studies during growth. *Behavior Genetics, 18,* 133–151.

Hewitt, J. K., & Heath, A. C. (1988). A note on computing the $chi^2$ non-centrality parameter for power analyses. *Behavior Genetics, 18,* 105–108.

Holliday, R. (1975). Testing the protein theory of aging. *Gerontologia, 21,* 64–68.

Holt, J. D., & Prentice, R. L. (1974). Survival analysis in twin studies and matched pairs experiments. *Biometrika, 61,* 17–30.

Hougaard, P. (1986). A class of multivariate failure time distributions. *Biometrika, 73,* 671–678.

Jardine, R. (1985). A twin study of personality, social attitudes and drinking behavior. Unpublished doctoral thesis, Australian National University, Canberra.
Jinks, J. L., & Fulker, D. W. (1970). A comparison of the biometrical genetical, MAVA and classical approaches to the analysis of human behavior. *Psychological Bulletin, 73,* 311–349.
Kendler, K. S., & Eaves, L. J. (1986). Models for the joint effect of genotype and environment on liability to psychiatric illness. *American Journal of Psychiatry, 143,* 279–289.
Kendler, K. S., Tsuang, M. T., & Hays, P. (1988). Age at onset in schizophrenia: A familial perspective. *Archives of General Psychiatry, 44,* 881–890.
Kirkwood, T. B. L. (1977). The evolution of ageing. *Nature, 270,* 301–304.
Lalouel, J. M., Rao, D. C., Morton, N. E., & Elston, R. C. (1983). A unified model for complex segregation analysis. *American Journal of Human Genetics, 35,* 816–826.
Lerner, I. M. (1954). *Genetic Homeostasis.* Edinburgh: Oliver and Boyd.
Martin, N. G., & Eaves, L. J. (1977). The genetic analysis of covariance structure. *Heredity, 38,* 79–95.
Martin, N. G., Eaves, L. J., & Eysenck, H. J. (1977). Genetical, environmental and personality factors influencing age of first sexual intercourse in twins. *Journal of Biosocial Science, 9,* 91–97.
Martin, N. G., Eaves, L. J., Kearsey, M. J., & Davies, P. (1978). The power of the classical twin study. *Heredity, 40,* 97–116.
Mather, K. (1953). Genetical control of stability in development. *Heredity, 7,* 297–336.
Mather, K. (1966). Variability and selection. *Proceedings of the Royal Society of London (B), 164,* 328–340.
Mather, K. (1967). Complementary and duplicate gene interactions in biometrical genetics. *Heredity, 22,* 97–103.
Mather, K. (1974). Non-allelic interactions in continuous variation of randomly breeding populations. *Heredity, 32,* 414–419.
Mather, K., & Jinks, J. L. (1982). *Biometrical Genetics: The Study of Continuous Variation* (3rd ed.). London: Chapman and Hall.
Meyer, J. M., & Eaves, L. J. (1988). Estimating genetic parameters in survival distributions: A multifactorial model. *Genetic Epidemiology, 5,* 265–275.
Morton, N. E. (1974). Analysis of family resemblance. I. Introduction. *American Journal of Human Genetics, 26,* 318–330.
Murphy, E. A., & Trojak, J. E. (1987). The bingo model of survivorship. III. Genetic principles. *American Journal of Medical Genetics, 26,* 667–681.
Murphy, E. A., Trojak, J. E., Berger, K. R., & Foster, E. C. (1987). The bingo model. IV. The statistics of survivorship in the Bingo-Gamma model. *American Journal of Medical Genetics, 28,* 1–12.
Murphy, E. A., Trojak, J. E., Hou, W., & Rohde, C. A. (1981). The bingo model of survivorship. I. Probabilistic aspects. *American Journal of Medical Genetics, 10,* 261–277.
Neale, M. C., Eaves, L. J., Hewitt, J. K., MacLean, C., Meyer, J. M., & Kendler, K. S. (1989). Analyzing the relationship between age of onset and risk to relatives. *American Journal of Human Genetics,* in press.
Neale, M. C., Eaves, L. J., Kendler, K. S., & Hewitt, J. K. (1989). Bias in correlations from selected samples of relatives: The effects of soft selection. *Behavior Genetics, 19,* 163–169.
Ott, J. (1985). *Analysis of Human Genetic Linkage.* Baltimore: John Hopkins University Press.
Plomin, R., & DeFries, J. C. (1985). *Origins of Individual Differences in Infancy: The Colorado Adoption Project.* Orlando, FL: Academic Press.
Province, M. A., & Rao, D. C. (1985). A new model for the resolution of cultural and biological

inheritance in the presence of temporal trends: Application to systolic blood pressure. *Genetic Epidemiology, 2,* 363–374.

Rao, D. C., Morton, N. E., & Yee, S. (1976). Resolution of biological and cultural inheritance by path analysis. *American Journal of Human Genetics, 26,* 331–359.

Rice, J., Cloninger, R. C., & Riech, T. (1978). Multifactorial inheritance with cultural transmission and assortative mating. I. Description and basic properties of unitary models. *American Journal of Human Genetics, 30,* 618–643.

Schieken, R. M., Eaves, L. J., Hewitt, J. K., Mosteller, M., Bodurtha, J. N., Moskowitz, W. B., & Nance, W. (1989). The univariate genetic analysis of blood pressure in children: The MCV twin study. Submitted for publication.

Williams, G. C. (1957). Pleiotropy, natural selection and the evolution of senescence. *Evolution, 11,* 398–411.

Wilson, R. S. (1983). The Louisville Twin Study: Developmental synchronies of behavior. *Child Development, 54,* 298–316.

Young, P. A., Eaves, L. J., & Eysenck, H. J. (1980). Intergenerational stability and change in the causes of variation in personality. *Personality and Individual Differences, 1,* 35–55.

# IV
# INTEGRATION THEMES AND FUTURE DIRECTIONS

JOHN K. HEWITT

A book of this kind, and the symposium that gave rise to it, leaves one impressed by the way that scientists studying behavioral development in such a diversity of settings, asking often quite different types of questions and using a variety of different methodologies, can recognize the value of each kind of approach. Of course there are disagreements, but there is not the kind of dismissiveness that often characterizes such exchanges. The reason for the productiveness of the current exchange of ideas is probably to be found in the shared assumption of most, if not all, of the contributors to this volume: that genetics will make an important contribution to our understanding of behavioral development.

Having established this large measure of agreement, we, the editors of this book, felt that different issues remained to be addressed or different features of the field of developmental behavior genetics ought to be highlighted. The different perspectives were difficult to reconcile in one concluding statement, and so we wrote three commentaries that pull together the conceptual themes of the book. In the first of these, Norman Henderson argues that we need to attend in detail to the definitions of the phenotypes we are studying. By their very nature, behavior genetic research strategies emphasize the care that must be taken in defining terms and methods of analysis, taking into account the generality of specificity of the phenotype, its ecological fidelity, the level of analysis of the phenotype, and its stability. Henderson emphasizes that, in general, we do not expect the genetic control measured under one set of circumstances to be the same as that measured under another. We have to recognize the different facets of an operational definition of the phenotype over which there may well be variability and, in the context of development, these specificities and instabilities, which may well be genetically influenced, are important aspects of study in their own right. In an evolutionary context, we have to be sensitive to the ecological fidelity of our operational phenotype, but here again genetic methods provide insights that other approaches cannot. In particular, whether the genetic architecture of a phenotype is that of a fitness character is often the best empirical guide we have to the ecological fidelity of the phenotype.

In the second commentary John Hewitt looks for the common framework that a genetic approach provides for the study of behavioral development. He is especially enthusiastic about the power of genetically informative research designs in the study of human behavioral development and argues that we are now in a position to identify a core design for this purpose. This is always a difficult exercise, since a precise blueprint is much more easily faulted than a vague call to good works, and of course different kinds of questions will inevitably require different approaches than the one offered. Recognizing this, Hewitt backs away from any such blueprint for studies of animal behavior, on the ground that the diversity of questions currently being addressed and the pace of theoretical and methodological advances make it impossible to identify any single approach as the best way to proceed.

The issue of research strategies is taken up again in the final commentary by Bob Benno and Marty Hahn, who return to the difficult questions they raised at the outset: Can productive links be formed between areas like the neurosciences and quantitative genetics? Are the resources required for integrative studies available? And are studies combining the different approaches taken in this volume an efficient way for the subject to proceed? They argue that, although there are considerable practical impediments, we can suggest an affirmative answer to the first of these questions, while the question of resources is seen not as one of dollars per se but rather as one of limitations—the limitations of training and human capacity. The answer to the last and most difficult question may have much to do with the politics of academic research. In the end, progress will depend on institutional support for faculty and researchers who seek to forge the necessary links across disciplines. Recent experience shows that this progress can be achieved, but it will require greater effort than that needed to carry out research within the more traditional delineations of academic disciplines.

# 14
# Quantitative Genetic Analysis of Neurobehavioral Phenotypes

NORMAN D. HENDERSON

Several chapters in this book have described strategies for estimating genetic influences on neurobehavioral traits, the role and importance of common genetic influences on different traits, and the relationship between the underlying genetic architecture of a trait and Darwinian fitness. How to estimate the influences of genetic and environmental factors and G × E interactions on phenotypic variation, the importance of studying genetic and environmental covariances among traits within and across ages, and the logic of evolutionary interpretations of different types of genetic influences on traits have been repeated themes. Although genetics provides a common theme in all chapters, Fuller's (1979) observation that "behavior geneticists, as a group, find unity in their concern with a special class of phenotypes rather than in a particular branch of genetics" continues to hold true. Fuller also argued that the choice of a phenotype can affect the form of genetic analysis and went on to expand his taxonomy of phenotypes with implications for genetic research. I would like to extend Fuller's arguments to show that even rather subtle aspects of a chosen neurobehavioral phenotype can alter the outcome of a genetic study and its interpretation.

## The Dimensionality of a Phenotype

When we choose a particular behavior or some sensory, motor, or neural capacity as a phenotype to be studied with quantitative genetic methods, we are not simply defining the phenotype in terms of measurement operations used to record an organism's response, such as errors in a maze, a score on a paper and pencil test, the number of aggressive responses observed, an ROC curve, or neural conduction velocity. Our phenotype is a joint product of the measurement operations used, the test environment, and the life history and genotype of the organism. In quantitative genetics the same measure taken in different environments, at different ages or following different experiences, is a priori assumed to reflect different genotypic characters until genetic correlations among the measures are shown to be substantial. A phenotype used in genetic research is thus operationally defined in terms of a measurement-environment-history (MEH) composite such as "latency of 11-day-old cage-reared housemice to return to their home nest when placed 12 cm from the nest during the dark cycle." Such specificity may seem uncomfortably narrow, but a change in any variable in the definition may result in a predictable difference in the genetic architecture of the new phenotype (e.g., Henderson, 1979, 1986; Chapter 11). Chapters in this volume by Eaves et al. (Chapter 13) and Corley and Fulker (Chapter 12) give parallel examples involving human development.

The same measurement operation in different test environments or following different life histories can result in a set of quite different genetic characters, a set of genetically correlated characters, or a single genetic character with high stability across environments and life history. Sometimes a construct assessed by using several different measures will be reflecting a single phenotype with high measurement stability in which the alternative measures all produce the same picture of genetic and environmental influences. Other times, constructs will not reflect a single phenotype with common genetic influences but instead encompass several different phenotypes, depending on the measurement operations, with the underlying genetic influences on each phenotype only moderately correlated with each other. The question of phenotype specificity or phenotype stability can only be addressed empirically, when multiple levels of the three MEH factors are examined in multivariate designs. Some designs may involve several different measurement operations, each thought to assess some aspect of a construct of interest (e.g., six different measures of maternal behavior), the same measures taken on individuals with different life histories (e.g., differences in expression of a maternal behavior as a result of the mother's prior maternal experience), or the same measures taken in different test environments (e.g., a maternal behavior as a function of offspring body temperature).

Although quantitative genetic research often involves only a single combination of MEH factors—examination of the sources of genetic variation contributing to individual differences in the phenotype defined by the MEH combination—most biobehavioral research consists of manipulating one of the three MEH factors and looking for resulting phenotypic differences. The study of measurement operations may involve comparisons among different types of assessment, issues of appropriate levels of measurement, or questions of aggregating behavioral responses to assess a broadly defined latent trait. The study of test environments may involve an analysis of behavioral responses to highly controlled stimuli in a psychologist's operant chamber or an examination of the effects of releasing stimuli on a feature-detection system by an ethologist. Life history encompasses all research involving developmental processes; changes resulting from learning, sensory stimulation, nutrition, and other experiences; and effects of chemical and physiological intervention on behavior. Most MEH levels are usually ignored—different training methods are compared across subjects without information on the individuals tested, repeated trials on an activity measure are summed into a single score for analysis, a study of genetic and family cultural influences on delinquency uses a composite measure based on community police records, and so on.

The focus on only one or two MEH combinations entails the risk of missing important interactions among factors, but it remains the common strategy of research—there is always likely to be a gap between prescriptions on how research should be designed to detect potential interactions and the realities of a data collection world. The term *interaction* is used here in its statistical sense, to mean that some of the variability in the measure is accounted for by unique combinations of two or more main effects not accounted for by the main effects alone. An alternate use of the term interaction, to describe the coaction of genes and normal environment through development, conveys little information, since gene–environment coaction is expected to underlie any phenotype. In contrast, large *statistical* genotype × environment interactions have been infrequent for measures assessed within the normal environmental ranges of the species being studied. Large G X E interactions are most typically found when environmental variables are manipulated to reach extreme levels, as in sensory isolation, hormone treatment, or central nervous system lesion experiments.

## Ecological Fidelity

For some measurement operations the test environment is an irrelevant factor; for others it is critical. Relative body weights of individuals should not change when the type of weighing scale is changed, but individual differences in conspecific aggression may be altered if the test arena is modified. Certain MEH combinations are also more relevant than others, and some combinations are meaningless, for example, the mating behavior of infants. Although phenotypes that represent absurd combinations are rarely studied, more subtle MEH mismatches can create gremlins in research data, particularly those combinations that are amenable to experimental procedures but produce phenotypic measures that lack biological relevance. This biological relevance, or *ecological fidelity,* of the phenotype involves the degree to which the measure reflects varying degrees of expression of a behavior exhibited by the organism in natural situations. We might expect, for example, that field observations would fall toward the high side of this ecological fidelity dimension and that measurements taken in highly controlled laboratory experiments would tend toward the low side. Similarly, one might argue that personality surveys and ability and performance tests tend toward lower ecological fidelity, since they substitute cognitive responses to hypothetical problems and situations for natural behavior in situations in which subjects may have elevated hypothalamic arousal. Ultimately, however, the ecological fidelity of a particular MEH combination is an empirical question. The laboratory experiment or the personality scale may or may not contain all relevant environmental cues necessary to elicit the phenotype found in natural situations.

When a measurement *does* lack ecological fidelity (e.g., it is done in an environment lacking in appropriate cues for the organism; it is done at an inappropriate age or requires a response that depends in part on prior experience not available to the organism; it is a poor assessment of the character of interest), most researchers are (properly) stymied by insignificant or uninterpretable experimental results. Unfortunately, this is not usually the case in developmental or genetic research because nearly any reliably measured phenotype, no matter how irrelevant or nonsensical from a developmental or evolutionary perspective, might show systematic changes with age or genetically influenced variation. With respect to genetic influences, as the ecological fidelity of a behavioral measure gradually drops owing to increasingly inappropriate MEH combinations, the behavior begins to approach that of a neutral trait with respect to Darwinian fitness. Such traits are characterized by a high degree of additive genetic variance and little genetic dominance (see Chapter 10). A design and measurement procedure that results in a reliable assessment of a biobehavioral character with low ecological fidelity will thus bias genetic results toward high heritabilities. The relatively high heritabilities often observed for behaviors in some animal laboratory experiments and some human traits assessed by paper and pencil methods may, for example, be the result of low ecological fidelity of the data collection methods.

## Level of Analysis

Any specific MEH combination helps characterize a measure not only in terms of its ecological fidelity but also in terms of its *level of analysis*. An agricultural geneticist interested in corn production might analyze one or more dependent variables such as the number of ears produced per plant, kernel size, number of kernels per ear, and disease resistance. Each contributes to yield and is amenable to quantitative genetic analysis. Alternatively, the geneticist could apply the same analysis to total yield per acre as a "bottom line" variable of interest. The genetic architecture of yield calculated in this manner would reflect both the genetic architecture of the various subcharacters that affect yield, each weighted according to its contribution, and the interactions and correlations among the subcharacters. Although the genetic information on the yield measure may be valuable for breeding corn to increase yield per acre, it reveals nothing about the various subcharacters that contribute to yield. If kernel size were the primary determinant of yield, a focus on this subcharacter alone might generate more precise and useful information than the yield measure. Yet, were the geneticist to study only kernel size as the dependent variable, the success in subsequent breeding for large kernels might have been offset by other factors, such as lower disease resistance or fewer kernels per ear, negating any net yield increase.

Adopting an idealistic research strategy to examine simultaneously all major subcharacters entering into the bottom line character of interest is usually impossible, due to a lack of resources and knowledge. Furthermore, one could argue that, in the corn example, yield per acre represents only one element of the true economic bottom line for the farmer—net profit. Genetic variation in such factors as growing time, production costs, shipping losses, and sugar content might also contribute to variation in profit, making the implementation of an ideal strategy even more formidable. Profit is influenced by characters such as yield per acre, production costs, and shipping qualities, each in turn influenced by many subcharacters such as kernel size and growth rate; yet even this considerable undertaking might have limited value without extensions to the different environmental conditions encountered in growing the corn.

These problems related to choosing the appropriate phenotypic level of analysis by the agricultural geneticist are not unlike those of the behavior geneticist, with two exceptions. The identification of subcharacters is often more difficult for behavioral phenotypes than for the well-studied morphological and physiological characteristics of food crops. More important, in the agricultural situation the supercharacter, such as yield or profit, is defined first and characters and subcharacters then follow in a "top down" manner. The reverse is often true with behavioral phenotypes, where latent trait "supercharacters" such as extraversion, conspecific aggression, and "g" are defined from the "bottom up" by a correlated cluster of specific behavioral

ANALYSIS OF NEUROBEHAVIORAL PHENOTYPES 287

responses. When development is a component in the behavioral study, developmental precursors also enter into the hierarchy of characters and subcharacters to be considered.

The hierarchy of possible phenotypes in a developmental genetic study is illustrated in Figure 14–1. The diagram is simplified in that it aggregates sublevels such as simple behaviors that are life history prerequisites to more complex behaviors and avoids complexities such as overlapping levels and feedback. At levels 1 and 2, developmental precursors establish constraints on further development and the range of gene expression at higher behavioral, physiological, and morphological levels (canalization). These pathways can also be responsible for positive genetic correlations among commonly influenced subcharacters through pleiotropic effects on early developmental processes. At this early stage, common gene influences (pleiotropy)

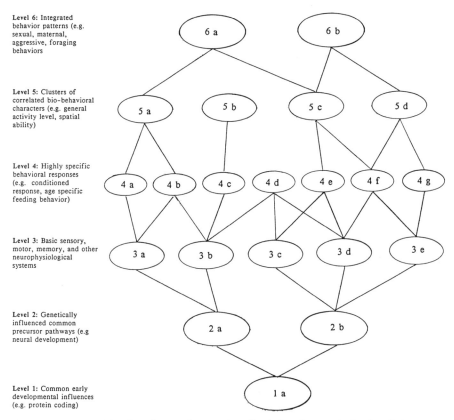

**Figure 14–1** Levels of analyses of operationally defined neurobehavioral phenotypes. Selection can occur at any level with implications for higher or lower levels. Phenotype 4d represents a highly specific behavior that plays no role in integrated behavior patterns of the organism subject to selection. It may nevertheless appear to be related to fitness because it shares influences of 3c and 3d with 4e, a behavioral response that is directly related to integrated behavior patterns. The "cluster" of measurements defining phenotype 5b reflect a single level 4 behavioral response.

can occur for characters that appear totally unrelated until the common mediating pathway is discovered—a pleiotropic gene in humans, for example, affects both male sterility and weakened pulmonary function, where the genetic influence is on the structure of scilia, common to both sperm tails and the hairs lining bronchial tubules. During early development a small number of recessive regulator genes can produce large effects on embryological development and subsequent viability. Natural selection at this stage of development thus is often acting on relatively few genes with large effects. In contrast, selection pressures at the behavioral levels, in the upper portion of Figure 14-1, will more often involve large numbers of background genes with small effects.

Neurophysiological phenotypes are represented at levels 2 and 3 of the diagram and narrowly defined (i.e., single MEH) behavioral phenotypes are represented at level 4. Level 5 represents clusters of level 4 phenotypes with genetic correlations approaching unity, indicating that the cluster represents various manifestations of a single genetic character. Natural selection is most likely to be operating on the integrated behavior patterns of the organism at level 6. These integrated behavior patterns may involve numerous genetic characters, often not highly correlated with each other, but each contributing to a common fitness-related objective. Maternal behavior of many mammals, for example, consists of a number of different behavioral elements such as pup licking, feeding, retrieval, foraging, and nest protection, all operating together to produce fitness—offspring surviving to breed—in a manner analogous to the yield-per-acre example in corn.

Normally only one or two subcharacters are examined that contribute to level 6 multidimensional behavioral repertoires, yet the genetic architecture of a narrowly defined level 2 through 5 phenotype depends on its ultimate role in all of the integrated behavior patterns at level 6. If, for example, high expression of a level-4 phenotype contributes positively, in terms of fitness, to several integrated behavior patterns of level 6, this should be reflected in the genetic architecture of the level-4 phenotype—genetic dominance favoring high expression of the character. Other phenotypes at levels 3 through 5 will have mixed effects on fitness of level 6 behaviors, contributing positively to some integrated behavior patterns and negatively to others. Thus there would be little or no net positive or negative relationship to fitness, resulting in genetic variance being primarily additive. With sufficient knowledge of how a set of behavioral elements are integrated into an adaptive level-6 strategy under appropriate environmental and life history conditions, one should be able to construct a multivariate composite phenotype from the relevant behavioral elements that exhibits a genetic architecture of a major component of fitness, even when individual behavioral elements show little or no such pattern.

In addition to integrated behavioral sequences at level 6, which are related to mating behavior, predator avoidance, foraging, and other functions, artificial composite phenotypes could be constructed which themselves have never been subject to selection. Genetic patterns may incorrectly suggest that such composites are fitness related simply because they tap a number of subcharacters that contribute to other phenotypes that have been subject to selection. Selection has not operated on a rodent's performance in a shuttlebox nor on a human's ability to program computers, yet both may appear to be major components of fitness. These are epiphenomena, a product of subcharacters that have been subjected to selection on the basis of their contributions to other level-6 characters. It should be evident from Figure 14–1 that certain phenotypes may exhibit a genetic architecture suggesting that the measured trait is correlated with fitness, even when selection has not operated on any phenotype even close to the MEH configuration studied. A behavior measured early in development may also appear to be fitness related, yet selection may be acting on a behavior pattern appearing much later in life for which the early behavior served as a prerequisite or is correlated with the adult behavior because of commonly shared lower level subcharacters.

Genetic correlations among phenotypes at levels 3 through 5 will be influenced by three opposing factors. First is the degree to which the different measurement and test environment combinations are just assessing various manifestations of the same phenotype (e.g., weight using two different scales). Such situations should produce high positive genetic and environmental correlations. Second is the degree that the genetic variance of two phenotypes is due to rare recessives having an influence on subcharacters shared by the phenotypes. Large influences on common subcharacters will obviously tend to create positive correlations. Counteracting these first two factors, which tend to produce positive genetic correlations, is the fact that selection is being applied simultaneously to phenotypes contributing to level-6 behaviors. Pleiotropic genes having positive effects with respect to fitness across situations will be brought rapidly to fixation; thus only pleiotropic genes having inconsistent effects on fitness across phenotypes will remain at intermediate frequencies and contribute to genetic variance and covariance of the phenotypes. As a result, low or even negative genetic correlations between fitness-related traits should occur. As a larger set of genetic correlations among phenotypes becomes available in the research literature, it would not be surprising to see a tendency for correlations among fitness-related phenotypes to shift from positive to negative as one progresses from the lower to the higher levels depicted in Figure 14–1.

The objective of many behavior genetic strategies involving single gene mutants and genetic mosaics is to trace gene influences from level 1 through level 6 phenotypes in what has become known as the *genetic dissection of behavior*. In this book Wimer (Chapter 5) describes some of these approaches with respect to brain development in housemice and Hall et al. (Chapter 6) describe them with respect to mutants interfering with various aspects of fruit fly courtship. The isolation and use of the *Drosophila* mutant strain dunce provides a good example of a genetic analysis extending from level 1 through level 5. Both male and female dunce mutants show marked deficits in effective courting patterns. Unlike wild-type adult males, dunce males continue to be distracted from their courting of females by the actions of immature males. Dunce females fail to benefit from preexposure to male courtship song in terms of enhanced receptivity to courtship and copulation (see Chapter 8). Reductions in the effectiveness of these level-5 sexual behaviors are the result of a decrement in acquisition in several learning situations (level 4), in turn due largely to a deficit in short-term memory (level 3). The deficit appears to be related to decreased heterosynaptic facilitation (level 2) brought about by the cyclic adenosine monophosphate (cAMP) cell-signaling response (level 1). It is now known that the dunce mutation, isolated behaviorally, affects the structural gene for cAMP-dependent phosphodiesterase (PDE II), mapped to chromomere 3D4 on the X chromosome (Tully, 1987).

## Ontogenetic Stability

A large class of life history variables that may influence the optimal expression of a set of phenotypes measured under the same measurement and test environment conditions are those associated with age changes. Changes in genetic influences on a particular behavior pattern during development (age $\times$ genotype interactions) are among the most studied in behavior genetics and the most developed from a theoretical standpoint (see, e.g., Chapters 2, 9, 11, and 13). I will arbitrarily label the degree of consistency of the underlying genetic influences of a set of similarly measured phenotypes assessed through development as the *ontogenetic stability* of the behavioral measure. Stability in this case does not refer to the absolute values of scores across age, since developmental trends are expected in nearly all behaviors, nor does it refer simply to high genetic correlations among the same measure taken at different ages. It refers to the consistency of the more complete picture of genetic architecture of the specific phenotype at different ages.

When a certain level of expression of a behavioral response tends to confer a fitness advantage to an organism at all ages or in all the changing ecological niches correlated with age, one should see a consistent genetic architecture throughout a major segment of the organism's development. It is not unreasonable to hypothesize, for example, that the ability to learn rapidly would confer a selective advantage to an organism throughout its life span. If so, one should expect to find genetic dominance favoring rapid learning across all ages and niches, assuming the cues and responses of the learning situation have some ecological relevance to the organism. In each specific age-test situation, however, the composite of sensory, motor, memory, and other nervous system functions involved in performance may differ from that required in other situations, resulting in low additive genetic correlations among learning measures taken at different ages. Similar instances are not difficult to find in human behavioral development. Young, Eaves, and Eysenck (1980), for example, found high genetic correlations for neuroticism between juvenile and adult measures but a relatively low across-age correlation for extraversion. Given the large difference in social cues and other age-related ecological niche factors that would contribute to a social behavior, it is not surprising that genes influencing social extraversion would differ with age.

The converse of this situation involves behaviors in which the optimum degree of expression of the character is highly niche or age dependent. In contrast to the case of learning, where more is usually better, in some situations high levels of locomotor activity are adaptive to an organism, but in other environments or at other developmental stages low or intermediate activity levels may be associated with fitness. Under such circumstances genetic architecture for the behavior can change radically across age, even when a substantial additive genetic correlation exists among the various measures of activity across the life span (e.g., Henderson, 1986). In this sense high additive genetic correlations and ontogenetic stability tend to be counterpoints of each other.

## Measurement Stability

Ontogenetic stability represents only one aspect of phenotypic stability. On a shorter time scale, individuals may vary considerably with respect to their phenotypic stability over microenvironmental variations during behavioral measurement. This *measurement stability* is variously defined as short-term measurement reliability, repeatability, or internal consistency. The degree of expression of some phenotypes is highly consistent across measures taken over a limited time span, whereas other phenotypes show little consistency on repeated measurement. Relative to the variation among individuals, the variation among short-term repeated measures of weight and other morphological characters within individuals is likely to be small. For many behavioral traits, however, the measurement process itself creates an environmental stimulus that can serve to modify the behavior. In addition, fluctuating test environment cues, often salient to the subject but unknown to the observer, can exist in behavioral tests. Finally, when multiple trials or measurements are used, each repeated measurement is based on a slightly changed life history of the organism. Any combination of these effects can result in low reliability or repeatability. Low reliability can be treated as simple measurement error, to be circumvented by averaging many observations of an individual into a single more reliable measure. There is some risk in aggregating what might be heterogeneous phenotypes due to life history differences, however, in terms of loss of understanding of biological mechanisms and underlying gene action. For example, the genetic architecture of performance during early trials in an avoidance learning task was found to differ substantially from that found on later trials for what appear to be sound biological and psychological reasons (Hewitt, Fulker, & Broadhurst, 1981). The long-running controversy in social/personality psychology concerning traits versus situational specificity has dealt with similar issues regarding the collapsing of individual behavioral responses into an aggregate behavioral measure (e.g., Rorer & Widiger, 1983).

The degree of short-term measurement stability of a trait will influence the relative size of genetic and environmental contributions to total trait variance, since low reliability necessarily means greater environmental variance. With a reliably measured morphological character such as body size, the difference between using a single measurement or the mean of several repeated measurements has little impact on heritability or other genetic estimates scaled as proportions of total phenotypic variance. In contrast, the number of measurements taken or the length of a behavioral observation period can have a substantial impact on heritability and related estimates derived from traits with low measurement stability. Some behavioral phenotypes are likely to be highly unstable because of their sensitivity to transient but highly salient environmental cues existing at the moment of measurement. An organism's genotype and life history may together influence a brief behavior, such as an orienting response to a stimulus, yet microenvironmental variations during the stimulus presentation may be far more potent factors in influencing the immediate behavior. To some degree, the amount of microenvironmental fluctuation is under control of the investigator—primarily through the degree of experimental control used and the size of the sample of behavior observed. Increased control reduces idiosyncratic environmental stimuli occurring during measurement, increased sampling helps average across random environmental perturbations. These design and analysis options have considerable impact on the proportion of environmental variance influencing an unstable behavioral phenotype, and hence on the size of genetic estimates. Figure 14-2 illustrates this effect by comparing the total percentage of phenotypic variance available for partitioning into genetic and life history environmental components for a highly stable (short-term reliability = .95) and unstable (reliability = .10) phenotype. When only a single measurement is taken, the upper-bound proportion of phenotypic variance that can be attributed to genetic influences differs considerably for the high- and low-reliability measures. As measures are averaged across increasing numbers of measurements or larger samples of behavior, these differences decrease substantially.

The confounding of heritability estimates and other genetic ratios by measurement reliability poses a problem in comparing research results. For example, Mousseau and Roff (1987) tabulated 1120 narrow heritability estimates of wild outbred populations collected from public records. The results indicated that heritability for life history characters such as fecundity, viability, and developmental rate is lower than for those of morphological measurements and that heritabilities of physiological and behavioral measures generally fall closer to the average of life history characters. While the results of this survey provide encouraging support for Fisher's (1958) fundamental theorem of natural selection concerning low heritabilities for high-fitness

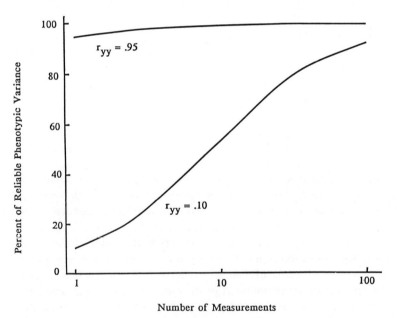

**Figure 14-2** Proportion of total phenotypic variance available for partitioning into genetic and among-subject environmental components as a function of the short-term stability of a single measure and the number of measurements taken. A small sample of a behavior highly sensitive to microenvironmental influences will necessarily show little genetic influence although the behavior may be highly heritable when viewed over a long period. Genetic results based on highly reliable morphological characters are little influenced by number of measurements.

characters, one is left to wonder how these results might change if measurement stability were taken into account. The use of contemporary meta analytical procedures (e.g., Hedges & Olkin, 1985) for such surveys would be useful in this regard.

Low short-term test stability of behavioral measures often leads to the *effect size fallacy,* the tendency to dismiss genetic influences on a phenotype as unimportant or not related to selection pressures because the genetic effects account for a relatively small proportion of total phenotypic variance. Test-retest reliabilities below .2 are probably not uncommon for measures of brief behavioral sequences, especially in social behaviors where variation in the behavior of the social partner is likely to contribute substantially to microenvironmental influences during testing. Gromko (1987), for example, reported coefficients in the .15 range for courtship displays in *Drosophila melanogaster.* If 85 percent of the total variance of a measured phenotype in a population is due to microenvironmental fluctuations during testing, the upper bound for genetic and systematic environmental influences on phenotypic variance is 15 percent, regardless of the underlying genetic character's relationship to fitness. The effect size fallacy involves the failure to recognize that a particular behavior sequence may be repeated hundreds of times in the lifetime of the organism. Despite trial to trial variation due to microenvironmental influences, there is some consistency in the behavior resulting from genetic and life history influences. Thus the reliability of a measure of the average lifetime expression of the phenotype would be extremely high. As a consequence, the relative genetic influence on individual differences in the lifetime expression of the behavior may be substantial, as may certain prior learning and other life history experiences.

We should also recognize that measurement stability may itself reflect an aspect of fitness for the organism. In many circumstances, an organism's response to the immediate environment is critical to its survival. In such circumstances test stability may be a useful index of the environmental monitoring and behavioral feedback of an organism, both of which may be subject to continuing selection pressure.

## General Stability

Both ontogenetic and measurement stability are specific aspects of phenotypic stability across MEH conditions. From a genetic standpoint they refer to the size of genotype by MEH interactions, where the latter were age and environmental factors. Large G $\times$ A or G $\times$ E interactions indicate that we are dealing with different genetic characters at different age levels or in different environments, and negligible interactions indicate phenotypic stability. These two forms of interactions do not encompass all situations of interest, however. A more general interactive form of genotype $\times$ organism interaction allows one to ask how consistent genetic architecture is for a particular trait across any set of conditions for the organism, including age, environment, and other genetically influenced effects. The general formulation is also useful for encompassing such concepts as range of reaction (e.g., Gottesman, 1963), phenotypic plasticity (e.g., Stearns & Koella, 1986), and developmental conversion (Smith-Gill, 1983) that deal with various aspects of G $\times$ O interactions.

Figure 14-3 is a contour diagram giving a topographical plot of relative fitness of an organism as a function of the degree of expression of character $X$ in conjunction with some other quality of the organism, such as an environmental gradient, age, or

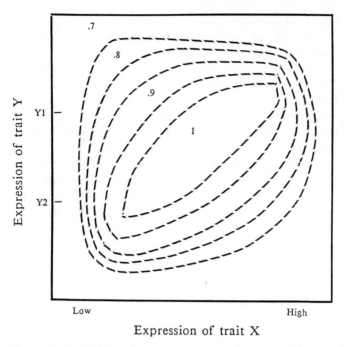

**Figure 14–3** Relative fitness contours as a function of the joint levels of expression of two phenotypes. The optimum level of expression of trait X (e.g., male conspecific aggression) depends on the level of expression of trait Y (e.g., body size or size of benefit). Had trait Y been body size of opponent, the contour lines would have sloped downward to the right.

other genetically influenced phenotype, on ordinate axis $Y$. In two dimensions these diagrams are analogous to topographical maps with optimal fitness equated with highest elevation. Figure 14–3 illustrates a topography in which the optimum degree of expression of $X$, with respect to fitness, is a positive function of some other characteristic of the organism $Y$.

Character $X$ might be level of conspecific aggression in males and variable $Y$ might be prior aggressive experience or age (life history variables), arena or opponent size (test environment variables), or a second genetically influenced phenotype such as body weight. Had one set out to study the genetic architecture of this aggressive behavior at a given age, with animals of a certain degree of aggressive experience or of a certain size, or with any other factor represented as level $Y$-$1$ in Figure 14–3, the optimum degree of expression of aggression and consequently the genetic architecture observed (dominance favoring high aggression) would have differed significantly from that obtained if one had studied this same response at level $Y$-$2$, where one would expect genetic dominance favoring low expression of aggression. Had variable $Y$ been ignored altogether, and one had simply studied the architecture of the aggressive response without regard to the organism's prior experience, age, weight, or other aspect of its phenotype represented on the ordinate of Figure 14–3, no single optimal level of expression of aggression would have been detected. The genetic architecture observed would likely have been one of substantial additive genetic variance and bidirectional dominance. It is these kinds of complexities that have made the study of social behaviors so challenging and progress slower than in other areas of developmental behavior genetics (e.g., Hewitt & Broadhurst, 1983; Chapters 3 and 4).

Obviously a considerable variety of fitness topographies is possible. Relationships between $X$ and $Y$ are often likely to be negative and possibly nonlinear, for example. An inverse relationship between expression of $X$ and level of $Y$ would be expected in compensatory relationships in which increased levels of $Y$ allow a relaxation in the degree of expression of $X$ necessary for optimum fitness. An example would be richness of food patch and amount of foraging activity. Energy expenditure and predator risks accompanying foraging suggest that low levels of foraging would be optimal when food was abundant. In contrast, costs associated with predator risk and energy use become necessary in sparse food patch situations, and thus high foraging levels may be associated with fitness in this situation.

When $Y$ consists of a second phenotype partly under genetic influence, Stearns' (1982) description of phenotypic plasticity is illustrated. Variables $X$ and $Y$ in this case may represent detection thresholds in two different sensory modalities, or maternal investment per offspring versus total offspring, or, as examined by Stearns and Koella (1986), age and body size at maturation. The likelihood of ignoring some aspect of the organism represented on the ordinate becomes particularly high when it involves a second behavioral or biological phenotype. One would be likely to study sensory modality $X$, ignoring sensory modality $Y$, or some aspect of maternal behavior $X$, independent of lifetime fecundity $Y$. As before, independent of $Y$ there may be no single optimum fitness value for $X$ in these situations.

The two-dimensional diagram of Figure 14-3 simplifies most natural situations, since several organismic characteristics are likely to interact simultaneously in establishing optimal expression of character $X$. Research that collapses across important $Y$ dimensions should still produce evidence of prior selection in the form of the genetic dominance. When studies do examine different levels of a relevant variable $Y$ in conjunction with character $X$, they should uncover dominance favoring different degrees of expression of $X$ of a function of $Y$ in accordance with the fitness topography. When $Y$ represents environment or age factors, this should result in predictable, or at least explainable, G × E or G × A interactions. When $Y$ represents a second genetically influenced phenotype, predictable patterns of genetic correlations should also emerge. In topographies similar to those in Figure 14-3, for example, genes pleiotropic for high or low expression of both characters should be favored over those pleiotropic in opposite directions for $X$ and $Y$. As a result, a positive genetic correlation should emerge and a better fit to this linear relationship should exist for heterozygotes than for homozygotes.

## Robust Behavioral Phenotypes

We occasionally encounter an extremely robust behavioral trait that possesses a high degree of general stability. A behavior pattern reflecting a consistent underlying trait is found to exist across a considerable range of MEH conditions. Such traits often can be traced to a common level 1 or level 2 influence (Figure 14-1), such as the dunce mutant described earlier wherein a common deficit in short-term memory was attributed to altered cAMP levels. Dunce performs poorly in a wide range of acquisition and retention situations, with varying degrees of ecological fidelity and across ages ranging from larvae to adult flies. Not surprisingly, robust behavioral phenotypes with substantial general stability, including human behavioral phenotypes such as general intelligence and neuropsychiatric disorders, have often been the focus of the most intense research efforts in behavior genetics. However, from the viewpoint of developmental behavior genetics, general stability is *not* a prerequisite for genetic analysis, and indeed the causes of instability, which may well themselves be genetic, should be approached by the same general genetically informed research strategies as the more robust traits.

## Conclusion

I have considered in some detail the issue of the definition of neurobehavioral phenotypes and how important it is to be precise about these definitions. In general, one should not expect the genetic control of a behavior measured under one set of circumstances to be the same as that measured under another set of circumstances. We have to recognize the different facets of an operational definition of the phenotype over which there may well be variability. In the context of development, these specificities and instabilities, which may well be under genetic control, are important aspects of study in their own right. In an evolutionary context we have to be sensitive to the ecological fidelity of our operational phenotype, but here again genetic methods provide insights that other approaches sometimes cannot. For example, whether the genetic architecture of a phenotype is that of a fitness character is often the best empirical guide we have to the ecological fidelity of the phenotype. We must be careful, however, not to think of a fitness character in terms of a static level of response. In some cases it is the plasticity of the response in different situations that results in increased fitness to the organism. A broad perspective, sensitive to details of measurement operations, the test environment, and the life history of the organism, remains essential when applying quantitative genetic methods to neurobehavioral phenotypes.

## REFERENCES

Fisher, R. A. (1958). *The Genetical Theory of Natural Selection* (2nd ed.). New York: Dover.
Fuller, J. L. (1979). The taxonomy of psychophenes. In J. R. Royce & L. P. Mos (Eds.), *Theoretical Advances in Behavior Genetics.* Alphen aan den Rijn: Sitjthoff & Noordhoff.
Gottesman, I. I. (1963). Genetic aspects of intelligent behavior. In N. Ellis (Ed.), *The Handbook of Mental Deficiency* (pp. 253–296). New York: McGraw-Hill.
Gromko, M. H. (1987) Genetic constraint on the evolution of courtship behavior in *Drosophila melanogaster. Heredity, 58,* 435–441.
Hedges, L. V., & Olkin, I. (1985). *Statistical Methods for Meta-Analysis.* New York: Academic Press.
Henderson, N. D. (1979). Adaptive significance of animal behavior: The role of gene-environment interaction. In J. L. Rovce & L. P. Mos (Eds.), *Theoretical Advances in Behavior Genetics.* Alphen aan den Rijn, Netherlands: Sitjthoff and Noordhoff.
Henderson, N. D. (1986). Predicting relationships between psychological constructs and

genetic characters: An analysis of changing genetic influences on activity in mice. *Behavior Genetics, 16,* 201-220.

Hewitt, J. K., & Broadhurst, P. L. (1983). Genetic architecture and the evolution of aggressive behavior. In E. C. Simmel, M. E. Hahn, & J. K. Walters (Eds.), *Aggressive Behavior: Genetic and Neural Approaches* (pp. 37-66). Hillsdale, NJ: Lawrence Erlbaum.

Hewitt, J. K., Fulker, D. W., & Broadhurst, P. L. (1981). Genetics of escape-avoidance conditioning in laboratory and wild populations of rats: A biometrical approach. *Behavior Genetics, 11,* 533-544.

Mousseau, T. A., & Roff, D. A. (1987). Natural selection and the heritability of fitness components. *Heredity, 59,* 181-197.

Rorer, L. G., & Widiger, T. A. (1983). Personality structure and assessment. *Annual Review of Psychology, 34,* 431-463.

Smith-Gill, S. J. (1983). Developmental plasticity: Developmental conversion versus phenotypic modulation. *American Zoologist, 23,* 47-55.

Stearns, S. C. (1982). The role of development in the evolution of life histories. In J. T. Bonner (Ed.), *Evolution and Development.* Berlin: Springer.

Stearns, S. C., & Koella, J. C. (1986). The evolution of phenotypic plasticity in life-history traits: Predictions of reaction norms for age and size at maturity. *Evolution, 40,* 893-913.

Tully, T. (1987). *Drosophila* learning and memory revisited. *Trends in Neuroscience, 10,* 330-335.

Young, P. A., Eaves, L. J., & Eysenck, H. J. (1980). Intergenerational stability and change in the causes of variation in personality. *Personality and Individual Differences, 1,* 35-55.

# 15

# Genetics as a Framework for the Study of Behavioral Development

JOHN K. HEWITT

Part of the reason for the current excitement about behavior genetics is that the basic questions about heritability that took up so much of our energy two decades ago have been superseded by new discoveries and new methodological advances. We are no longer asking *whether* population variation in this or that behavior is influenced by genes but *how* genes exert their influence. In particular, we want to know how genetic and environmental factors operate during development, and developmental questions that can be given clear expression within a genetic framework are inaccessible to other approaches. A genetic approach helps to define the questions to be asked, suggests the methods we should use to answer them, and leads to quantitative analyses of the scope and limitations of these methods. In short, the genetic approach leads to the formulation of testable hypotheses and answerable questions in a way that is often the envy of psychologists, ecologists, and epidemiologists.

## THE GENETIC APPROACH TO HUMAN BEHAVIORAL DEVELOPMENT

A genetically informative longitudinal study (e.g., of twins or of adopted and biological children) properly designed, executed, and analyzed can provide tests of a range of hypotheses that are central to our understanding of human development. We can test whether the same genes influence behavior in the same way at different ages. Are the genetic influences on conduct disorder (to take an example from child psychiatry) the same before and after puberty, or are new kinds of genetic influence "switched on" at puberty? Are the genetic influences on intelligence essentially the same from birth to adulthood, or are different genetic factors at work during the different cognitive stages? The answers to such questions come from our ability to estimate both the genetic effects at two or more ages *and the genetic correlation between the effects at one age and those at another.* Different sets of genes acting at different ages will result in a low genetic correlation. We can take our analysis of developmental continuity a step further and test whether the effects of genes, manifest at one age, *persist* in the organism at older ages, or whether continuity of genetic influences is a consequence of continuous de novo activity of the genes. For example, do genetic influences on conduct disorder manifested early in childhood have persistent consequences for later adolescence, or are the genetic influences on later adolescent behavior unaffected by earlier expression? This test, discussed in Chapters 11 and 13, is provided by the different patterns of variances and covariances predicted for longitudinal twin data by the different mechanisms that give rise to age to age correlations. The practical importance of the distinction between the two mechanisms is that in the absence of persistence, intervention to prevent the expression of genes at one age will have no benefits at later ages unless the intervention is continued. As a further extension of our analysis of development, methods are now being researched that allow us to test whether the genetic control of the *timing* of behavioral development is independent of the genetic control of the level of behavior (Neale, Eaves, Hewitt, MacLean, Meyer, & Kendler, 1989; Meyer & Eaves, 1988). For example, do genes that endow greater mature verbal ability also entail earlier development of verbal skills? Does greater liability to psychiatric disorder lead to earlier onset of the disorder? Are liability and age of onset controlled by different genetic systems?

As well as analyzing the causes of continuity and change during development for any particular phenotype, we can test whether observed correlations between different phenotypes are genetic rather than environmental. For example, is the co-morbidity of behavioral disorders genetic in origin? In a twin study this question is answered by whether the pattern of cross-twin–cross-trait correlations for monozygotic and dizygotic twins is consistent with genetic or with environmental predictions. These cross-twin–cross-trait correlations are even more informative because, provided that each of two traits has a different genetic and environmental "architecture," the pattern of correlations will be determined by, and hence reveal to us, the *causal* relationships between the variables. For example, do stressful life events cause psychological distress? Does lack of psychological well-being lead to the experience of stressful life events? Are they reciprocally causal? Finally, are both life events and psychological distress produced by a third causal variable (Heath, Neale, Hewitt, Eaves, & Fulker, 1989)? In determining the relationships between variables during development, we can further test whether the same genes influence the same traits in males and females and whether they have the same magnitude of effect. For example, do genes predisposing to hyperactivity in boys manifest themselves in some other disorder in girls? Are genetic influences on cognitive ability the same in boys and girls? We can answer these questions in a study of twins by a comparison of the opposite-sex–dizygotic-twin correlation with those of the male and female same-sex dizygotic twin pairs.

These kinds of questions are central to the study of human behavioral development, and they all yield to an appropriately designed, executed, and analyzed behavior genetic study. All the questions and hypotheses have been posed in terms of *genetic* influences but, as long as these are taken into account, the remaining environmental sources of variation and covariation can be analyzed in the same way without the need for untested ad hoc assumptions. Thus the behavior genetic study tells us more about *environmental* influences than does a nongenetic study; a study of individuals (in the absence of randomized experimentation), or a study of sets of individuals for which genetic and environmental resemblance are confounded, cannot reveal anything about the genetic or environmental origins of the variation or covariation that we observe. It is, for example, no more than blind faith to hold to the belief that a correlation between parents' behavior and that of their children necessarily implies a causal path from parent's to child's phenotype. Having acknowledged this, the next stage is to consider the consequences for the patterns of observations on parents and children of alternative models, including genetic and cultural inheritance. For a reasonably simple system, the inclusion in a research design of monozygotic (MZ) and dizygotic (DZ) twin children and their parents, as well as of MZ and DZ adults (and their children), enables us to separate genetic and environmental influences on both the children's and the parents' behavior, to estimate the genetic correlation between the two and, in turn, to determine the predominant direction and magnitude of any phenotype-to-phenotype causal path.

## OBSTACLES TO TRADITIONAL LONGITUDINAL STUDIES

The strengths and elegance of this genetic approach to the study of human behavioral development are only just beginning to be exploited. Major longitudinal studies, like the Louisville Twin Study reported by Matheny in Chapter 2 and the Colorado Adoption Project reported by Corley and Fulker in Chapter 12, are providing the kinds of data bases we need, but their duration and expense are obstacles. The duration of the traditional longitudinal study not only entails an institutional and investigator commitment that spans decades but, perhaps more significantly, requires a commitment to a particular focus of investigation and particular measurements grounded in the theory and hypotheses of pressing interest at the start of the study. Twenty years later we might wish to have looked at different variables in different ways. A further problem for these traditional longitudinal genetic studies is that of sample attrition. A 5-percent annual attrition would leave us with less than half our original sample after 16 years, and the subsample that remains will be genetically and environmentally unrepresentative for many variables of interest. In addition to these difficulties, we should note that studies of adopted children start out with a restricted range of environments provided by prescreened adoptive parents and what is probably a restricted range of genotypes represented in the biological parents and adoptees. Faced with these kinds of obstacles, it is not surprising that only a few research groups have undertaken the daunting challenge of a traditional genetically informed longitudinal study.

## OVERLAPPING COHORT LONGITUDINAL TWIN STUDY: A DESIGN FOR HUMAN DEVELOPMENTAL BEHAVIOR GENETICS

A design that overcomes these difficulties is the overlapping cohort longitudinal twin design (Heath & Eaves, 1986). This will form the basis of a new study of adolescent behavioral development planned to begin in 1990 (Hewitt et al., 1988). In this study, samples of twins will be recruited at each of the ages spanning the period of interest; in the particular case of the study of Hewitt et al., the age range is 8 through 16. Thus if 100 pairs of twins are recruited for each year of age, there will be 900 pairs initially. These twins and their parents will be assessed annually on four occasions or until the twins reach age 17, when they leave the study. In the second, third, and fourth year of the study additional samples of 8-year-olds will be recruited and followed for the remainder of the study. If for each cohort we recruit sufficient families to yield 100 at the fourth occasion of measurement, after allowing for attrition the total study will yield 1200 families for which there will be complete data. The sample sizes available for statistical calculation depend on the particular cross-sectional and cross-temporal comparisons. There will be 400 families for a cross-sectional analysis at any given age, 300 for any given one-year interval, 200 for any given 2-year interval, and 100 for any given three-year interval (i.e., four annual measurements). With this design a nine-year developmental period is spanned in only four years of data collection. The price we pay is in our inability to estimate wide interval correlations for particular cohorts. However, providing that the correlations for intervals of greater than three years are largely predictable from those for lesser intervals, the models and hypotheses we have discussed can be tested with the information available. If random samples of twins are recruited, monozygotic twin pairs, same-sex dizygotic twin pairs, and opposite-sex dizygotic twin pairs will be represented in approximately equal proportions.

Among the strengths of this design are that, because the target developmental period is spanned in only four years of data collection, the institutional and investigator commitment is a manageable period during which the important continuity of the investigating team can be maintained. Further, the theoretical and measurement focus remains current throughout the project. Equal power for cross-sectional analyses at every age is obtained, and information from each age is available at the same time. The ravages of attrition do leave the analyses of the older age groups irrevocably weakened, and we do not need to wait for the passage of time to collect and analyze data at the older ages. The use of twins rather than adoptees avoids the problems of unrepresentativeness of biological and adopting parents and the difficulty of obtaining sufficient sample sizes with adequate age structure. The inclusion of male, female, and opposite-sex (dizygotic) twin pairs permits the analysis of the control of sex differences during development. The inclusion of parents provides, from our genetic perspective, our best indication of the adult end point of the developmental trajectory: Parent-offspring correlations and their patterns of change over time provide information about whether the same genes influence parents and children and how close adolescents are to their developmental plateaus. The parents also provide indices of environmental circumstances that we expect may be relevant to the course of their children's development; further, as observers, they can furnish ratings of their children and information about their children's background and history that no other adult is as well placed to provide.

Thus, not only does a genetic approach ask questions that are answerable *in principle,* but it provides a method of study that can be carried through in practice to yield real progress in our understanding of human behavioral development.

## GENETIC APPROACH TO ANIMAL BEHAVIORAL DEVELOPMENT

If the study of human behavioral development yields so nicely to the genetic approach, the more so should the study of animal behavior and neural development (Chapter 7). Here we have available not only controlled biometrical crossbreeding and selection studies but a whole range of techniques from molecular and physiological genetics. Hall et al. (Chapter 6) have discussed the application of molecular genetic techniques in understanding a biological clock mechanism in *D. melanogaster,* and Wimer (Chapter 5) has outlined the use of chimeric mice which, like other forms of mosaic such as the genetically created gynandromorphs in *Drosophila,* open up the possibility of developmental "fate mapping" (tracing the adult phenotype to its developmental origins in the early zygote) as well as linking the control of behavior to particular neuroanatomical structures. The special advantage of studying genetic mosaics is that we can observe the effects of the genetic variant on particular structures set against a "normal" background, as distinct from the study of nonmosaic genetic mutants in which the genetic variant under study is ubiquitous and its site of action therefore difficult to localize in the absence of prior knowledge of the gene product.

Despite the rapid progress to date and obvious promise of these genetic dissection techniques for the future, they have been applied to an as yet relatively restricted range of examples of interest to behavior genetics. The examples include neuroanatomical phenotypes in mice, with or without gross behavioral consequences such as the motor disturbances of staggerer or weaver mice, and behavioral phenotypes of *Drosophila* courtship, conditioning, and periodicity.

Part of the reason for resistance to the wholesale adoption of molecular genetic approaches in animal behavior is that although biochemical and neural mechanisms will be illuminated by the molecular genetic approach more rapidly and successfully than by any other, many questions about behavior are essentially evolutionary questions. As Arnold (Chapter 9), Cairns and Gariépy (Chapter 3), and Leamy (Chapter 8) showed, these questions require answers in terms of the characteristics of natural selection and the properties of population genetic variation and correlations. To understand the physical basis for how a sequence of behaviors might follow one from another or how some behaviors are more malleable than other behaviors or at some times rather than others is not to answer what seems the more fundamental question of *why* this should be so. A second part of the reason for resistance is that in animals, as in humans, most of the phenotypic variation we are interested in is polygenically determined and we have to be especially careful to establish what the relationship is between the mechanisms revealed by disrupting a complex system and the control of normal variation in that system.

Thus, whereas one feels able to make quite specific suggestions about the favored approach to studying human behavioral development, even to the extent of laying out a particular experimental design appropriate to the current level of knowledge and kinds of questions to be answered, this does not seem to be a useful exercise for animal behavior genetics. Different classes of questions require different kinds of answers, and the pace of new theoretical and methodological advances makes it foolhardy to suggest that this or that approach is the way to go.

## REFERENCES

Heath, A. C., & Eaves, L. J. (1986). Research strategies for the resolution of developmental genetic models. *Fifth International Congress on Twin Studies,* Amsterdam. September (Abstract).

Heath, A. C., Neale, M. C., Hewitt, J. K., Eaves, L. J., & Fulker, D. W. (1989). Testing structural equation models for twin data using LISREL. *Behavior Genetics, 19,* 9–35.

Hewitt, J. K., Eaves, L. J., Heath, A. C., Neale, M. C., Kendler, K. S., Erickson, M., Rutter, M., & Simonoff, E. (1988). *Adolescent behavioral development: Twin study.* Unpublished grant proposal, MH45268.

Meyer, J. M., & Eaves, L. J. (1988). Estimating genetic parameters of survival distributions: A multifactorial model. *Genetic Epidemiology, 5,* 265–275.

Neale, M. C., Eaves, L. J., Hewitt, J. K., MacLean, C., Meyer, J. M., & Kendler, K. S. (1989). Analyzing the relationship between age of onset and risk to relatives. *American Journal of Human Genetics, 45,* 226–239.

# 16
# Issues of Integration

MARTIN E. HAHN AND ROBERT H. BENNO

We now return to the issue with which we began this book—integrated approaches to the study of behavior. In the Introduction, we posed three practical questions about integrated approaches:

Would a diverse group of single approaches yield more understanding of behavior than complex integrated ones?

Would integrated studies of behavior be so expensive in resources that, even though desirable, they could never be accomplished?

Is it possible to forge productive links between such research areas as the neurosciences and quantitative genetics?

Before we attempt to answer these questions in the context of the preceding chapters, we ask the reader to please examine Figure 16-1. Figure 16-1 presents a schematic of a completely integrated approach to the understanding of behavior, an approach that includes the elements contained in this book: behavior, development, ecological/evolutionary setting, neural bases of behavior, and genetic analysis. What is necessary to perform such integrated research is, first, a detailed description of the behavior as it is performed by the organism during the ages of interest; second, a description of the ecological setting in which the behavior occurs during development; third, isolation of the neural substrate of the behavior; and finally, a developmental genetic analysis of the behavior and its neural substrate across the ages of interest. Such an integrated analysis would seem to be the ultimate goal of an investigator interested in behavior and would satisfy Lorenz's (1981) guidelines about understanding the system of which behavior is a part.

## Would a Diverse Group of Single Approaches Yield More Understanding of Behavior Than Complex Integrated Ones?

In our opinion, the linked knowledge gained from integrated strategies yielding descriptions not only of the parts of the system but also of the relationships among the parts would be more comprehensive than the knowledge gained by a diverse group of single approaches. The search for order among the facts about behavior would be greatly enhanced, for example, by understanding of the ecological relevance of a developing behavior, as shown by Arnold (Chapter 9), by quantification of the

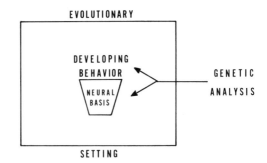

Figure 16-1 Schematic of a completely integrated approach to the understanding of behavior.

interplay between ontogeny and phylogeny for mouse aggression, as done by Cairns and Gariépy (Chapter 3), or by the detailed modeling of influences on developing cognitive abilities in children, as carried out by Hewitt (Chapter 11) or Corley and Fulker (Chapter 12).

## Would Integrated Studies Be So Expensive in Resources that, Though Desirable, They Could Never Be Accomplished?

We see two potential problems associated with resources used in integrated studies. Studies integrating, for example, behavior, development, and genetic analysis can involve large amounts of investigator time. As an example, one of us (Hahn) is currently analyzing the results of a developmental genetic investigation of ultrasonic call production in young mice. The study involved recording 16-second samples of the vocalizations of about 600 infant mice on each day from the ages of 2 through 12 days. The recordings fill over 80 500-meter reels of audiotape and, with analysis at the rate of about eight hours per tape, over 600 hours will be spent in quantifying the characteristics of those ultrasonic calls. At that pace, a completely integrated investigation of infant mouse vocalizations, including identification of the neural networks responsible, identification of the effects of the calls on adult mice, and describing the stimulus circumstances under which calls are emitted, might easily consume the entire scientific lifetime of one investigator. However, since the amount of time consumed by single investigators working piecemeal on the same behavior would probably be at least as great, an integrated study would not draw a penalty for taking too much time.

Of greater concern may be the diversity of training and expertise required of a single investigator or the difficulty of collaboration among a group of investigators from diverse backgrounds. We suggest that collaborative efforts will be the more successful approach, and we hope that this book may alert some readers to the power of such efforts.

## Is It Possible to Forge Productive Links Between Such Research Areas as the Neurosciences and Quantitative Genetics?

Clearly, an understanding of genetic and environmental contributions to the structure and function of the nervous system is critical to an understanding of individual differences in behavior as well as of the mechanisms underlying species-typical behaviors. Yet a survey of abstracts of the Society for Neuroscience's or of the Behavior Genetics Association's annual meetings over the last several years shows relatively few that relate genetics and the neurosciences. It is especially clear that few productive links between the neurosciences and quantitative genetics have been formed. In his introduction to Part II of this volume, Benno delineated some concepts coming from the work of Stent (1981) that could provide a foundation for more productive linkages. Stent described an ideological and an instrumental view of the relationship between the genes and the nervous system. In the ideological view, genes are seen to specify the assembly of the nervous system in complete detail. One experimental-genetic method used to explore this view is a slight variation on Mill's method of similarities (1950) in which individuals of genetically homogeneous populations (e.g., inbred lines) are screened for differences in neurobiological characteristics to determine the precision with which genes code for those characteristics during development. In the instrumental view, there is little concern for how genes specify nervous system structures; instead the focus is the causal pathway between different genotypes and different nervous system phenotypes. The experimental-genetic approach applied here is Mill's method of differences (1950), in which animals from a heterogeneous population are compared. Often, the form of this experiment is a comparison of normal animals with conspecifics who have isolated, known mutations. Such work is highly productive, but may be criticized as having little application to achieving understanding of the normal range of behavioral variation.

## Successful Integrations

In the spirit of facilitating integrative approaches to the study of behavior, we conclude this chapter with brief descriptions of programs of research that have integrated genetics and studies of the nervous system. Not only do data coming from these programs validate the interdisciplinary approach, but because of their interdisciplinary basis the programs have sometimes displayed some interesting innovations in methodology. Continuing to work within the classification system of Stent (1981) we present studies from the instrumental and ideological approaches.

### *The Ideological Approach*

One intriguing and widely observed phenomenon is broad variability in phenotypes among individuals of the same species during development. Such variability has been observed in a number of traits in presumably isogenic populations and is sometimes attributed to "developmental noise" (Waddington, 1957). Collins (1968), for example, has demonstrated differences in the strength of behavioral laterality within inbred strains of mice, and Wahlsten (1982a,b) has consistently observed variability in the size of the corpus callosum in the inbred Balb/cJ mouse strain. An investigator utilizing isogenic populations of organisms or pairs of identical twins may be able to describe the genetic and environmental interplay that is the source of observed phenotypic variability.

The cerebellum of the mouse brain lends itself to study of the role of genes and environments in development because it develops postnatally. Inouye and Oda (1980) documented distinct branching patterns in the cerebellum at the third folia just rostral to the primary fissure in inbred mouse strains. To study this phenomenon

# ISSUES OF INTEGRATION

further, Cooper, Hahn, Hewitt, and Benno (1987) examined the cerebellar foliation patterns of four inbred mouse strains and the 12 $F_1$ hybrids produced by a 4 × 4 diallel cross. We observed two distinct patterns. In the first, the arbor vitae of the third primary branch of the cerebellum split into two foliate branches divided by a sulcus, a mouthlike pattern we called Type I. In the second pattern, the third primary branch of the cerebellum ended in a single, fingerlike projection (Type II). Figure 16-2 shows the distribution of the foliation patterns as expressed in the percentage of Type I. An examination of that figure indicates that mice of the DBA/2J strain (D) consistently expressed the Type-I pattern and members of the SJL/J strain (S) mostly displayed the Type-II pattern. Mice of the C57BL/10J (C) and Balb/cJ (B)

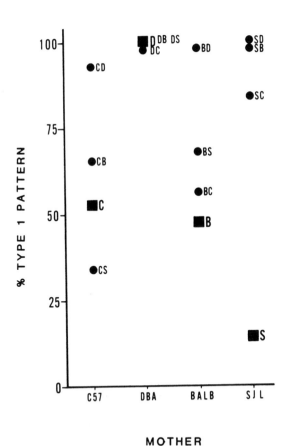

**Figure 16-2** Genotypic distribution of cerebellar foliation pattern in inbred strains of mice. The patterns that are described in the text are determined from creryl violet stained mid-sagittal sections of mice brains produced by a 4 × 4 diallel cross.

strains were intermediate in their anatomy. The hybrids, genetically identical and perfectly heterozygous within each group, displayed the Type-I pattern if they came from a DBA parent but expressed both foliation patterns to differing degrees without DBA parentage. Further examination of the data with support from a biometrical genetic analysis indicates dominance in the inheritance of this foliation pattern. The DBA pattern is expressed almost completely regardless of the strain of the other parent, and the biometrical analysis (a Hayman analysis of variance modified by John Hewitt) reveals significant additive, directional dominance and maternal components. This combination of the ideological approach and biometrical genetic analysis provides entree into the intriguing problem of individual differences within a species during development.

## *The Instrumental Approach*

One way the power of the instrumental approach can be appreciated is in studies of single gene mutations in *Drosophila*. A number of single gene mutations are characterized as altering neuronal cell differentiation and pattern formation. Pairs of strains of *Drosophila* can be produced that differ at only one locus—one strain carrying the mutant gene, the other carrying the "normal" gene. Behavioral differences between the strains can be carefully quantified and correlated with nervous system morphological differences. Thus, the investigator is able to determine which neural circuits are important in controlling the behavior of interest. Hall (Chapter 6) has effectively used this single-gene approach to define the neural circuitry involved in *Drosophila* courtship. In his studies, Hall describes two x-chromosome mutations in male *Drosophila* that affect the courtship song: the *cac* mutation, which affects "tone pulses," and the *per* mutation, which affects song rhythm. His evidence from genetic mosaics of *Drosophila* shows that the *cac* mutation may perturb either the development of a hypothetical "song circuit" in the thoracic nervous system of young flies or, alternatively, act only in adults to alter the physiology of the putative pattern generator in the thoracic region. Hall suggests that by using molecular biological approaches to clone the *cac* gene it should be possible to distinguish between these two possible timings of $cac^+$ gene expression. The strength of using the molecular biological approach in tandem with neurogenetics is also apparent in Hall's work on the *per* gene in *Drosophila*. Early studies with the *per* mutation showed that the *per* gene not only codes for song rhythm in *Drosophila* but also is important in the generation of circadian rhythms. Although it is not clear at this time what part of the nervous system is altered by the *per* mutation, Hall has recently been able to sequence the *per* gene, demonstrating the importance of the threonine-glycine repeat region of this gene in the normal expression of *Drosophila* courtship. Although this type of neurogenetic approach to the understanding of the development, structure, and function of neural networks is a relatively new field of biology, it is clear that the use of the techniques of molecular biology will allow identification of the specific sequences of DNA that are important in the control of behaviors of interest.

A very different but equally powerful example of the instrumental approach comes from Crusio and collaborators (1986, 1987a,b). They combined quantitative genetic analysis and naturally occurring variation in the size of portions of the hippocampus to study learning and open-field habituation in the mouse. They began by noting the evidence that the circuitry of the hippocampus plays a critical role in learning and memory tasks, though most of the evidence of that role has been gathered by lesion studies of the hippocampus. They argued that naturally occurring variation in hippocampal size would be a better means of assessing the role of the hippocampus in behavior than lesion-induced variation, since lesions produce a variably damaged and incomplete hippocampal system. On the other hand, when naturally occurring variation is used to study the hippocampus, a complete and normally functioning hippocampus is the object of study. Employing eight inbred strains of mice which displayed a range of areas of the mossy fiber hippocampal terminal fields, they observed a strong negative relationship between hippocampal size and performance errors in a radial arm maze (Crusio, Schwegler, & Lipp, 1987b), thus adding support to the putative role of the hippocampus in learning and memory. Seeking to further clarify that role they employed a diallel cross genetic design and biometrical genetic analysis to estimate the size of additive, dominance, and other genetic components of variation to the size and cell distribution patterns in regions of the hippocampus. Those analyses allowed them to conclude that some terminal field regions of the hippocampus had undergone stabilizing selection for size and that one field, the suprapyramidal terminal field, had undergone strong directional selection for increased size in the evolutionary past of the mouse. That is, mice with a larger suprapyramidal terminal field had a reproductive advantage over mice with a smaller field. Thus, biometrical genetic analysis is effective in answering "how" questions about the nervous system and is useful in answering the very elusive "why" questions as well.

## CONCLUSION

The contributions to this volume have illustrated how questions about behavioral development can be approached from neural, biometrical, and evolutionary perspectives. We have seen "case studies" in which scientists from varied backgrounds have approached a variety of problems; and we have examined the different kinds of questions they have asked as well as the different methods and techniques they used to answer them. Clearly, an overall understanding of behavior is not just a matter of understanding its evolutionary context. Nor is it sufficient to understand the mechanisms that lead from genes to neural structure. A complete understanding of behavior requires the genetic, neural, developmental, and ecological/evolutionary perspectives.

However, this is not to advocate an approach that is multidisciplinary in the usual sense. What sets the contributions to this book apart is that they all share a common foundation in using the genetic properties of organisms in an attempt to discover why and how the organisms behave as they do. Clearly there is an excitement about the power of a genetic approach that is now beginning to be realized and that is leading to the emergence of developmental behavior genetics as a discipline. Many problems remain to be resolved, and progress in some areas is more rapid than that in others, but there are few if any questions about behavioral development that will not be answered more fully by a genetic approach than by one that ignores genetics.

## REFERENCES

Collins, R. (1968). On the inheritance of handedness. I. Laterality in inbred mice. *Journal of Heredity, 59,* 9–12.
Cooper, P., Hahn, M. E., Hewitt, J. K., & Benno, R. (1987). Biometrical genetic analysis of a cerebellar foliation pattern. *Neuroscience Abstracts, Vol. 13,* p. 254.
Crusio, W. E., Genthner-Grimm, G., & Schwegler, H. (1986). A quantitative-genetic analysis of hippocampal variation in the mouse. *Journal of Neurogenetics, 3,* 203–214.
Crusio, W. E., & Schwegler, H. (1987a). Hippocampal mossy fiber distribution covaries with open-field habituation in the mouse. *Behavioural Brain Research, 26,* 153–158.
Crusio, W. E., Schwegler, H., & Lipp, H. (1987b). Radial maze performance and structural variation of the hippocampus in mice: A correlation with mossy fibre distribution. *Brain Research, 425,* 182–185.
Inouye, M., & Oda, S. (1980). Strain-specific variations in the folia pattern of the mouse cerebellum. *Journal of Comparative Neurology, 190,* 357–362.
Lorenz, K. (1981). *The Foundations of Ethology.* New York: Simon and Schuster.
Mill, J. S. (1950). *A System of Logic.* New York: Hafner (Hafner Library of Classics).
Stent, G. S. (1981). Strength and weakness of the genetic approach to the development of the nervous system. *Annual Review of Neuroscience, 4,* 163–169.
Waddington, C. H., (1957). *The Strategy of the Genes.* London: Allen and Unwin.
Wahlsten, D. (1982a). Deficiency of corpus callosum varies with strain and supplier of the mice. *Brain Research, 239,* 329–347.
Wahlsten, D. (1982b). Mode of inheritance of deficient corpus callosum in mice. *Journal of Heredity, 73,* 281–285.

# Index

Acetylcholine cells, ix, 126
Adaptive significance
  of normal versus mutant behaviors, 100
  of the *per* gene in *Drosophila,* 107
Additive genetic action, 161
Additive genetic variance, 161, 169, 191, 285
  in developing cognitive ability, 247
  in the diallel cross, 194
Adoption, allowing a comparison of rearing and genetic influences, 236
Adoption studies, 164
  assessment of development and genotype in, 237
  criticisms of, 237
  criticisms of classic studies, 240
  elements of the ideal design in, 236
  findings of classic studies, 239–40
  findings of recent studies, 241
  logic of, 236
  longitudinal type of, 237
  parent–offspring data and, 237
  selective placement and, 237, 239
  transracial, 240
Age-related changes
  and Huntington's disease, 267
  lack of models for, 266
  and schizophrenia, 267
Age-related changes in personality, and the effects of genes $\times$ age, 267
Age-specific trait, 167
Aggression, 5, 40, 64. *See also* Agonistic behavior
  age-dependent differences in, 43, 45–47, 75
  in female mice, 54
  in male mice, 40–59, 74, 294
Aging process
  as the accumulation of genetic and environmental changes, 266
  genetic and environmental determinants of, 267
  genetic test of the "homeostatic" model for, 272
  as the gradual failure of homeostasis, 272

  theories of, 266
  various models of, 267
Agonistic behavior. *See also* Aggression
  defined, 42, 67
Allometry
  formula for, 144, 147
  intraspecific and interspecific, 144
  prediction of, 148
American sunfish, 179
Animal learning, genetics of, 164, 218
Animal learning and human development, similarities between, 225
*Aplysia,* 133
Arousal-reactivity, as a mediator of aggression, 43
Artificial selection, 192
  for aggression in mice, 43, 74
  bidirectional, 44
  for brain weight in mice, 134–39
  and correlated characters, 44, 52
  rate of, 44
  results of, for aggression, 45*f,* 45–57
Attachment, 77

Behavior clusters, 69
*Behavior Genetics,* 18, 55
Behavior genetics, 164
  animal studies in, 217
  as defined by Fuller and Thompson, 7
  established as a field, 18
  human studies in, 217
  progress in, 19
  and understanding the development of intelligence, 239
Behavioral change, related to a diet shift, 178
Behavioral inhibition, 48, 50
  and selection for aggression, 51
Behavioral laterality, 306
Behavioral–neurobiological systems, and evolution, 55
Behavioral phenotypes
  as age-independent, 290

## INDEX

Behavioral phenotypes (*continued*)
  as age-specific, 290
  choice of, for behavior genetic research, 283
  developmental hierarchy of, 287
  ecological fidelity of, 281, 285. *See also* Biological systems as entireties
  level of analysis of, 286
  measurement–environment–history (MEH) for, 283–86
  measurement stability and the estimation of genetic parameters in, 293
  measurement stability of, 284, 291
  method of testing measurement stability in, 284
  multivariate approaches and, 288
  as niche-independent, 290
  as niche-specific, 290
  as operationally defined, 283
  problem of poor ecological fidelity in, 285
  as robust, 295
  as "supercharacters," 286
*Behaviorism,* 14–15
Behaviorism, 4, 14, 18
Bidirectional genetic dominance effects, 191. *See also* Stabilizing selection
Biological systems, as entireties, 62–63, 73, 185, 302, 304
Biometrical genetic approach, 161
Biometrical genetics, xiii, 4, 9, 20, 161, 217
Biometry, defined, 161
Birdsong, 177
  age-specific changes in, 182
  comparative studies of, 181
Bluegill sunfish, 179
Body size
  early and late growth of, 149
  regression of brain size on, 153
  scaling of neonatal to adult for, 150
  selection for, 145, 148–49
Brain size
  artificial selection for, 157
  as correlated response to selection for body size, 145
  as a fitness character, 148
  genetic regression of body size on, 149
  maternal influences on, 83, 146–47, 150, 156–57
  prediction of, 145
  and relationship to metabolism, 145–46, 150, 157
  scaling, with body size, 144–46, 148
  selection for, 148
Brain and body size
  correlation at different ages for, 92, 149
  correlation and regression in random bred mice, 151–54, 153$t$
  correlation and regression in inbred/hybrid mice, 154–56, 155$t$
  evolution of, 83, 144, 150
  genetic models of, 92, 148–51
  in phylogeny, 118
  thyroid hormone effects on, 136–37, 137$f$, 138$f$, 139$t$

*Caenorhabditis elegans,* 116, 132
Cell adhesion molecules, 86, 126
Cellular site of gene action, 85, 90, 100
Cerebellar development, 85–91, 86$t$, 125, 135–39
  anatomical and biochemical measures of, 135–36
  correlation with onset of behavior, 135–39
Cerebellar mutations in the mouse
  lurcher, 90
  Purkinje cell degeneration, 90
  reeler, 88
  staggerer, 86–87
  weaver, 87–88
Cerebellum, and foliation patterns in the mouse, 306
Change and continuity in development, environmental contribution to, 223
Chimeric mice, 124, 302
  technique of producing, 87
  used in genetic dissection of behavior, 289
Circadian rhythms in *Drosophila. See* Neurological mutants in *Drosophila per*
Classic cross-breeding design, 94
Clonal development in the nervous system, 89
Cloning of genes, 102
  behavioral genes in *Drosophila,* 103, 105–10
Colorado Adoption Project, 236–65, 300
  described, 242
  parental and proband IQ correlations in, 243, 244$t$
Component failure model, of blood pressure change during aging, 232$f$
Conditioning
  instrumental, 11
  Pavlovian, 11
Continuity and discontinuity in behavior, x, 4, 41, 73
  genetic contribution to, 232
  path model of, 225, 227$f$
  role of genes and the environment in, 231
Continuously varying characteristic, 161
Corpus callosum defects, 92, 306
Correlation between physical and mental development, 29

# INDEX

Covariation, patterns of, 161
Critical periods
  in behavioral development, 72
  and birdsong, 182
  for socialization, 72
  and theory of behavioral development, 70
Cultural transmission, and developing cognitive ability, 247

Darwin, Charles, 7
Darwinian evolution, 7
Darwinian fitness, 104, 190
  of behaviors as "just so stories," 190
  and the expression of behaviors, 191
Developing cognitive behavior, 305. *See also* General cognitive ability; Infant intelligence; Mental development
Developing motor behaviors
  effects of thyroid hormone on, 136–37, 137$f$, 138$f$, 139$t$
  in mice, 134–39
Development. *See also* Ontogeny
  biometrical approach to, 233
  causal links between early and later events in, 63, 233, 261, 289
  contrast of genetic and environmental influences on, 217
  evolutionary setting and, 63
  general model of, 226, 227$f$
  general model of blood pressure changes in humans in, 226
  general model of changes in blood pressure variance in, 228$f$
  as integrated into the study of behavior genetics, xi
  spurts and lags in cognitive, 267
  spurts and plateaus in, ix, 5. *See also* Periods of acceleration and lag
  stability and change in, ix, 92, 299
Developmental behavior genetics
  as a new discipline, 26, 164, 167, 281, 309
  progress in, 20
Developmental noise, 81, 306
Developmental perspective, 40–41
Developmental social behavior genetics, 61
  agenda for, 71
Developmental stability, 191
Developmental trajectory
  and aggression, 52
  and behavioral inhibition, 52
Diallel cross-breeding design, 69, 94, 150, 163, 214
  accuracy of genetic estimates in, 210
  alternative analyses for, 200
  and behavior, 192
  bidirectional dominance variance in, 195
  as compared to artificial selection, 192
  as compared to the triple test cross, 192
  defined, 193
  effect of a strain substitution in, 202–4
  fixed model type, 199
  Griffing analysis of, 199
  Hayman analysis of, 199
  hypothetical worked examples of, 193
  maternal effects in, 194
  methods of analysis for, 199
  reciprocal effects in, 194
  and sample size, 201
  utility of the fixed effects model in, 211–12
  and variance–covariance plots, 196–98, 197$f$
  various models for analysis of, 199
Directional dominance variance, 194
Directional selection
  for aggression in the mouse, 40–59
  for brain size, 148
  and the evolution of ontogenies, 176
Discretely varying characteristics, 161
Dogs, 5, 70, 72
  breed differences in, 72
  critical periods in, 72
Dominance genetic variation, 161, 190, 285
*Drosophila*, xii, 302, 308
  Darwinian fitness characters in, 219
  eclosion in, 104
  learning in, 219
  *Melanogaster*, 103–7
  metamorphosis in, 100, 104
  *Simulans*, 103–7
  stabilizing selection in, 192
  utility of linking laboratory and ecological data in, 211
*Drosophila* courtship behavior, 83, 100–105, 219, 289, 293, 308
  enriched environment effects and, 104
  as example of genetic dissection of behavior, 289
  as experience dependent, 104
  kinetics of, 101

Ecological context, 185, 302
  and development, 63
Ecological niche, 177
  ontogenetic dimension of, 177, 206
Environmentalism, 14
Epigenesis, 81
  definition of, 114
  influences on, 114–15, 117–18, 117$f$, 118$t$, 119, 134

Epistasis
  defined, 190
  as regulating homeostasis, 271–72
Eugenics, 4, 9
  criticisms of, 13
  goal of, 9
  negative, 13
  scientific aspects of, 9
  social/political aspects of, 9, 12–13
Evolution
  of behavioral ontogenies, 163, 167, 186. *See also* Ontogeny and phylogeny
  of behaviors, ix, xii, 9
  of ecological niche ontogeny, 179
  human, 9
  of ontogenies, ix
  and psychology, 8, 11
  synthetic theory of, 16
Evolution and behavior, 42
Evolution and development, linkages between, 41–42, 54
Evolution of behavior
  coupled case (high genetic correlation across age), 172, 173*f*
  uncoupled case (low genetic correlation across age), 172, 172*f*
Evolution of genetic systems, alternate views, 218
Evolutionary history of stabilizing selection, 94
Evolutionary significance
  of *Drosophila* mutant studies, 100
  usefulness of laboratory induced mutants and, 110
Evolutionary trajectory, 173
  effects of genetic correlation on, 175

Facultative shifts in behavior, 181, 185
Function–structure bidirectionality, 42
Functional validation of neurons, 124

Galton, Francis, 9
Gene action, 167
Gene × age interaction, 290, 293
Gene × environment interaction, 37, 92, 114, 220, 293. *See also* Epigenesis
  ideological view of, 306
  instructional type of, ix
  instrumental view of, 306
Gene × environment interactionism, 14
Gene expression
  *per* gene and, 105
  staggerer gene and, 86

General cognitive ability. *See also* Developing cognitive behavior; Infant intelligence; Mental development
  as measured in the Colorado Adoption Project, 243
  parent–offspring correlations and, 261
  sibling resemblance in, 261
  trends in genetic correlation during development of, 262
  trends in heritability during development of, 261
General intelligence, 237. *See also* General cognitive ability
  correlations between twins across ages in, 255
  longitudinal results of, 238
  reasons for studying, 238
Generalized numerical optimization package (MINUIT), 258
Genes and chromosomes, 16
Genetic architecture
  of behavior, xiii, 62
  changes across development in, 164
  changes across development in mouse activity in, 221
  changes across trials in rat learning in, 221
  and correlated characters in selection, 213
  of developing behavior in the mouse, 207
  and escape/avoidance conditioning in rodents, 95, 220
  and the evolutionary history of a trait, 190, 214
  and fitness characters, 176, 207, 281
  interpretation of, 195
  of learning in animals, 218
  of a trait, 162, 206, 286, 288
Genetic basis of behavior, as a taboo topic, 15
Genetic correlation
  in Darwinian fitness characters, 213
  as derived from genetic covariance, 170
  and the development of intelligence, 239
  and the evolution of behavioral ontogenies, 172, 186
  evolutionary importance of, 186
  for morphological traits measured at different ages, 175
  and pleiotropy, 175
  as responsive to selection, 163
  between traits, 163, 213
Genetic coupling, in the development of the striped swamp snake, 178
Genetic covariance, 170
Genetic epidemiology, 266–80
Genetic linkage, 161
Genetic mosaics, 82, 302. *See also* Chimeric mice

# INDEX

Genetic polymorphism, 181
Genetic variance, 168
　components of, 191
*The Genetical Theory of Natural Selection,* 161
Genetically associated variations in brain morphology, 95
Genetics and psychology, separation of, 15
Growth signature, 26
Gynandromorphs, 102, 102*f*

Hawaii Family Study of Cognition, 242
Heritability, 169, 176, 186
　broad, 222
　changes across learning trials in rats, 223
　changes in during learning, 224
　of developing cognitive ability, 261
　and estimates of brain size, 148, 150, 151–52, 151*t*
　as a function of age, 169, 175
　narrow, 222
Heterochrony, in development, 43, 74
Hippocampus
　development of, 94–95, 94*f*
　genetic "lesions" of, 309
　hippocampal lamination defect (Hld) in the mouse, 94
Homeostatic mechanisms in aging
　as regulated by gene × environment interaction, 271–72
　as regulated by genetic systems, 271–72
"How to" manual, xiii
Huntington's disease, 267

Ideological approach to neural development, 81–82, 85, 306–8
In vitro, ix
　conversion of adrenergic to cholinergic neurons, 127
　mutagenesis of the *per* gene in *Drosophila,* 109
In vivo, ix
　conversion of adrenergic to cholinergic neurons, 126–27
Inbred-hybrid differences
　in learning, 218
　meaning of, 190
Inbreeding depression, 191
Inclusive fitness, 19
Incomplete penetrance. *See* Variable expressivity
Individual differences
　between-family factors and, 77
　changes in etiology across learning trials, 223
　etiology of, 3, 7, 10, 37, 54, 73, 75–76, 162
　etiology of, in cognition, 238
　within-family factors and, 77
Infant intelligence. *See also* Developing cognitive behavior; General cognitive ability; Mental development
　as an age-specific trait, 238
　and failure to predict later IQ, 238
　measurement problems in, 238
Infanticide, 60–61
　inheritance of, 60–61
　as a reproductive strategy, 60
"Inside out" migration of neurons, 94, 122, 124
Instincts, ix
Instrumental approach to neural development, 81–82, 308–9
Integrated approaches to behavior, xiv, 18, 282
Intelligence testing
　beginnings of, 14
　controversy and, 14
Interdisciplinary approaches, xi
Intermale fighting, 67, *See also* Aggression; Agonistic behavior
Isogenic, 82

Kin selection, 19

Learning impairments in *Drosophila. See* Neurological mutants, amnesiac, dunce, rutabaga
Linear time series model
　as applied to cognitive development in twins, 270
　assumptions of, 271
　components of, 268
　failures of to model accurately with, 270
　of longitudinal measures on a trait, 269*f*
　questions addressed by, 268
Longitudinal studies, 5, 20
　and adoption designs, 237
　of aggression in mice, 46
　cosibial design in, 45
　and developmental social behavior genetics, 69
　importance of, 185
　of mental ability in twins, 28
　of physical and behavioral development in twins, 5, 26
　of social behavior, 41
　of temperament, 35
　of twins, 25
　use of physiological markers in, 70
　use of theory in, 70
Louisville Twin Study, 5, 25–39, 243, 261, 300

Male mating tactics
  alternate forms of, 177, 180
  evolution of, 179
  in fish, 179
  ontogeny of, 179–80
Maternal influences, 161, 193. *See also* Brain size, maternal influences on
Mathematical models of learning, all-or-none vs. linear, 223
Mendel, Gregor, rediscovery of the work of, 12
Mendelian genetics, 7, 9, 168, 181. *See also* Particulate inheritance
Mental development, ix. *See also* Developing cognitive behavior; General cognitive ability; Infant intelligence
  stability and change in, 28
  tests of, 29
Methodological issues in the study of developing behavior
  economy in design, 7
  genotypic sampling, 208
  measurement drift, 208
  sample-population relationship, 209
  seasonal factors in sampling, 208
Methods to study developing behaviors
  adoption studies, 236–63
  comparative approach, 177, 185
  and environmental influences, 299–300
  genetic approach, 298
  genetic reference populations, 209
  integrated studies, 304
  motor development in mice, 135–39
  obstacles to longitudinal studies, 300
  overlapping cohort longitudinal twin study, 300
  potential for longitudinal genetic study, 298
  power of longitudinal adoption design, 262
  use of cross-twin-cross-trait correlations, 299
Methods to study development
  linear time series models, 268–71
  use of age and genotype as predictor variables, 273
Methods to study developmental social behavior
  agenda for, 71
  cross-fostering, 74, 168
  examples of, 56, 64, 72–77
  experimental analysis, 61
  Gottlieb's views, 63
  homogeneous set design, 64, 76
  Lorenz's views, 62, 304
  normative description, 61
  panel-of-testers design, 64
  song simulator in *Drosophila*, 103
  standard tester design, 64, 76

Methods to study heritabilities of brain and body size, 151–56
Mice, xii, 5, 40–59, 67, 75, 175, 308
  aggression in, 40–59
  Fuller BWS, 134–39
  inbred/hybrid, 146
  infant behaviors in, 206
  random bred, 146, 151
  recombinant inbred strains, 95
  utility of inbred strains of, 214
  wild-type, 94–95
Mill's methods
  of differences, 306
  of similarities, 306
Model. *See also* Linear time series model; Multivariate path model; Survival analysis model; Univariate path model
  for age-related change in blood pressure, 230
  of genetic influences on animal learning, 225
Modeling, and the developmental process, 164, 277
Modeling disease risk
  with age and genotype as predictors, 273, 274$f$, 275$t$
  results in monozygotic twins of, 275
Modeling the developmental process
  principles and problems of, 277
Model systems for neuronal development
  Fuller BWS mice, 134–39
  inbred mice, 133
  mutant mice, 134
  selective breeding, 134
Molecular genetics
  and rhythmic singing behavior in *Drosophila*, 105–10
  rise of, 16
Mosaic development, 102, 102$f$, 116
Multidisciplinary approach, xi
Multivariate analysis
  and development, 164
  origins of, 20
Multivariate measurement of behavior, significance of, 193
Multivariate path model, 251
  applications of, 257
  and assortative mating, 250
  assumptions of, 255–56
  correlation estimates for ages 1–4 and 7 years with, 259
  extensions for varying family groupings with, 253
  rules for deriving correlations in, 249
  test statistics for model fitting with, 260–61
Mutation, 176
  and evolution by saltation, 181

# INDEX

Natural selection, 190
   acting on the average phenotype, 162
   acting on the pattern of genetic control, 162
   and behavior, xi, 214, 217, 219
   and bidirectional dominance variance, 195
   and Darwinian fitness, 201
   and developing organisms, xi
   and genetic parameters, 176
   and homeostatic mechanisms of aging, 271
   and human behavior, 77
   and the intermediate as the optima, 192
   and man, 8
   as the mechanism of evolution, 8
   operating on integrated behavioral phenotypes, 288
   related to age, 170, 206
Nature-nurture question, 15, 17, 40, 55, 63
Neo-Darwinism, 3, 16
Nerve growth factor, 130, 133
Nervous system, biochemical maturation of, 135
Nervous tissue, determination of, 120
Neural circuits
   cerebellar, 85–88, 86*f*, 135–37
   complex, ix
   development of and relationship to behavior, 135
   effects of genes and environment on, 81
   hippocampal, 82, 93–95, 94*f*
   phylogenetic/ontogenetic view of, 119
   song circuit, 102
Neural crest cell development, 124–27
Neural induction, 121, 121*f*
Neural tube, 122–24
Neurological mutants in *Drosophila*
   amnesiac, dunce, rutabaga, 103–4
   biochemical abnormalities in, 104
   cacophony *(cac)*, 100–102
   isolation of, 100
   period *(per)*, 103–9, 107*f*
Neurological mutants in mice, 85–89
Neuronal modifiability, 93, 117–18, 130–31
Neuronal development
   ecosystem view of, 120
   cell lineages and, 89
   critical periods in, 93, 95, 115
   genetic vs. epigenetic control of, 81, 114
   historical viewpoints of, 114–15
   ideological approach to, 81–82, 85
   instrumental approach to, 81–82
   phlogeny and, 83, 116, 118
   preprogrammed, 115–16
   thyroid hormone effects on, 136–37, 137*f*, 138*f*, 139*t*
   trophic substances role in, 120, 128, 130–31, 131*f*, 133

Neuronal differentiation
   aggregation into cellular masses and, 125–26
   axonal outgrowth and, 127–28
   changes in cell shape in, 127
   dendritic maturation in, 130–31
   disturbances in, 100
   expression of neurotransmitter in, 126–27
   histogenic cell death and, 89–90, 131–33, 132*f*
   migration and, 86, 88, 94, 122–24, 125*f*, 126*f*
   proliferation and, 89, 122, 123*f*
   specific classes of neurons and, 123
   synaptic modifiability and, 130, 133. See also Neuronal modifiability
   synaptogenesis and, 128, 130, 133
Neurons
   cell birthdays of, 119
   cell lineages of, 116
   cerebellar granule cells, 86–88, 86*f*, 124–25, 125*t*, 131
   cerebellar Purkinje cells, 85–91, 131, 135–39
   cortical pyramidal cell orientation in, 130
   hippocampal mossy fibers, 95
   hippocampal pyramidal cells, 93–95
   Mauthner's cells, 127–28, 129*f*
   Type I, Type II cells, 118–19, 118*t*
Nonadditive genetic variation, 190. See also Dominance genetic variation; Epistasis
Norepinephrine cells, ix, 126–27

Odor cues, and predatory patterns, 178
*On the Origin of Species*, ix, 7–10
Ontogenetic stability, as consistent genetic architecture across age, 290
Ontogeny, xii, 15. See also Development
Ontogeny and phylogeny, relationships between, 40–42, 55, 74, 76, 185, 305
"Open" and "closed" behaviors, 42
Overdominance, 190

Particulate inheritance, 3, 11–12, 161. See also Mendelian genetics
Particulate vs. continuous inheritance, 12
Path analysis, 217. See also Multivariate path model; Univariate path model
   origins of, 20
Periods of acceleration and lag, 26. See also Development, spurts and plateaus in
Phenotypic variability
   in brain morphology, 81, 85, 91*f*, 92, 146
   in *Drosophila*, 107
Phenylketonuria, 16
Phylogeny, xiii

Physical development, stability and change in, 28
Plasticity. *See* Neuronal modifiability
Pleiotropy, 161, 289
Polygenic control, 217
Polygenic hypothesis, 161
Polygenic system, 201
Polygenic traits, 176
Psychology
  child, 15
  comparative, 41
  experimental, 10–11
  scientific, 7
Psychometrics, 13
  and the measurement of cognition, 237
Pumpkinseed sunfish, 179

Quantitative genetic approach, 192
Quantitative genetic perspective, 179
Quantitative genetics, 146–47, 161
  and analysis of behavioral ontogenies, 185
  and links to the neurosciences, 306
  rise of, 17

Rats
  learning in, 220
  wild-trapped, 220
Recessive modes of inheritance, 93
Regulative development. *See* Mosaic development
Rock bass, 179
Rodents. *See also* Mice; Rats
  learning and Darwinian fitness, 219

Schizophrenia, 267
Selection
  and correlated response, 170–72
  correlated response at two ages, 170, 171$f$
  direct response to, 170
Selective placement in adoption, model treatment of, 246
Single gene approach, to genetic dissection of behavior, 30, 289
Snake, water (striped swamp snake), 63, 177
Social behaviors
  as "closed," 55
  defined, 65
  genetic and developmental influences on, 54–55, 60
  as increasers/decreasers of gene transmission, 68
  as influenced by natural selection, 60, 294, 294$f$, 295
  lists of, 64, 67
  ontogeny and phylogeny of, 61–62
  origins of, 61
  potential for multivariate approach and, 68
  Scott's categories of, 67
  sequences of, 68
  Tinbergen's hierarchical structure of, 67–68
Social dominance, 69
Social interactions
  cue function in, 65, 71
  defined, 65
  ontogeny and phylogeny of, 40
  study of, 56
  Tinbergen's view of 65, 66$f$
Social relationships
  defined, 66
  intermale fighting and, 67
  as an organism's environment, 64, 76
  Scott's views on, 64
  and stability and change in behavior, 66. *See also* Continuity and discontinuity in behavior
  stabilizing nature of, 67
*Sociobiology*, 3
Sociobiology, 19, 60
  origins of, 19
  philosophical and political attacks on, 19
Song pulses, 101$f$
  rhythmic fluctuation in, 103–8, 106–7, 107$f$
Species-typical behavior, 42, 47, 75, 305–6
Stabilizing selection, 191, 202
Survival analysis model, and the aging process, 275–77
Synchronized growth patterns, in twins, 26

Temperament
  measurement of, 34
  synchronized changes in twins, 36
Triple test cross, 192
Twins studies
  behavioral development in, 25
  of blood pressure, 228
  concordance for height and weight in, 27
  concordance for temperament in, 35
  home environments and, 34
  mental development in, 26
  mental differences in monozygotic in, ix
  model fitting for blood pressure in, 229
  MZ-DZ differences in, 9
  physical development of, 25–26

social behavior of, 76
temperament in, 34
trend analysis in, 31–32
Typological vs. populational thinking, 10

Ultrasonic vocalizations in mice, 75, 305
species × age effects, 75

Univariate path model. *See also* Path analysis
elements of, 245
extended to adoptive and nonadoptive families, 246–48, 247*f*

Variable expressivity, 91*f*–92

Zygosity determination, 27